AS Geography
for Edexcel

Digby ▸ Hurst ▸ Chapman ▸ King ▸ Owen

OXFORD

OXFORD
UNIVERSITY PRESS

Great Clarendon Street, Oxford OX2 6DP

Oxford University Press is a department of the University of Oxford.
It furthers the University's objective of excellence in research,
scholarship, and education by publishing worldwide in

Oxford New York

Auckland Cape Town Dar es Salaam Hong Kong Karachi
Kuala Lumpur Madrid Melbourne Mexico City Nairobi
New Delhi Shanghai Taipei Toronto

With offices in

Argentina Austria Brazil Chile Czech Republic France Greece
Guatemala Hungary Italy Japan Poland Portugal Singapore
South Korea Switzerland Thailand Turkey Ukraine Vietnam

Oxford is a registered trade mark of Oxford University Press
in the UK and in certain other countries

© Oxford University Press 2008

Authors: Bob Digby, Catherine Hurst, Russell Chapman, Anna King,
Andy Owen

The moral rights of the authors have been asserted

Database right Oxford University Press (maker)

First published 2008

British Library Cataloguing in Publication Data

Data available
ISBN 978-0-19-913482-3

10 9 8 7 6 5 4 3

Printed in Spain by Cayfosa Quebecor

Acknowledgements

The publisher and authors would like to thank the following for permission to use
photographs and other copyright material:

P6: ASSOCIATED PRESS (top right); p8: Galen Rowell/Mountain Light/Alamy;
P15: SHEHZAD NOORANI/Still Pictures (top left); p15: ASSOCIATED PRESS (top right);
P15: Dinodia Images/Alamy (bottom right); p17: David Rydevik; p18: U.S. Air Force photo
by Master Sgt. Val Gempis (bottom right); p20: ASSOCIATED PRESS (top right); p22:
ASSOCIATED PRESS (top right); p23: Photo by Mike Hewitt/Getty Images (top right); p25:
National Geographic (top right); p26: ASSOCIATED PRESS (top right); p27: David Parker/
Science Photo Library; p28: Copyright 2007 photolibrary.com (bottom right); p34: NASA
Images (all); p37: Colin Harris/Light Touch Images; p39: Copyright 2002-2007 Nature
Picture Library (top right); p40: Copyright 2007 photolibrary.com; p41: Chris Jackson/Getty
Images (bottom right); p43: Helene Rogers/Alamy (bottom left); p44: Broker|Dreamstime.
com; p45: Ian Murray/Geography photos; p46: sinopictures/viewchina/Still Pictures
(bottom right); p47: www.renet.at (top left); p50: Greenpeace/Morgan (top right); p50:
The Carbon Neutral Company Limited (bottom right); p50 Coldplay (middle right); p51:
MARTIN BOND/Still Pictures (middle); p51: Copyright 2007 photolibrary.com(top right
spot image); p54: TIMOTHY A. CLARY/AFP/Getty Images; p56: Mark Sykes / Alamy(bottom
left); p56: Jeannette Daugbjerg Hansen Photography(bottom right); p57: Copyright 2007
photolibrary.com (middle right); p59: Hans De Keulenaer/Leonardo energy (middle left);
p62: NASA(all); pg63: NASA (top right); p66: Walt Disneyworld News; p68: Alamy (top
right); p70: Photolibrary; p71: Alamy (top right); p76: Christian Aid 2007; p80: Spectrum
Photofile Inc. (middle right); p80: Still Pictures (bottom right); p81: Link Picture Library;
p84: Associated Press; p85: Photolibrary; p86: Still Pictures (top right); p87: Photolibrary
(top right); p88: Still Pictures (top); p88: Australian Image Originals (bottom); p90: Alamy
(bottom left); p100: Panos Pictures (top right); p101: Reuters (top right); p103: CAMERA
PRESS (top right); p107: Rex Features(top); p107: Trevor Leighton (middle top); p107:
Associated Press (middle bottom); p107: Rex Features (bottom); p113: Photolibrary (top
left); p113: Getty Images (bottom right); p114: Photolibrary (top right); p114: Associated
Press (middle right); p114: Alamy (bottom left); p116: India Picture (top); p118: Adrian fisk
(left); p118: Reuters (right); p119: Still Pictures; p120: Photolibrary; p121: Associated Press
(left); p121: Dinodia (right); p122: Alamy; p124: Caribbean Colours (top right); p124: Samir
Hussein Photography (bottom); p125: Associated Press (top right); p125: Alamy (middle);

p127: Reuters; p128: 2007 Marks and Spencer plc.; P132: Associated Press (bottom right);
p136: MAURICIO LIMA/AFP/Getty Images (bottom right); p137 Associated Press (top right);
p137: JORGEN SCHYTTE/Still Pictures (middle right, left); p137: Margie Politzer/Lonely
Planet Images (middle right, centre), p137: Nick Turner/Alamy (middle right, right); p138:
Apex news and Pictures(middle bottom); p139: Copyright 2007 photolibrary.com (bottom
right);p140: Wayne Grundy (top right); p144: EUMETSAT (middle right); p146: Copyright
2007 photolibrary.com ; p147: Environment Agency (middle right); p147: Metoffice(bottom
right); p148: Michael Semmens Photography (top right);p148:OUP (middle right); p149:
Benjamin Evans(top right);p150: FEMA (top right); p151:NASA (bottom right); p152:FEMA
(top right); p152: Copyright 2007 photolibrary.com (bottom middle);p153: Jim Wark/Still
Pictures; p154: NASA 9top right); p156: Mario Tama/Getty Images (bottom left); p157: NASA
(middle, all); p158: Associated Press (top right); p159: Ron Giling/Lineair/Still Pictures 9top
left); p159: PAUL GLENDELL / Still Pictures(top right); p160: Howard Birnstihl Photography
(top right); p160: Ian Waldie/Getty Images (bottom right); p161: Spectrum Photofile Inc.;
p167: Bill Bachman / Alamy (top right); p167: Christine Osborne Pictures (bottom right);
P174: Alamy(top); p174: Alamy (bottom); p175: The Photolibrary Wales Ltd (top); p175:
Photolibrary(top right); p175: Chris Fairclough Worldwide (bottom right); p176: Airview
Aerial Photography(bottom right); p179:Alamy (top right); p180: Gavin Gough; p182:
Alamy(bottom right); p183: Lonely Planet Images(top); p183: Getty Images (bottom left);
p183: Alamy (bottom right);p184: Nation Trust Photo Library (bottom);p185: Brenton West;
p186: Southampton Container Terminals Ltd(bottom); p187: Alamy(bottom left); p187:
Getty Images(bottom right); p188 Alamy(top right); p189: Alamy (middle left); p189: Naturepl
(middle right); p190: Friends of the Earth Southampton (top right); p190: Derek Lovejoy
& Partners (bottom left); P192: A. Logie(bottom right); p193: National Gas Archive(top)
p193: Alamy (middle right); p193 Alamy (bottom left);p194:Alamy; p195: A. Logie(bottom
right);p196: Shutterstock;p197: Andrew Stacey (top right);p198: Creative Commons;p199:
Andrew Stacey (top); p199: KeyTERRA-FIRMA Ltd (bottom);p200: John Hardy; p201:
Special Projects Engineer (top); p201:alamy (bottom);p204: Photolibrary; p206: Gordon
Langsbury, rspb-images.com; P212: Richard Klune/Corbis (bottom left); p212: Photofusion
Picture Library /Alamy (bottom right); p214: Darrin Jenkins / Alamy (top right); p216: Liz
Somerville/Photofusion/Photographers Direct (bottom left);p216: Travel and Places/Alamy
(bottom right); p217: Robert Estall photo agency/Alamy (top right); p220: Neil Hunt/Alamy
(middle right); p220: Sandra Young /Photographers Direct (bottom right); p221: Justin
Kase/Alamy(top right); p223: Roger G Howard Photography/Photographers Direct (bottom
right); p224: David Hobard Photography/Photographers Direct (bottom right); p225: Aidan
Semmens Photography/ Photographers Direct (top right); p226: geogphotos / Alamy (bottom
right); p228: 2007 Snoasis Concern Ltd; p230: Karin Duthie/Alamy (top left); p232: Karin
Duthie /Alamy; p233: natavillage.org (all); p234: Karin Duthie/Alamy (bottom left); p235:
David Reed/CORBIS(all); p236: Karin Duthie Illustrative Options images(bottom right);
p238: YOAV LEMMER/AFP/Getty Images (bottom right); p239: Mark Newman/Lonely Planet
Images(top right); p240: Andrew parker/Alamy (top right); p240: Darren Johnson/ Lazyfruit/
Photographers Direct (middle right); p242: Homer Sykes/Alamy (bottom right);p242:
David Hoffman Photo Library/Alamy(bottom); p243: Mpix(top right); p243: MPX Stock
Images/ Photographers Direct (middle right);p245: Neil Setchfield/Alamy(top left); p245:
Eric Hands Photography/ Photographers Direct(top right); p246: Simmons Aerofilms/Hulton
Archive/Getty Images (top right); p247: Design for manufacture/englishpartnerships.co.uk;
P252: POPPERFOTO/Alamy (bottom left); p252: Christopher Pillitz/Alamy (bottom right);
p253: supershoot/Alamy (bottom left); p255: David Hoffman/ DHPL/Photographers Direct(top
right); p256: ASSOCIATED PRESS(top right);p257: 2007 The London Organising Committee
of the Olympic Games and Paralympic Games Limited (bottom right); p258: ASSOCIATED
PRESS(top right); p258: LondonPhotos – Homer Sykes/Alamy(bottom left);p260: Elizabeth
Molineux ; p261: London Development Agency; p262: Copyright 2007 photolibrary.com;
p263: ASSOCIATED PRESS; p264: WWW.AIRVIEWONLINE.COM.AU (bottom left); p266:
Joanne Moyes / Alamy (top right); p266:webbaviation.co.uk(middle right); p266: Elizabeth
Molineux(bottom right); p267: Andy Marshall/Alamy; p268: Putmans photographers(top
left); p268: Yvonne White/ WhiteGoldImages/Photographers Direct(top right); p268: Robert
Clare/ Photographers Direct(bottom right); p270: Peter Barritt /Alamy (top right); p274:
Copyright 2007 photolibrary.com; p276: Daniel Gale/Shuttersock (top right); p276: Buzz
Pictures/Alamy (bottom left); p278: Geoff Tydeman Freelance Photography/ Photographers
Direct(top); p278: Simon Vine/Alamy(middle); p279: Paul Bradforth Photography/
Photographers Direct(top right); p282: Dennis Johnson/Lonely Planet images (Middle left);
P287: Apex News & Pictures (bottom left); p287: Alamy (bottom right);p289: Alamy (all).

Cover photo: Oxford University Press

The Ordnance Survey map extract on pp.138 is reproduced with the permission of the
Controller of Her Majesty's Stationery Office © Crown Copyright.

Illustrations are by Barking Dog Art.

**Bob Digby would like to thank Andrew Chapman, Max Nye, and Alex Taylor for
helping with research**

Every effort has been made to contact copyright holders of material reproduced in this
book. Any omissions will be rectified in subsequent printings if notice is given to the
publisher.

Contents

1 **The world at risk** 5

1.1 What has happened to Tebua? 6
1.2 Hazards and disasters 8
1.3 Global hazards 10
1.4 Hazards and vulnerability 14
1.5 Bushfire! 20
1.6 Why are some places hazard prone? 22
1.7 Hazard hotspots 24
1.8 The causes of climate change 30
1.9 The effects of climate change 34
1.10 The impacts of global warming 38
1.11 Dealing with climate change: mitigation and adaptation 44
1.12 Global warming: which way now? 46
1.13 Strategies for dealing with climate change 50
1.14 Energy use, efficiency and conservation 56
1.15 Tackling risk and vulnerability in Bangladesh 60
1.16 Tackling risk and vulnerability in Indonesia 62

2 **Going global** 65

2.1 Taking the Mickey to the world? 66
2.2 What is globalisation? 68
2.3 Getting connected: colonialism 70
2.4 Global differences 74
2.5 Staying connected: global groupings 76
2.6 Strained connections 78
2.7 Corporate connections 82
2.8 Global remix: winners and losers 86
2.9 World on the move 92
2.10 Migration into Spain 94
2.11 Causes and consequences of European migration 96
2.12 Analysing population change in the UK 102
2.13 Megacities: Los Angeles 110
2.14 Megacities: Mumbai 116
2.15 Environmental consequences of globalisation 122
2.16 Social consequences of globalisation 124
2.17 Managing change for a better world 126

3 **Extreme weather** 131

3.1 Extreme heat 132
3.2 From one extreme to another 134
3.3 Increasing the risk 136
3.4 One extreme storm 138
3.5 What caused the Boscastle flood? 142
3.6 Managing flooding 146
3.7 Hurricane Katrina, August 2005 150
3.8 All about hurricanes 154
3.9 Managing hurricanes 156
3.10 The Big Dry 158
3.11 What caused the drought? 162
3.12 Managing Australia's water 165
3.13 Gloucestershire under water 168

4 **Crowded coasts** 173

4.1 The coast: not all sand and sea 174
4.2 The shift to the coast 176
4.3 Crowded coasts: Bournemouth 178
4.4 Conflict on the coast 182
4.5 Industry on the coast: Southampton Water 186
4.6 Protest on the coast: Dibden Bay 190
4.7 The Holderness coast … going …going … 192
4.8 Managing erosion on the Holderness coast 198
4.9 Thames estuary at risk 204
4.10 Can coasts be managed sustainably? 208

5 **Unequal spaces** 211

5.1 An unequal nation 212
5.2 East Anglia: emerging inequalities 214
5.3 Managing inequalities in rural East Anglia 218
5.4 Can rural communities be sustainable? 222
5.5 Planning for the future 228
5.6 Botswana: a different kind of inequality 230
5.7 The marginalised Batswana 232
5.8 Divided Botswana 238
5.9 Capital inequalities: London 240
5.10 A tale of two boroughs 242
5.11 Sustainable urban communities 246

6 **Rebranding places** 249

6.1 Rebranding places 250
6.2 Regeneration in East London 252
6.3 London and the 2012 Olympics 256
6.4 Transport regeneration in East London 260
6.5 Sydney's Olympic story 262
6.6 Re-imaging Manchester 266
6.7 Can Walton-on-the-Naze be resuscitated? 268
6.8 Cornwall in crisis? 270
6.9 Cornwall's Garden of Eden? 274
6.10 Rebranding 'cool' Cornwall 276
6.11 Rebranding farming 280
6.12 Can rebranding apply to LEDCs? 282

Exams: how to be successful 287

Glossary 297
Index 302

About the questions in this book

● Over to you

'Over to you' questions mostly provide you with opportunities for collaboration, for pair or group work.

● On your own

'On your own' questions mostly provide you with opportunities for independent work.

Exam question: There are 'exam-style' questions, with marks allocated. These are clearly identified. They are included in certain 'On your own' questions, and on the final page for each chapter. You will encounter questions like these in your exam, so they give you the chance of valuable exam practice.

The exam questions are based on Edexcel examination questions; they have not been taken from past exam papers, and they have not been provided by Edexcel.

Should the people of Kiribati sue the countries emitting the most greenhouse gases?

What do you think ?

There are also 'What do you think?' boxes. These ask questions about controversial issues, and will challenge your critical thinking. Questions of this type will not appear in exam papers.

Answers

The *AS Geography for Edexcel Activities & Planning OxBox CD-ROM* provides:
* answer guidance for the 'Over to you', 'On your own', and 'What do you think?' questions
* answer information and mark schemes for the exam questions.

Edexcel accepts no responsibility whatsoever for the accuracy or method of working in the answers given.

What do I have to know?

This topic is about how some parts of the world are at risk from hazards (e.g. volcanoes, earthquakes), and others from climate change. It has seven guiding questions.

1 Global hazards: What are the main types of physical risks facing the world and how big a threat are they?

What you need to learn
- Define – hazard and hazard types, disaster, hazard risk, disaster risk equation
- Why some people are vulnerable to hazards
- Why global warming is a major hazard

Although this is introductory and requires generalized knowledge, the examples used (e.g. Kiribati) can be used later, e.g. global warming.

2 Global hazard trends: How and why are natural hazards now becoming seen as an increasing global threat?

What you need to learn
- Magnitude and frequency of different hazards, and hazard trends
- The impacts of these upon lives, property, infrastructure and GDP
- Why some natural disasters are increasing

You need to know examples of why some hazards are increasing, and different trends (e.g. lives lost versus damage caused).

3 Global hazard patterns: Why are some places more hazardous and disaster-prone than others?

What you need to learn
- How to assess hazard risk in your local area.
- The global distribution of major natural hazards
- Disaster hotspots: the California coast and the Philippines

You need to know how to research hazard risk in your own area, how different hazards have different global distributions, and case study knowledge of California's coast and the Philippines.

4 Climate change and its causes: Is global warming a recent short-term phenomenon, or should it be seen as part of longer-term climate change?

What you need to learn
- Global warming is part of on-going climate change
- The causes of this may be natural and/or human
- Recent climate change is unprecedented

This section of the book is about process, not case studies.

5 The impacts of global warming: What are the impacts of climate change and why should we be concerned?

What you need to learn
- The impacts of climate change on the Arctic and Africa
- The potential impacts of a global sea level rise, and how these affect places disproportionately
- Why climate change is difficult to predict
- The 'tipping point' and its significance

This demands case study knowledge about why some places may suffer from more impacts of global warming than others.

6 Coping with climate change: What are the strategies for dealing with climate change?

What you need to learn
- Mitigation and adaptation strategies, with examples
- Why views about climate change differ, with examples
- How carbon footprints can be reduced
- Why global agreements to limit carbon emissions are difficult to reach
- How 'act local, think global' could affect climate change

This section demands detailed knowledge of examples to deal with climate change in different parts of the world.

7 The challenge of global hazards for the future: How should we tackle the global challenges of increasing risk and vulnerability in a more hazardous world?

What you need to learn
- How global warming can impact on other global issues, e.g. conflict, famine
- Some countries face a greater challenge from global warming; examples of strategies and possible solutions

This section requires knowledge of why some countries face a difficult future with global warming, together with how carbon emissions can be reduced.

1.1 What has happened to Tebua?

In this unit you'll investigate the risks facing some of the world's most remote island nations.

Gone for good?

Early in 2002, writer Curtis A. Moore went in search of Tebua, an island in the Pacific. Local legend said that it had existed for thousands of years. 'But now,' he wrote, 'I've been told it is gone – swallowed by the sea. Its fate, some say, was triggered by global warming – the unnatural increase in the Earth's temperature caused by air pollutants that trap solar heat.'

Tebua lay off Tarawa Atoll, on the extreme western edge of the nation of Kiribati, a group of islands stretching across the Pacific in an area nearly as wide as the USA. Kiribati consists of very low-lying sand and mangrove islands, one metre or less above sea level in most places. To visiting westerners, it seems like paradise. It covers 2.5 million square miles of ocean, an area 27 times that of the UK. But its land area is a mere speck – just 720 km², one quarter the size of London. Its population is only 92 500. Economically, Kiribati is even smaller. Its total GNP (Gross National Product) is equal to that of about 3000 average Americans. Its exports consist mostly of coconut flesh, used to make soaps and oils.

Kiribati is disappearing. Most of its population live on Tarawa, the most densely populated chain of small islands, surrounding a central lagoon. Its beaches are flat and barely 500 metres wide, and have become so eroded by storms that sand has been imported from Australia to maintain them. Many families have already moved, dismantling their wooden homes and rebuilding them further inland. Gradually people are being squeezed into a narrow strip of land between the lagoon and the Pacific. More people each year are leaving, becoming the world's first **environmental refugees**.

● **Environmental refugees** are people forced to migrate as a result of changes to the environment.

Tarawa Atoll, Kiribati ▲▶

Kiribati's nearest large neighbours are Australia and New Zealand ▶

Where will Kiribati's population go?

Kiribati's president, Anote Tong, warned Australia and New Zealand – the two developed countries in the region – to prepare for a mass exodus within the next decade. Speaking at the annual South Pacific Forum in Fiji, Mr Tong said that rising sea levels would create countless environmental refugees. 'If we are talking about our island states submerging in 10 years' time, we simply have to find somewhere else to go,' he said. He warned of a flood of refugees – pointing to increased levels of migration already from South Pacific countries. About 17 000 islanders have applied for residence in New Zealand in the past two years, compared with 4000 in 2003. While New Zealand has been generous so far, and already has a sizeable Pacific population, accepting large numbers of refugees could be a political problem.

Adapted from the *Independent*, October 2006

Why did Tebua disappear?

Tebua's disappearance is one of many signs that global warming is occurring. Researchers have been predicting for years that rising sea levels would soon cover the islands of the South Pacific. Since 1920, temperatures in the southwest Pacific have risen by about 1 °C, although the world as a whole has warmed by 0.6 °C in the same period. Including Australia and New Zealand, nearly 30 countries stretch over 20 000 km^2 across the Pacific. Five of these have warmed by more than 1 °C in the last century. In one, Micronesia, sea level has risen 21.4 mm every year since 2001. Increasing numbers of tropical storms in the area have created a region that suffers from **multiple hazards**, including the threat of tsunami.

Global warming is happening, although whether it is natural, or caused by the effects of human activity and pollution, is still being debated. Many scientists believe that increased temperatures and the melting of the ice caps are being caused by emissions of **greenhouse gases** from the world's wealthiest countries. One country alone – the USA – creates 25% of global greenhouse gas emissions. The problem is that the worst effects of rising sea level will be felt by the world's poorest countries, which make only the tiniest contribution to global warming and are the least equipped to cope with it.

> ● **Greenhouse gases** are those gases which are said to retain heat within the Earth's atmosphere and contribute to global warming, e.g. carbon dioxide, nitrous oxide and methane.

As if global warming wasn't enough …

Not only is Kiribati suffering from the impacts of sea level rise, but it has also been subject to many tsunami warnings; there were two major alerts in the first three months of 2007:

- On 13 January 2007, an earthquake off the coast of Japan, which measured 8.2 on the Richter scale, triggered a tsunami alert in Kiribati – 3000 miles away.
- On 2 April 2007, an earthquake off the Solomon Islands, also measuring 8.2, triggered a further alert.

Although neither of these actually caused damage in Kiribati, they show that places like Kiribati are at risk from multiple hazards.

> *Should the people of Kiribati sue the countries emitting the most greenhouse gases?*

● Over to you

1. Using the World Bank website (www.worldbank.org), compile a data file showing Kiribati's economy (e.g. GNP, exports) and population data (e.g. birth and death rates, infant mortality).
2. In pairs, decide whether or not it matters if Kiribati survives as a nation. Present your ideas to the class.
3. How willing do you think Australia and New Zealand might be to accept environmental refugees from Kiribati, and why?

● On your own

4. Define these terms from the text: global warming, environmental refugees, multiple hazards, greenhouse gases
5. Using Google, or similar, research recent hazard threats in Kiribati.
6. How far is Kiribati a nation facing 'multiple hazards'?

What do you think?

1.2 Hazards and disasters

In this unit you'll find out what hazards and disasters are and how they affect people.

What are hazards?

Every year, many events happen around the world which may be described as **natural hazards**. The box on the right lists ten natural events, but which of them are hazards? For a natural event to become a *hazard*, it has to involve people. It is the way people, social systems and environments link together that determines whether an event becomes a hazard. What may be just a natural event in an uninhabited location may be a severe hazard somewhere else if people are involved.

Background

Two types of hazard

The hazards in this chapter are of two types:

A **Hydro-meteorological** – those caused by running water and its processes (hydro) and those associated with or caused by weather patterns (meteorological). They include:

- floods, debris and mud flows
- hurricanes (also known as tropical cyclones), coastal storm surges, thunder and hailstorms, rain and wind storms (including tornadoes), blizzards and other severe storms
- drought, bushfires, temperature extremes, sand and dust storms.

B **Geophysical** – those caused by earth processes. These are of two types:

- internal earth processes of **tectonic** origin, e.g. earthquakes, tsunami, and volcanic activity
- external earth processes of **geomorphological** origin involving mass movements, e.g. landslides, rockslides, rock falls.

These types sometimes overlap, e.g. a snow avalanche may be hydro-meteorological in origin, but geophysical as an event.

Can global warming be seen as a natural hazard?

What do you think ?

- A **natural hazard** is a natural event or process which affects people, e.g. causing loss of life or injury, economic damage, disruption to people's lives or environmental degradation.

10 hazardous events

1. A cyclone affecting Hong Kong
2. A hurricane passing over a remote unpopulated island
3. A flood in a rural area which floods the roads, but does not affect any houses
4. A volcano erupting on a remote unpopulated island
5. An avalanche in a ski resort
6. An avalanche high on mountainous slopes remote from any settlement
7. A tsunami wave 50 centimetres high off the coast of Japan
8. An earthquake in Kashmir, northern Pakistan
9. A drought in Australia's outback
10. A drought in the south-east of England

▼ *An avalanche – but is it a hazard?*

What are disasters?

When does a **natural hazard** become a **disaster**? Dregg's model of defining disasters shows how some kind of overlap is required before a hazard becomes a disaster. A disaster is a matter of scale; it is simply bigger than a natural hazard. However, it is difficult to define precisely. Insurance companies – who do a lot of research into global hazards – attempt to define disasters. In 1990, Swiss Re defined a disaster as an event in which at least 20 people died, or insured damage of over US$16 million value was caused. But values and currencies change – so would this definition be appropriate now, or should it be changed?

Disasters and vulnerable populations

Whether a hazard becomes a disaster or not can depend on how **vulnerable** the people who are exposed to it are. An increasing proportion of the world's population lives in areas which are exposed to hazards. Examples include:

- people in Bangladesh who are threatened by floods and cyclones
- people who live on steep slopes where landslides may be common, such as the favelas (shanty towns) in many Brazilian cities.

The greater the scale of a natural hazard, and the more exposed people are, the greater a disaster is likely to be.

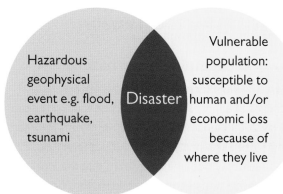

▲ *Dregg's model of defining disasters. M. Dregg is a geographer who has written a lot about the study of hazards.*

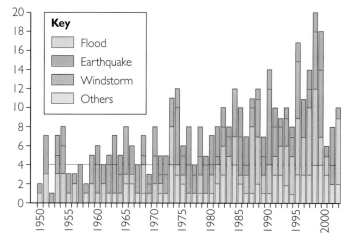

▲ *Number of major natural disasters causing over 1000 deaths and/or economic damage costing US$ 1 billion (using 2002 figures), 1950-2002*

Who studies hazards?

People study hazards from different perspectives, such as:

- scientists, e.g. geomorphologists (who study landform processes), geologists (rocks) and hydrologists (water)
- those who study societies, e.g. economists, sociologists, and psychologists
- applied scientists, e.g. civil, structural and hydrological engineers
- private companies, e.g. insurers
- public bodies, e.g. national and local governments
- cultural organizations and individuals, e.g. writers, photographers and even musicians.

It seems as though most people and organisations have an interest! Geography's unique role is to bring together – or synthesise – information from different studies to provide an overview of the issues.

● Over to you

1 Classify the ten listed events on page 8 into those which are hazards and those that are not. Explain your reasoning briefly in each case.
2 Now draw a table to classify these ten events into those that are hydrological, meteorological, geomorphological, and tectonic.
3 Using the graph above, explain which type of disaster is most frequent, and suggest why.
4 From your own knowledge and research, name three examples of events in each disaster type.

● On your own

5 Define the terms from the text: natural hazard, disaster, hydro-meteorological hazards, hydrological, meteorological, geophysical hazards, tectonic, geomorphological, vulnerable populations.
6 Explain, using examples, whether you think hazards are really 'natural'.

In this unit you'll learn how hazards increasingly threaten people's lives and property.

Is the world becoming more hazardous?

The media is always keen to report on disasters. As a result, images of events such as Hurricane Katrina, or the Boxing Day 2004 tsunami, stay in people's minds for years. Watching TV reports on the USA's hurricane season, or Bangladesh's latest flood, makes it seem that the world is a hazardous place – and becoming more so. But, how far is this actually true?

The two graphs on the right show that the number of reported natural disasters rose sharply between 1930 and 2006, particularly floods and windstorms. You might expect that this would lead to greater casualties and economic losses. However, many disaster events can now be predicted much more accurately, because of satellite imagery and increased hazard awareness. As a result, early prediction can reduce the effects of disasters – disaster deaths have fallen, although financial damage has rocketed (see the graphs below). Single disasters can dramatically affect these figures – the Kobe earthquake (1995) and Hurricane Katrina (2005) are the two biggest disasters economically, while the 2004 Asian tsunami killed far more people.

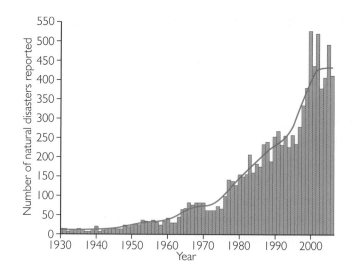

▲ Global trends in the number of natural disasters reported, 1930-2006

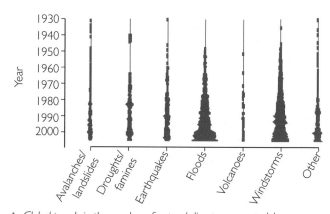

▲ Global trends in the number of natural disasters reported, by type, 1930-2006

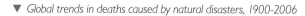

▼ Global trends in deaths caused by natural disasters, 1900-2006

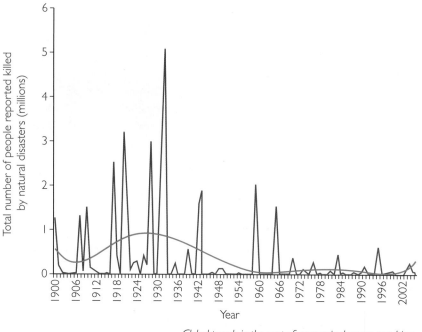

Global trends in the cost of economic damage resulting from natural disasters, 1950-2006 ▶

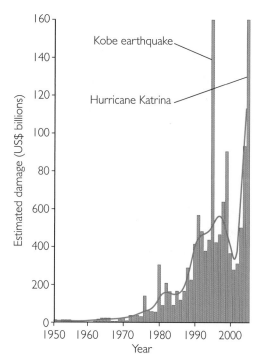

The effects of natural disasters on people

The number of people affected by natural disasters has risen sharply. The rise in global population, from just over 1 billion in 1900 to about 6.5 billion in 2007 has meant that some areas which were once uninhabited are now being developed and built up. To put it simply, more people are living in hazardous places where they might be exposed to natural disasters. Examples include:

- coastal cities in which harbourside or beachside houses and flats are exposed to hurricanes
- housing developments on river flood plains which, by their nature, are likely to flood.

Who suffers most from disasters?

There are some general trends, but the impact of disasters varies considerably between regions of the world and between countries. The two world maps below show the number of natural disasters by country between 1976 and 2005, and the numbers of people affected in each country per 100 000 population. Study the maps carefully and consider how far they show patterns that you would (a) expect and (b) not expect at all. You might get some idea about why these areas are affected so much by studying the graph on the next page, which shows the most common types of natural disaster. Globally, floods and windstorms (which include hurricanes, cyclones, tornadoes and storms) account for 75% of natural disasters.

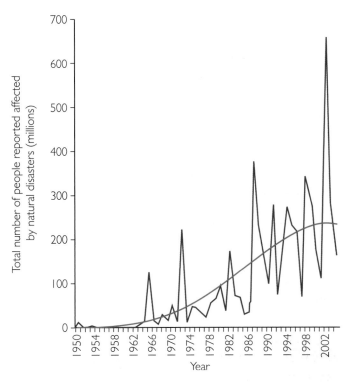

▲ Global trends in the number of people affected by natural disasters, 1950-2006

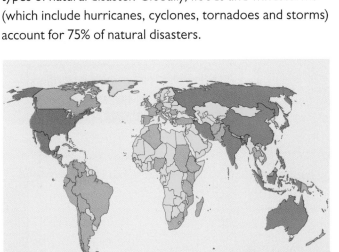

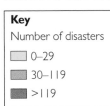

Key
Number of disasters

- ☐ 0–29
- ☐ 30–119
- ☐ >119

▲ The number of natural disasters by country, 1976-2005

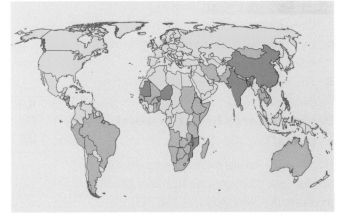

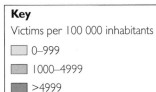

Key
Victims per 100 000 inhabitants

- ☐ 0–999
- ☐ 1000–4999
- ☐ >4999

▲ The number of people in each country affected by natural disasters, per 100 000 population, 1976-2005

Global hazards - 2

Investigating the world's worst hazards

Which are the world's worst hazards? The answer depends on the year. 2004-5 was bad for disasters, beginning just before the New Year with the 2004 Boxing Day tsunami. Between December 2004 and December 2005, 300 000 deaths occurred globally – one of the worst 12 months on record (it compared to 40 000 in 2003-4, and 21 000 in 2005-6). On average, 77 000 people were killed annually in natural disasters between 2000 and 2005, although that figure is lower if the 2004 tsunami is removed from the calculation. Care should be taken when comparing years directly, because the total number of deaths can be hugely affected by a single major disaster like this. Most deaths from disasters occur in Asia.

Which hazards are most common?

The bar chart below shows the total numbers of the most common natural disasters which occurred in 2006, and the average yearly numbers for 2000-2005.

- Floods were easily the most common event.
- Windstorms (e.g. hurricanes) were second.

Together these accounted for nearly three-quarters of all disasters globally.

- Drought and extreme temperatures combined were third. Their causes and impacts are different, but they are created by similar weather conditions.
- Earthquakes were fourth, but the numbers were much lower than the top three events.
- Landslides (including avalanches) were relatively few in fifth place.
- Volcanoes were lower still in sixth place.

Which hazards have the worst impacts?

Floods and windstorms may be greatest in number, but do they cause the most deaths or create the most damage? The data are complex but patterns do stand out:

- Earthquakes cause occasional major damage, but there is no upward trend.
- Damaging floods are increasing, but not consistently so.
- Damaging windstorms are also increasing, though again not always consistently.

2005

Type of disaster	Month	Main location	Deaths
Earthquake	October	Pakistan	73 338
Hurricane Stan	October	Guatemala	1513
Hurricane Katrina	August	USA	1322
Earthquake	October	India	1309
Flood	July	India	1200
Earthquake	March	Indonesia	915
Flood	June	China	771
Earthquake	February	Iran	612

2006

Type of disaster	Month	Main location	Deaths
Earthquake	May	Indonesia	5778
Typhoon Durian	December	Philippines	1399
Landslide	February	Philippines	1112
Heat wave	July	Netherlands	1000
Heat wave	July	Belgium	940
Typhoon Bilis	July	China	820
Tsunami	July	Indonesia	802
Cold wave	January	Ukraine	801
Flash flood	August	Ethiopia	498
Typhoon Samoai	August	China	373

▲ The worst natural disasters by number of deaths, 2005 and 2006

▼ The total numbers of the most common natural disasters which occurred in 2006, and the average yearly numbers for 2000-2005

Is the world becoming more hazardous?

What do you think?

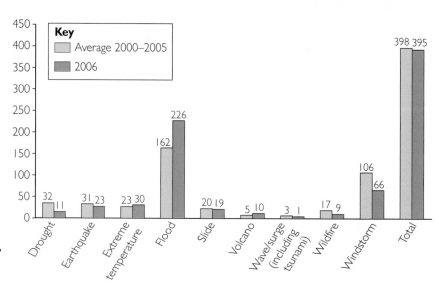

Key
- Average 2000–2005
- 2006

Why are floods and windstorms increasing?

The media almost always say that it is due to global warming. The theory is that the:

● increased warming of the earth causes warm air to rise, creating convection cells
   – which form hurricanes

● increasing temperatures increase evaporation, which in turn leads to increasing rainfall
   – and therefore greater flooding.

Or is it part of a natural cycle? Research shows that the Atlantic Ocean – where many windstorms begin – appears to work in a cycle of peaks and troughs, as the graph on the right demonstrates. The period around 1930–1935 showed increased storm activity, with major falls and increases occurring in cycles since then. So, although there has been an increase in hurricane activity since the mid 1990s, there were previous increases in the 1950s and 1970s.

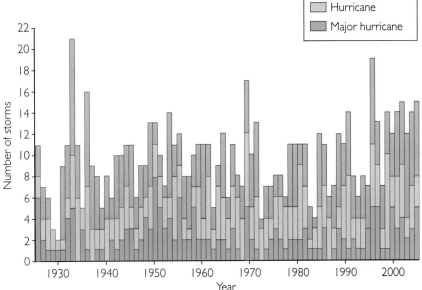

Key
- Tropical storm
- Hurricane
- Major hurricane

Trends in the numbers of Atlantic storms occurring since 1925 ▶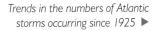

● Over to you

1 Use the four graphs on page 10 to outline the trends in global disasters.
2 Suggest why the number of disasters has increased, but the number of deaths has fallen.
3 Why should **a** costs of disasters and **b** numbers of people affected have risen sharply?
4 Study the two world maps on page 11 and complete the table at the bottom.
5 Suggest why Asia has the greatest number of deaths from natural disasters.

● On your own

6 Research the following hazards web sites to update recent disaster trends:
   - CRED – (Centre for Research on the Epidemiology of Disasters)
   - The United Nations International Strategy for Disaster Reduction (UNISDR).
7 Survey the most recent natural disasters in terms of **a** location of the worst events, **b** deaths and people affected, **c** value of damage caused, and **d** which types of hazards cause the most damage.
8 Write 500 words about the ways in which your findings are **a** similar to and **b** different from this chapter.

Pattern	Examples of countries and hazards that they face	Explanation for this pattern
The number of disasters is high and the number of victims per 100 000 population is also high		
The number of disasters is high but the number of victims per 100 000 population is low		
The number of disasters is low but the number of victims per 100 000 population is high		

Hazards and vulnerability - 1

In this unit you'll look at the idea of vulnerability and why hazards affect people and places in such different ways.

How significant are natural hazards?

There are no data for deaths from hazard events globally, only for those events which are large enough to be called disasters. Although numbers vary considerably from year to year, on average fewer than 100 000 deaths are recorded each year from natural disasters worldwide. This is:

- 30 times fewer than the number who die from HIV/AIDS
- 35 times fewer than the number of road deaths
- 50 times fewer than the number of smoking-related deaths.

Perhaps we are too concerned about hazards and disasters and the threats they represent.

However, what makes disasters interesting for geographers is that their impacts vary enormously between places. Asia dominates by far. This suggests that some places are more vulnerable than others to natural disasters.

People at risk – hazard vulnerability

While disasters may result from natural processes, they vary hugely in the ways in which they affect people – some people are more vulnerable to their effects than others. In 1976, an earthquake occurred near Guatemala City. Upper- and middle-class areas of the city were unaffected. Slum areas, by contrast, were destroyed. In total, 22 000 people died, almost all of them urban poor. This illustrates the concept of a 'class-quake', i.e. that hazards can be unequal in the way that they affect people.

Richer countries have vulnerable populations too …

In 1995, the city of Kobe suffered Japan's worst earthquake for 70 years. By global standards, its death toll was low – about 5500 – but it showed how vulnerable some of Japan's people were to hazards, as the panel on the right explains.

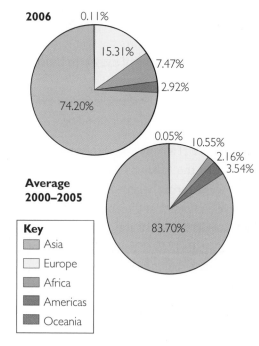

2006

0.11%
15.31%
7.47%
2.92%
74.20%

0.05% 10.55%
2.16%
3.54%
83.70%

Average 2000–2005

Key
- Asia
- Europe
- Africa
- Americas
- Oceania

▲ The percentage of people killed by natural disasters by continent in 2006, and as an average for 2000-2005 inclusive

---

Nagata is the home of the ***burakumin***, Japan's untouchable caste. It is packed with grimy rubber factories and iron works, and the houses are almost all old, timber and wattle constructions. Of 10 000 homes destroyed, over 3000 were burned to the ground. Nishinomiya was also an area of desolation. Few buildings were not a victim of collapse or fire.

A. Norton, writing about the Kobe earthquake

● Over to you

1. In pairs, research one of the pairs of earthquakes opposite to find out how and why the impacts were so different. Produce a display or PowerPoint to illustrate your findings.
2. How far does your research illustrate the disaster risk equation?
3. In class, discuss whether disasters are ever natural, or whether they are almost always human.

● On your own

4. Define these terms in your own words: hazard vulnerability, the disaster risk equation.

Exam question: Why do similar kinds of hazards have different impacts in different places? Use examples in your answer. (10 marks)

Similar hazards have different impacts in different places. Does it matter?

What do you think?

Comparing four earthquakes

Location	Armenia, 1988	San Francisco, 1989
Earthquake strength	Both 6.9 on Richter scale	
Number killed	25 000	63
Number injured	300 000	3500

Location	Bam (Iran), 2003	Central California, 2003
Earthquake strength	Both 6.5 on Richter scale	
Number killed	26 000	2 killed
Number injured	250 000	very few

▲ The Bam earthquake in Iran destroyed the city

▲ The central California earthquake in the USA destroyed several buildings

Look at the two tables above, which contrast four earthquakes. Each table illustrates how, when disasters occur, the effects are most felt by vulnerable people. The people of Armenia and Iran were more vulnerable than those in California, because less money was available to spend in those countries on protecting people from poorly constructed collapsing buildings – the biggest cause of death.

Hazard researchers measure **hazard vulnerability**, using this disaster risk formula:

$$\text{Disaster risk (D)} = \frac{\text{Hazard (H) x Vulnerability (V)}}{\text{Capacity (C)}}$$

For any hazard event, its impacts are the result of people's vulnerability, not the strength of the hazard itself. Vulnerability is based on 3 factors:

- Vulnerable people do not live in dangerous places because they want to; they do so because they have to. Land may be cheaper there, or unwanted by others.
- Vulnerable people cannot afford to build well, even if building regulations are enforced. In Kashmir's 2005 earthquake, lack of building regulations was not the cause of building collapse, but too few inspectors were employed to enforce them. Buildings that did comply – in the wealthiest areas – were less affected.
- Rapid urbanisation has forced the poor into high-risk areas of cities, particularly in developing countries.

● **Hazard vulnerability** is the capacity of a person or group to anticipate, cope with, resist and recover from the impact of a natural hazard.

▲ A shanty town in Mumbai, India, during the monsoon. Rising land values have forced the poorest people into high-risk environments like this.

Hazards and vulnerability - 2

The Boxing Day tsunami, 2004

The impacts of the 2004 Boxing Day tsunami

3 India

The south-east coast of the mainland, especially Tamil Nadu, was worst affected. Up to 140 000 people were displaced. In the Andaman and Nicobar Islands, salt water contaminated freshwater sources and destroyed arable land. Most of the Islands' jetties were also destroyed.

Countries affected by the 2004 Asian tsunami ▼

5 Somalia

This was the worst-hit African country, with damage concentrated in the tip of the Horn of Africa. Homes and boats were destroyed and freshwater wells and reservoirs were contaminated. Up to 30 000 people were displaced.

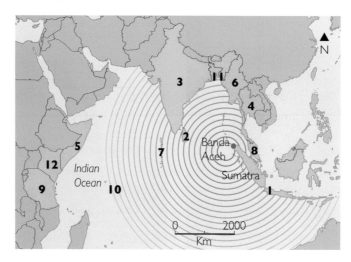

4 Thailand

The west coast was severely hit, including islands and tourist resorts near Phuket, so the death toll included 1700 foreigners from 36 countries.

1 Indonesia

Western Sumatra, the closest inhabited area to the earthquake's epicentre, was devastated by the tsunami. Up to 70% of some coastal populations were killed or missing. Up to 400 000 people were displaced.

12 Kenya

Kenya was one of the few affected countries to be warned and to take some action, so many people were able to escape the coastal areas as a result – and only one person was killed.

7 Maldives

Of the 199 inhabited islands, 20 were destroyed. The shallowness of the water limited the tsunami's destructive power but flooding was extensive. A sea wall protecting the capital Male prevented half the city being destroyed. Many luxury tourist resorts were badly damaged – affecting the economy of the country, which depends on tourism.

2 Sri Lanka

The second worst affected country – the southern and eastern coastlines were ravaged, with homes, crops and fishing boast destroyed. 400 000 people lost their jobs.

The distribution of dead and missing ▶

	Country affected	Dead and missing
1	Indonesia	236 169
2	Sri Lanka	31 147
3	India*	16 513
4	Thailand	5395
5	Somalia	150
6	Burma	61
7	Maldives	82
8	Malaysia	68
9	Tanzania	10
10	Seychelles	3
11	Bangladesh	2
12	Kenya	1
	Total	**289 601**

* including the Andaman and Nicobar Islands

The tsunami horror

Most people remember when they first saw film of the 2004 Boxing Day tsunami, showing holidaymakers and locals watching the sea retreating an abnormal distance – and then its return towards the shore as a wall of water – realizing only at a late stage how huge the returning waves were. It rates as one of the world's worst disasters. Its scale was rare – of a kind that occurs about once in every 100 years.

Tsunami occur where:
- earthquakes measure more than 6.5 on the Richter scale
- the earthquake's focus is shallow beneath the Earth's surface
- the focus is also beneath the ocean.

The earthquake that caused the Boxing Day tsunami was estimated at between 9.0 and 9.3 on the Richter scale, and was over 100 times stronger than the one which caused the Kobe earthquake in 1995. The thrust heaved the floor of the Indian Ocean towards Indonesia by about 15 metres, and, in so doing, sent out shock waves. Once started, these radiated out in a series of 'ripples', moving almost unnoticed across oceans until they hit land. The longer and shallower the coastal approach, the more the ripples built up height. The waves that struck the shallow coastline near Banda Aceh (only 15 minutes from their origin), and parts of Sri Lanka, were nearly 17 metres high on impact. By contrast, islands in the Maldives experienced a four-metre high sea swell, rather than a crashing wall of water.

▼ *The tsunami wall of water hitting Ao Nang, Thailand*

Sri Lanka – who died in the 2004 tsunami?

Sri Lanka was the second most seriously affected country after Indonesia, with over 30 000 deaths, 5700 people missing, and 861 000 people displaced.

One survey carried out in March 2005 in part of Ampara (an eastern coastal district of Sri Lanka), by Nagasaki University of Japan, found that the most vulnerable people had suffered the most. This area had previously experienced rapid coastal urbanisation. Its economy was also based on tourism and subsistence fishing, which left it vulnerable to the tsunami.

In this part of Ampara, out of a population of 3533 (living in 859 households), 12.9% died. Of these:

- most deaths occurred during and immediately after the disaster
- more than double the number of women died, compared to men
- 56% of victims were children
- the elderly and disabled were more likely to die than young, healthy adults; 15% of deaths were of people aged over 50.

Other factors which increased people's vulnerability were:

- whether they were indoors at the time of the tsunami (13.8% of casualties). Women and children were more likely to be inside on the morning of the tsunami. Even compared with those on the beach or in the sea, people at home were more likely to die.
- the quality of the building they were in, either in terms of its structure or its location and exposure to the full force of the waves. 14% of deaths occurred in buildings which were completely destroyed, compared to 5% which occurred in buildings that held up well or withstood the initial impact.

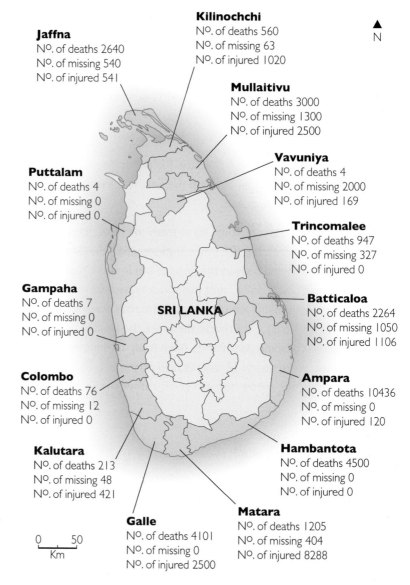

N

Jaffna
No. of deaths 2640
No. of missing 540
No. of injured 541

Kilinochchi
No. of deaths 560
No. of missing 63
No. of injured 1020

Mullaitivu
No. of deaths 3000
No. of missing 1300
No. of injured 2500

Puttalam
No. of deaths 4
No. of missing 0
No. of injured 0

Vavuniya
No. of deaths 4
No. of missing 2000
No. of injured 169

Trincomalee
No. of deaths 947
No. of missing 327
No. of injured 0

Gampaha
No. of deaths 7
No. of missing 0
No. of injured 0

SRI LANKA

Batticaloa
No. of deaths 2264
No. of missing 1050
No. of injured 1106

Colombo
No. of deaths 76
No. of missing 12
No. of injured 0

Ampara
No. of deaths 10436
No. of missing 0
No. of injured 120

Kalutara
No. of deaths 213
No. of missing 48
No. of injured 421

Hambantota
No. of deaths 4500
No. of missing 0
No. of injured 0

Galle
No. of deaths 4101
No. of missing 0
No. of injured 2500

Matara
No. of deaths 1205
No. of missing 404
No. of injured 8288

0   50
Km

▲ *The number of people affected by the 2004 tsunami in Sri Lanka*

▼ *A couple stand in the remains of their home in Ampara, Sri Lanka. They have lost everything they own to the 17-metre high tsunami waves*

- whether they belonged to a fishing family (15% of deaths).
- whether they had lower educational qualifications. Those with higher educational qualifications were 20% less likely to die if educated to secondary level, and 60% less likely if educated to university level. University educated people earn more and could afford to live away from high-risk locations.
- whether they earned lower incomes. In Ampara, 15 000 rupees (US$150) per month is a high wage. Most deaths occurred in households earning 1–2999 and 3000–5999 rupees, with few deaths in the highest earnings category.

▲ *Replacement housing on the coast of Ampara nearly three years after the 2004 tsunami*

How do other disasters compare?

A similar mortality pattern has been reported in other disasters to that found in the tsunami.

- In both the 1999 Taiwan and the 1988 Armenian earthquakes, there was a high mortality rate among women and young children.
- In the 1995 Kobe earthquake in Japan, a high mortality rate was reported among the elderly, especially those with physical disabilities.

However, previous studies on floods have revealed different mortality patterns, and there seems to be vulnerable groups unique to each type of disaster.

- In a study of flood-related deaths in Europe and the USA, middle-aged men were found to be most vulnerable.
- An Australian study reported that 80% of flood deaths were men. Risky behaviour, such as trying to swim across rivers or using cars to flee, caused increased mortality.

Environmental change and the tsunami

One clear factor has emerged from several countries affected by the tsunami – the countries which suffered the most were those where the tourism industry has grown rapidly in recent years. Many coastal areas of Thailand and Sri Lanka have been cleared of coastal mangrove swamps to make way for hotels and resorts. Mangroves act as a natural barrier, absorbing wave power and creating a natural coastal buffer zone. Damage from the tsunami was noticeably reduced in coastal areas which had maintained their mangrove swamps, beach forest and coral reefs.

Were some of the impacts of the 2004 tsunami avoidable?

What do you think?

● Over to you

1. Use the map to describe and explain the pattern of those affected by the tsunami in Sri Lanka.
2. Why was the east coast of Sri Lanka so much more vulnerable than the west?
3. Explain why the following groups were so much more vulnerable to the effects of the tsunami than the population as a whole:
   - women
   - children, the elderly and disabled
   - those who were in buildings
   - those on low incomes and those in fishing occupations
   - those with low educational qualifications.
4. In pairs, devise a plan to educate coastal inhabitants about tsunami, so that future disasters can be avoided.
5. How would you persuade Sri Lanka's government of the need to maintain – and not clear – its mangrove forests?

● On your own

6. Research how mangrove clearance affected the impacts of the 2004 tsunami in Thailand.
7. How and why is education as important as tsunami early warning systems in preventing similar impacts in future?
8. Write a 500-word essay on 'Those who suffer most in disasters are the most vulnerable'.

1.5 Bushfire!

In this unit you'll find out how different strategies reduce some of the impacts of bushfires in Australia.

Australia's bushfires

Bushfires are part of Australian summers. Every summer, warnings appear in Australia's national parks, and stories in the media show the effects of bushfires from previous years to emphasise their seriousness. While bushfires cause significant property damage, with an average of 84 homes destroyed every year, there are far fewer deaths now – averaging only 5 each year.

▲ *A bushfire raging in the Blue Mountains near Sydney in November 2006*

Background

How do bushfires start?

Bushfires always attract attention because people think they are started deliberately. This is true in less than 10% of cases. The most frequent causes are:

- carelessness, e.g. an outdoor barbecue or cigarette end
- lightning strikes.

However, once started, bushfires can be difficult to stop. They spread in one of three ways:

- Ground fires – dry leaf litter and twigs catch light and the fire spreads through the undergrowth.
- Crown fires – where fire spreads through the treetops. These are common in Australia's eucalypt (or 'gum') trees, the resin of which is highly inflammable. This can actually benefit the tree; many species of gum tree regenerate quickly once fire has been through. Some produce seeds which only crack and germinate after a bushfire.
- Spot fires – caused by burning embers landing away from the main fire and starting new fires in their own right.

▼ *How fire spreads. Note the three different ways – ground fires, crown fires and spot fires.*

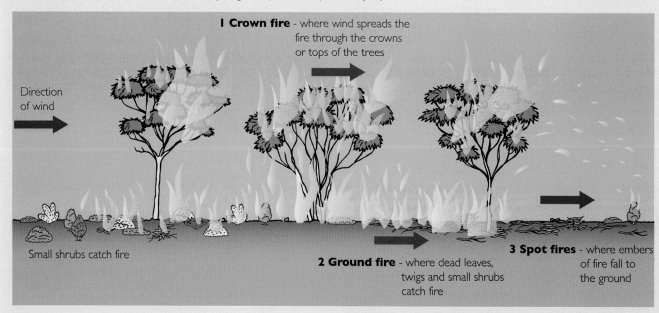

Direction of wind

I Crown fire - where wind spreads the fire through the crowns or tops of the trees

Small shrubs catch fire

2 Ground fire - where dead leaves, twigs and small shrubs catch fire

3 Spot fires - where embers of fire fall to the ground

How have bushfire deaths been reduced?

The number of bushfire deaths has fallen steadily. The 'Ash Wednesday' bushfires in 1983 killed 75 people and covered 400 000 hectares. Yet in the bushfires of 2006-7, only four people were killed. This follows a pattern in which even the worst bushfires, covering twice the area of those in 1983, kill only a few people and, in some cases, none at all. How has this been achieved?

Two methods have been used to cut deaths and increase safety:

Controlled burning

Population increase in areas at risk of bushfires has given rise to a need for controlled burning. For instance there has been a steady flow of people moving out of Melbourne to smaller country towns outside the city. Macedon, 70 km to the north west of Melbourne, is one such town. The area contains Mount Macedon National Park, an area of outstanding scenic beauty. The area is almost entirely forested with eucalypt species.

Now, local councils minimise the fire risk by controlled burning – regular burning of leaf litter to reduce the fuel for the bushfires. This controlled burning is done every year.

> *Should native forest be left to regenerate, even if there is a fire risk to life and property?*
>
> **What do you think?**

Over to you

1 In pairs, produce a PowerPoint of images of Australian bushfires, using images from Google. Identify the areas affected, the causes, and the impacts.
2 Should people move to areas where bushfires are a threat? Or should there be planning controls to stop this? Debate this in class.

On your own

4 Conduct research into El Niño and how it affects bushfires in Australia. Find out **a** what El Niño is (Unit 3.11 will help with this), and **b** the link between El Niño and bushfires.
5 Consider what the implications of controlled burning are for **a** plant growth, **b** local councils. Draw a table showing what the benefits and problems of controlled burning might be.

Fire Officers inspect properties in bushland areas to assess the risk and advise residents about burning. However, this strategy has its opponents. Environmental groups claim that the frequent burning does not allow new seedlings to grow sufficiently tall and strong to survive a bushfire. Eventually, they say, the forests will decline and die out.

Education programmes

People survive when they stay with their house during a bushfire. Now, instead of mass evacuation, people are educated about what to do in a bushfire, knowing that evacuation is a last resort. Victoria's Fire Authorities issued a poster to households and communities to educate people about how to keep their properties at low risk from fire.

People can also install protective measures for themselves. In 1983, the Macedon area was devastated by the Ash Wednesday bushfires. The house pictured below was built recently from timber left after the fires. The owners of the house want safety from future bushfires and have adopted several forms of protection, such as the sprinkler system.

▼ *Fire protection installations on a house near Macedon, Victoria*

Why are some places hazard prone?

In this unit you'll assess hazard risks in your own area, using London as an example.

Tornadoes in the capital

We often think of hazards as things that happen somewhere else. Tornadoes are associated with dramatic weather conditions in the American Midwest. Yet there are over 70 tornado events in the UK every year. In December 2006, North London was hit by a tornado that was so localized that it affected just a few streets in Kensal Rise. In just a few minutes, roofs were ripped off houses, trees brought down, and cars damaged or crushed by falling debris. Amazingly – because it happened at lunchtime – no one was killed and there were a few minor injuries.

▲ *The North London tornado on 7 December 2006, ripped the side off this house in Kensal Rise*

Background

What causes the UK's storms and local tornadoes?

Tornadoes are uncommon in the UK, but the conditions that bring them are not. Storms are brought by depressions or low pressure areas, formed by circulating air around an area of low pressure. These draw together cold winds from polar regions to the north, and tropical winds from the south.

Tornadoes are more likely to form in the UK when cold air moves rapidly south over warmer land or sea, and when there is a large pressure difference over a small area, dragging the air in more quickly. As the air moves south, the following sequence of events occurs:

- As the air warms at the surface, it becomes lighter and starts to rise, making it unstable.
- This sets up small convection currents, which rise through the atmosphere, like gusts of wind.
- The Earth's rotation causes the rising air to spin.
- The faster the spin, the more the air is likely to 'touch down' on the Earth's surface. This forms the tornado.

▼ *The weather conditions that led to the North London tornado*

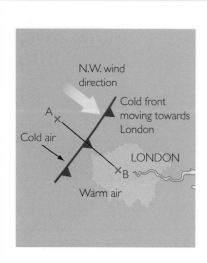

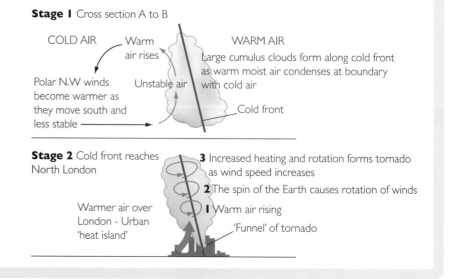

Assessing London's flood risk

Tornadoes apart, London's biggest hazard by far is flooding. The worst-affected areas are close to the Thames, but others are also at risk (see the map below). London's flooding has four main causes:

- Prolonged rainfall, when surface run-off flows over impermeable tarmac and concrete surfaces into drains, which lead to sewers and then to the river.
- Sudden storms. London's summer temperatures are the highest in the UK. Evaporation from the Thames causes huge amounts of moisture to be transported to the upper atmosphere – where it condenses and falls as rain, often in torrential thunderstorms. This causes a backlog of water in drains, which, in turn, causes flooding.
- **Storm surges**. These occur when tides are highest, bringing incoming tides up the River Thames to the point where they meet out-going river flow.
- Exceptionally low pressure air, reducing the 'weight' on the sea and causing it to rise above its normal level.

When these causes coincide, as happened in 1953, flooding can be extensive and damaging.

Looking to a hazardous future?

The map shows London's flood risk areas, although the risk has now been reduced by the Thames Flood Barrier. Its purpose is to prevent incoming tides from flooding the city when storm water may be flowing away. The Barrier was completed in 1982, but is now too small and needs replacing as sea levels continue to rise due to global warming. A replacement would be expensive – £30 billion or more – but the consequences of flooding a major global financial centre such as London (earning 40% of the UK's GDP) would be unthinkable.

▲ The Thames Flood Barrier, protecting the UK's financial heart

> *Does London deserve its own 'hazards' budget for events such as flooding?*
>
> **What do you think?**

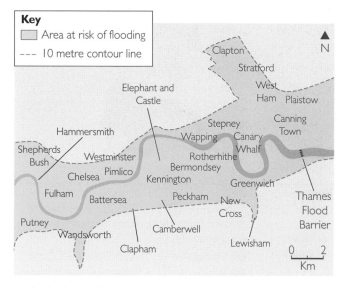

Key
- ☐ Area at risk of flooding
- --- 10 metre contour line

▲ Flood risk map of London

● Over to you

1 In groups, produce a report about hazard risks in your area. Work as follows:
   a Use websites such as www.environment-agency.gov.uk to assess the flood risk for local postcodes.
   b Report on short-term weather events, e.g. floods, fog.
   c Research geomorphic hazards, e.g. landslips, slumping.
   d Assess tectonic hazards, e.g. local land disturbances.
   e Assess any threats from global warming.
   f Prepare a hazards risk map for your area and assess the social, economic, and environmental impacts.

● On your own

2 Research the causes and impacts of the 1953 flood in London and eastern England. Does it prove that London needs special flood protection?

In this unit you'll find out about multiple hazards in two hazard hotspots: the Philippines and California.

Most countries face some kind of hazard, but six countries stand out as being the most hazard-prone in the world: the Philippines, Japan, India, Bangladesh, China and Indonesia. The first half of this unit focuses on the Philippines and how its exposure to a range of hazards makes it an extremely high-risk environment:

- It sits across a major plate boundary, so it faces significant risks from volcanoes and earthquakes.

- Its northern and eastern coasts face the Pacific, the world's most tsunami-prone ocean.
- It lies within South-East Asia's major typhoon belts. In most years, it is affected by 15 typhoons and struck by 5 or 6 of them.
- Landslides are common in mountain districts.

Pages 24-26 focus on how great a threat these hazards pose to the Philippines, their impact, and how well the country copes.

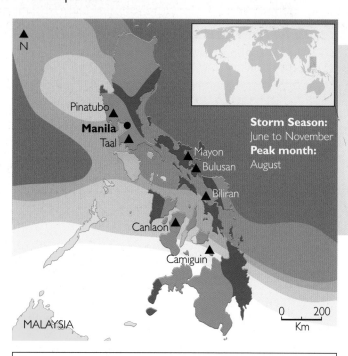

Storm Season: June to November
Peak month: August

Area: the Philippines consists of about 7000 islands, and is 25% bigger than the UK
Population: 91 million in 2007
Wealth: GDP in 2006 was US$5000 per capita; a middle-income country according to the World Bank
Landscape: mostly mountainous, with coastal lowlands; many people live and work on steeply sloping land

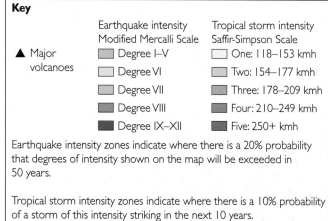

Key

▲ Major volcanoes

Earthquake intensity Modified Mercalli Scale	Tropical storm intensity Saffir-Simpson Scale
Degree I–V	One: 118–153 kmh
Degree VI	Two: 154–177 kmh
Degree VII	Three: 178–209 kmh
Degree VIII	Four: 210–249 kmh
Degree IX–XII	Five: 250+ kmh

Earthquake intensity zones indicate where there is a 20% probability that degrees of intensity shown on the map will be exceeded in 50 years.

Tropical storm intensity zones indicate where there is a 10% probability of a storm of this intensity striking in the next 10 years.

Mount Pinatubo's volcanic eruption in June 1991

Mount Pinatubo's eruption was the biggest the world had seen for over 50 years. The volcano showed signs of eruption in April 1991, with steam explosions and minor earthquakes. A 10-km exclusion zone was set up around Pinatubo by government advisers, who eventually extended the zone to 30 km – evacuating more people each time it was extended. Two weeks before the blast, they produced a video outlining the risks of **pyroclastic flows** and **lahars**.

By 9 June 1991, 58 000 people had been evacuated, reaching 200 000 by 12 June (when the first eruption sent a cloud of ash 20 km into the atmosphere, spreading over South-East Asia within three days). The second eruption, on 15 June, was cataclysmic; a dome on the side of the volcano collapsed, creating a pyroclastic blast and causing huge lahars. However, effective monitoring and management reduced Pinatubo's death and injury toll to just over 4300 people.

- 350 people died, including 77 in the lahars that occurred.
- Some evacuees died in camps, where they were exposed to disease.
- 80 000 hectares of farmland were buried beneath ash, disrupting the livelihoods of 500 000 farmers and their family members.
- Economic losses were US$710 million, mainly agriculture and property.

Background

Tectonic hazards

The Philippines lies on the boundary between two **tectonic plates**, the Philippine and Eurasian. The Eurasian Plate is forced beneath the Philippine, creating the deep Manila Ocean Trench, to the west. The plates move in a series of 'jerks', producing an earthquake each time they do so. This kind of plate boundary is known as a **destructive plate boundary**, because one plate is destroyed beneath another.

The most destructive volcanic eruptions occur along destructive plate boundaries. As the **subducted** Eurasian plate melts in the Earth's mantle, the molten rock – or **magma** – forms a **magma chamber**. Some magma becomes explosive, especially where it combines with gases within the chamber. Because these rocks have only just melted, they cool easily wherever they come close to the Earth's surface. Little lava is produced; instead, the magma solidifies just below the Earth's surface and produces a solid cap over the volcano's outlet or **vent**. Pressure grows beneath it so that, when a volcanic blast does happen, it does so with huge force.

▲ *Mount Pinatubo erupting in 1991, showing the massive ash cloud from the first eruption*

80% of the world's volcanoes occur along destructive plate boundaries. Their explosive, damaging nature brings three risks: the blast itself, pyroclastic flows, and lahars. Pyroclastic flows consist of intensely hot blasts of solid rock and ash, travelling at high speeds at temperatures of hundreds of degrees that burn and destroy on impact. Lahars are huge mudflows generated when lakes within the volcanic crater are released by the blast, or when the heat instantly melts snowfields around the peaks.

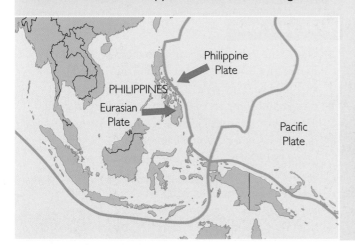

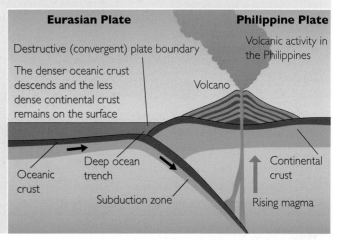

Hazard hotspots – the Philippines

Other hazard risks

Some hazard risks in the Philippines are complex because they have multiple effects. One earthquake in 2006:

- killed 15 people, injured 100 and damaged or destroyed 800 buildings
- generated a local tsunami 3 metres high
- triggered landslides which breached the crater wall of Parker Volcano, and then fell into Maughan Lake …
- … creating a flood which washed away houses.

The Guinsaugon landslide

Guinsaugon was a village in the central Philippines. In February 2006, a mudslide completely engulfed the village and its land, covering 3 km² and killing about 1150 people (see right). It was not unusual – a series of storms in December 2004 killed 1800 people in the north-eastern Philippines. In 2003, 200 people were killed in landslides. Typhoons and storms kill several thousand people there every decade.

How serious is each hazard risk? A summary of hazard impacts affecting the Philippines, 2000–2006 ▼

▲ *The Guinsaugon landslide in 2006. Filipino soldiers are recovering the dead. The hillside in the background shows where the landslide came from.*

The *physical causes:*

- There was unseasonable torrential rain; 2000mm of rain fell in 10 days in February – normally the dry season.
- La Niña – a cyclic ocean and wind current affecting South-East Asia – was probably the cause of the rainfall.
- A 2.6 magnitude earthquake struck just before the slide and may have triggered it.

The *human causes* included:

- deforestation of native forest cover protecting the soil. In 50 years, logging has reduced several million hectares of forest to about 600 000 today.
- replacement of native forest by shallow-rooted trees, such as coconuts, further reducing soil protection.

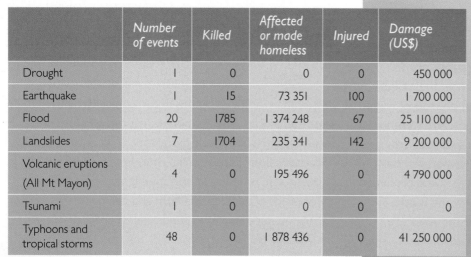

	Number of events	Killed	Affected or made homeless	Injured	Damage (US$)
Drought	1	0	0	0	450 000
Earthquake	1	15	73 351	100	1 700 000
Flood	20	1785	1 374 248	67	25 110 000
Landslides	7	1704	235 341	142	9 200 000
Volcanic eruptions (All Mt Mayon)	4	0	195 496	0	4 790 000
Tsunami	1	0	0	0	0
Typhoons and tropical storms	48	0	1 878 436	0	41 250 000

● Over to you

1. In pairs, **a** research a factfile of about 15 socio-economic indicators (e.g. GNP, literacy rates) for the Philippines from the World Bank website, **b** decide how developed the Philippines is.
2. Draw a Venn diagram with three circles overlapping, labelled 'social', 'economic' and 'environmental'. In pairs, classify the impacts of the Mount Pinatubo eruption.
3. Explain how successfully you think government agencies managed the threat of the eruption.
4. Study the hazard table and, in pairs, decide which is the most serious hazard that the Philippines faces, and why.

● On your own

5. Define the following words from the text: tectonic, destructive plate boundary, subducted, magma chamber, vent, pyroclastic flows, lahars.
6. Research the impact of two hazards in the Philippines, e.g. a typhoon and a tsunami. Classify these in a Venn diagram, as you did with Pinatubo.
7. Using the 2006 earthquake and Guinsaugon landslide as evidence, explain why it is difficult to classify the causes and effects of hazards.
8. Write 500 words to show whether damage done by hazards in the Philippines is mainly social or economic.

Hazard hotspots - California

Ever since 1849, when gold was discovered, California has been one of the most desirable places to live in the USA. It is wealthy; its economy is the world's 6th largest, bigger than France or Italy! 25 Californian counties have per capita incomes of over US$65 000 (about £35 000) per year, making them amongst the world's wealthiest places. Yet the risk map on the right, showing the likelihood of hazard occurrences, identifies California as the USA's most hazardous state. There are two reasons for this: plate tectonics and climatic patterns, particularly those related to El Niño and La Niña.

Plate tectonics and California

San Francisco seems like a city living on the brink of disaster. Its residents know that it lies along the San Andreas fault, where the Pacific Plate moves north-westwards past the North American Plate. The two move in the same direction but the Pacific Plate moves more quickly, thus creating friction. This is called a **conservative plate boundary**. The San Andreas fault is the fracture – or **fault line** – between them. It runs along the Californian coast from Los Angeles north to San Francisco. Other fault lines run parallel to the major fault; San Francisco is built over two of them.

These faults move regularly, causing earthquakes. In 1906, San Francisco was destroyed in an earthquake measuring 8.2 on the Richter Scale. It fractured gas pipes (which caused explosions and fires) and water mains (which could have prevented the spread of the fires). A further earthquake, of magnitude 7.1, occurred in 1989. With its epicentre at Loma Prieta, it caused major damage and deaths. Some buildings collapsed, while others were badly shaken. Five years later, a further earthquake shook Northridge in Los Angeles (see the next page for details of both earthquakes).

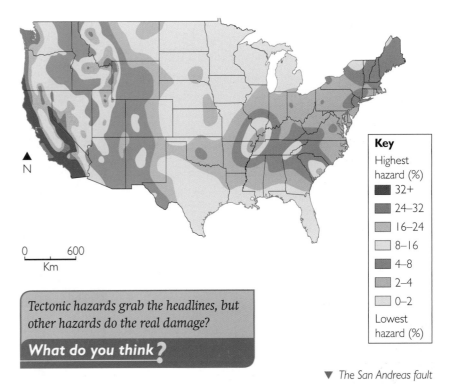

N

0   600
Km

Key

Highest hazard (%)

- 32+
- 24–32
- 16–24
- 8–16
- 4–8
- 2–4
- 0–2

Lowest hazard (%)

Tectonic hazards grab the headlines, but other hazards do the real damage?

What do you think?

▼ *The San Andreas fault*

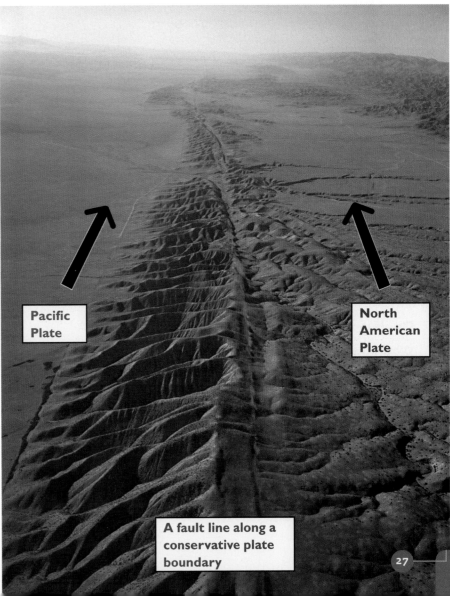

Pacific Plate

North American Plate

A fault line along a conservative plate boundary

Hazard hotspots - California

- Date and time: 5:04 pm, 17 October 1989
- Magnitude and location: 7.1; epicentre – Loma Prieta in the Santa Cruz mountains
- A magnitude 5.2 aftershock struck the region 37 minutes after the main earthquake
- 63 people died and 13 757 were injured (most were killed when a freeway collapsed)
- 1018 homes were destroyed and 23 408 damaged
- 366 businesses were destroyed and 3530 damaged
- The damage cost US$6 billion

The 1989 Loma Prieta earthquake in San Francisco ▶

▼ *The 1994 Northridge earthquake in Los Angeles*

- Date and time: 4:31 am, 17 January 1994
- Magnitude and location: 6.7; striking the densely populated San Fernando Valley in northern Los Angeles
- There were many thousands of aftershocks (mostly of magnitude 4.0–5.0) during the following weeks, causing further damage
- 57 people died and over 1500 were seriously injured
- 12 500 buildings were damaged; 25% suffered severe-to-moderate damage
- 9000 homes and businesses were without electricity for several days (20 000 without gas), and 48 500 people were without water
- There was damage to several freeways serving Los Angeles – choking traffic for 30 km

Dealing with earthquake threats

What if there was another major earthquake? The panels above show that in wealthier countries the economic damage can be great, whereas the impacts in developing economies tend to affect people. To protect themselves, most Californians insure their property against earthquake damage. After the Loma Prieta and Northridge earthquakes, demand for insurance rose sharply. But, by 1996, it had dropped to below 1989 levels, and has declined further since. Many people avoid the cost of taking out insurance. Insurers despair about this – asking what people would do in the event of another earthquake. In San Francisco, people say that no earthquake occurred between 1906 and 1989, so why bother with odds that low?

Why do wealthy countries suffer more economic damage from hazards, while poorer countries suffer more deaths and injuries

What do you think?

Climatic patterns and California

California has a reputation as a state where the weather is always perfect, but it suffers periodic changes which can be hazardous. Sometimes drought occurs and forest fires threaten, while at other times floods and landslides provide headline news. Flood risks in California vary, but they coincide with El Niño; forest fires and drought coincide with La Niña.

In Australia, bushfires (see page 20-21) coincide with El Niño. How does El Niño have such different effects? Read more about El Niño on page 163 to understand this better.

Background

El Niño is part of a cycle of approximately 7 years, in which 1-2 years occur as El Niño years and 1-2 occur as La Niña (see the table below). Each brings distinctive threats:

- During El Niño years, air currents move eastwards across the Pacific, bringing moist air to South America and the eastern Pacific. The Californian coast is also affected by this current, which brings torrential rains – causing flooding and landslides.
- During La Niña years, the current reverses to move across the Pacific towards Australia, bringing moist air to Australia and the western Pacific. During this time, warm dry air blows over California from the American deserts, bringing drought and the threat of forest fires to California.

El Niño years (1-2 years in 7)	La Niña years (1-2 years in 7)
• On-shore sub-tropical winds blow towards California from the Pacific Ocean • On-shore moist winds • Increased sub-tropical storms bring heavy rain • Increased landslides result from heavy rain	• Off-shore winds blow towards California from the deserts of Arizona and New Mexico • Dry warm air with little rain • Increased drought • Increased risk of bushfires (called wildfires in the USA)

◄ *Comparing the hazardous effects of El Niño and La Niña in California*

● Over to you

1 In pairs, **a** research a factfile of about 15 socio-economic indicators for California (e.g. average incomes, car ownership); **b** decide how developed California is.
2 Draw a Venn diagram, with 3 circles overlapping, labelled 'social', 'economic' and 'environmental'. Classify the impacts of the Loma Prieta and Northridge earthquakes.
3 How do these impacts compare with those of Pinatubo (see page 25)?
4 On a world map, draw different currents to show El Niño and La Niña, and annotate the map to show their impacts on California.

● On your own

5 Define the following words from the text: conservative plate boundary, fault line.
6 Research the impacts of two hazard events in California, e.g. a landslide and flooding. Classify these in a Venn diagram – social, economic and environmental.
7 In 750 words, compare **a** the causes and **b** the impacts of hazards in California and the Philippines.

Exam question: Choose either the Philippines or California and explain why it is considered a disaster hotspot. (10 marks)

In this unit you'll investigate the causes of climate change and global warming. Some argue that it's natural and long-term, while others blame recent human activities.

Is climate change natural or human induced?

The three warmest years on record have all occurred since 1998, and 19 of the warmest 20 years since 1980.

The problem with global warming is that everyone agrees that it is happening, but there is fierce debate about how and why.

- Not all areas of the world are warming at the same rate; rates vary between the two hemispheres.
- Some areas of the world have not warmed at all in recent decades, while other areas have.
- Industrial areas are not warming as fast as more remote areas!
- Are the changes natural or caused by human activity?

The natural greenhouse effect

The driving force for weather and climate is the sun and its **radiation**. The Earth and its atmosphere intercept this radiation – and about 30% is reflected back into space by clouds and by reflective surfaces such as snow. If this was the only factor at work, the Earth's surface would be 33 °C cooler than it is now and human life would be impossible. But some radiation is absorbed by the Earth's surface, and some is re-emitted by gases known as greenhouse gases, which warm the Earth's atmosphere by 33 °C. This process is called the **greenhouse effect**, and is a completely natural process.

▼ *The greenhouse effect*

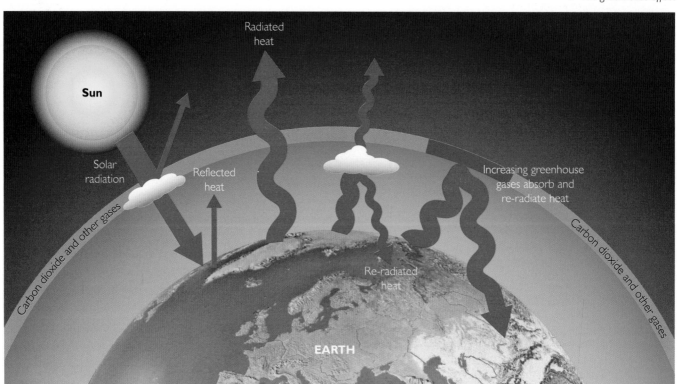

Greenhouse gases and their contribution to global warming

Carbon dioxide

- Although only making up 0.04% of the Earth's atmosphere, CO2 is a major greenhouse gas.
- It is given off when carbon-based fuels are burned, e.g. coal and oil. Fossil fuels have increased atmospheric CO2 by 25%.

Chlorofluorocarbons (CFCs)

- They were first used in the 1960s as propellants in spray cans, foam plastics, and refrigerant fluids.
- CFCs absorb solar radiation and contribute to global warming.
- In the 1980s, UK scientists discovered thinning in the ozone layer between 10 and 25 km above Antarctica, probably caused by CFCs (in laboratory tests, CFCs were seen to destroy ozone).

Methane

- A minor gas but very effective in retaining heat. Since 1950, annual emissions have increased four times faster than the increase in CO2.
- Causes include rice production, burning vegetation, coal mining – and livestock flatulence from intensive cattle rearing!

- The **enhanced greenhouse effect** is the increase in the natural greenhouse effect, said to be caused by human activities which increase the quantity of greenhouse gases in the atmosphere.

Nitrous oxide (N$_2$O)

- Agricultural fertilisers, burning fossil fuels, and production of synthetic chemicals (e.g. nylon) release N2O into the atmosphere.
- N2O traps infrared radiation in the atmosphere, changing to nitric oxide which destroys ozone and thereby allows harmful ultraviolet rays into the Earth's atmosphere.

Ozone

- Found mostly in the troposphere – the height of which in the atmosphere varies between 10 and 25 km. It acts like a greenhouse gas but plays a vital role in dispersing harmful ultraviolet rays, so that fewer enter the Earth's atmosphere.
- The warming effect of ozone is greatest at 12 km height, where most aircraft operate and where pollution from N2O is greatest.

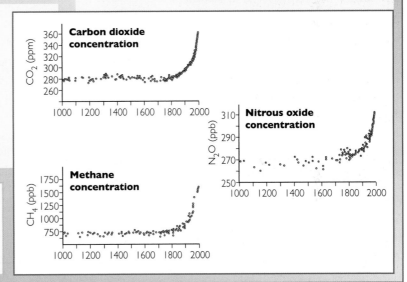

The enhanced greenhouse effect

The quantities of several greenhouse gases in the atmosphere have increased by 25% since 1750, when industrialisation began in the UK. They have grown enormously in the past 250 years, and are growing faster and faster now. Since the 1980s, 75% of CO$_2$ emissions have come from burning fossil fuels. Most climate researchers believe that this is leading to increasing global temperatures. The term **enhanced greenhouse effect** is used to describe this situation.

Human activities, such as burning fossil fuels and deforestation, release natural stores of carbon and nitrogen, which then combine with oxygen to form greenhouse gases:

- carbon combines with oxygen to form CO$_2$ (carbon dioxide)
- nitrogen combines with oxygen to form N$_2$O (nitrous oxide)

As global temperatures increase, so too does the main greenhouse gas – water vapour. Increasing global temperatures lead to greater evaporation of water, which then leads to greater condensation. This causes increased cloud cover, which then traps heat in the atmosphere.

The causes of climate change - 2

Global warming – a long-term process?

Global warming is part of a natural cycle of climate change which has taken place at least over the last 18 000 years (as the graph on the right demonstrates). Research shows that climate since the Pleistocene period, which began 1.8 million years ago, has alternated between cold glacial periods – or ice ages – and warmer periods. Possible explanations for why this might have happened are: changes in solar output, variations in the Earth's orbit, cosmic collision and volcanic emissions.

Changes in solar output

Variations in solar energy have been known to occur for centuries. Solar energy – called 'Sun spots' – works in cycles, so that the Sun's energy varies over short periods. It is quite possible that these affect global temperature. But a 2006 study showed that there has been no net increase in solar output since the mid 1970s (during which time global warming has speeded up), and that changes in solar output within the past 400 years are unlikely to have played a major part in global warming.

Variations in the Earth's orbit

According to one theory, known as the **Milankovitch cycles**, the Earth's orbit varies every 100 000 years, with changes in the Earth's axis every 41 000 years. These variations are thought to be the main factors affecting glaciation cycles. Think of the Earth like a wobbling top, so that every so often the axis changes, together with the position of the Earth in relation to the Sun. In the past, this may have altered solar energy over some parts of the Earth by 25%. The evidence from **deep sea core**

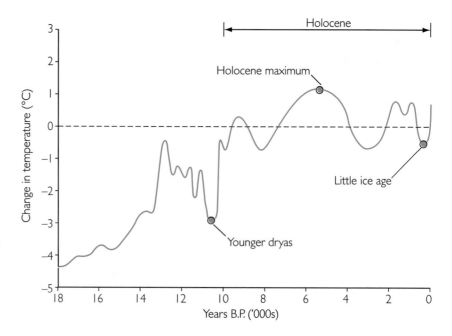

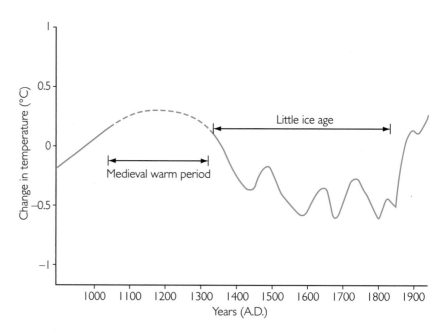

▲ Variations in global temperature in the past 18 000 years (top) and in the past 1000 years

samples, which have laid undisturbed for millions of years, is that the Earth has had eight glacial-interglacial cycles over 800 000 years, and that changes can occur quite suddenly. Within these broad changes are smaller cycles – some lasting centuries while some only last for decades – which might explain the variations in the graph.

> *Does it matter whether the causes of global warming are natural or human?*
>
> **What do you think ?**

Cosmic collision and volcanic emissions

A number of theories show the impact that **volcanic emissions** can have on global climate, together with the impacts of meteorites or larger objects from space. The eruptions of Mount St Helens in 1980 and Mount Pinatubo in 1991 released huge quantities of volcanic dust which blocked out sunlight and also caused increases in cloud cover and rainfall for 1-2 years after the events. But these events are irregular, and relatively small; the US Geological Survey estimates that, by comparison, human activities produce 150 times the amount of CO_2 emitted by volcanoes.

Accuracy of weather recording

Most weather data have been collected only in recent times. In some parts of the world, records go back as far as the 1700s – but weather science and data collection have developed hugely in the last 50 years. The recording of weather data is now much more accurate, together with more accurate forecasting. In addition, evidence from weather observers and farmers shows that:

- the European growing season – during which plants grow when temperatures exceed 6 °C – has increased by 11 days since 1960
- spring begins 6 days earlier than it did then
- there are more frost-free days
- butterflies have extended their range north.

The same is true of Canada, where the most northerly limits of tree growth have extended further north. You will explore more about the effects of climate change in the next two units.

▲ *The eruption of Mount St Helens in 1980 released huge volumes of dust and CO_2 into the atmosphere*

● **Over to you**

1 Draw an A4-size copy of the greenhouse effect diagram, and on it draw and add labels to show **a** greenhouse gases, **b** where they come from, and **c** which parts of the diagram they affect.
2 Explain how your diagram shows an enhanced greenhouse effect.
3 In a table, summarise in two columns a evidence that climate change is long-term b how likely it is that this evidence explains current climate change.

● **On your own**

4 Define the terms from the text: radiation, greenhouse gases, greenhouse effect, enhanced greenhouse effect, Milankovitch cycles, deep sea core samples, volcanic emissions.
5 How and why is it important to have accurate weather data?

1.9 The effects of climate change - 1

In this unit you'll explore the effects of climate change on Greenland, look at changing global temperatures, and discover how the world's oceans may be affected as a result.

Warmer weather is good … isn't it?

'If you're a fisherman in Greenland,' reported Melbourne's daily paper in 2007, 'global warming is doing wonders for your business because the harbour no longer freezes over.' The article reported a boost for tourism, offering the 56 000 Inuit inhabitants economic growth. Air Greenland started a direct flight from Baltimore, USA, in 2007 to attract American visitors. One critic, thinking of the CO_2 emissions that this would cause, named it 'eco-suicide tourism'.

Since 1975, the area of Greenland's ice sheets melting each summer has increased by 30%. Now, Greenland's ice cap is losing 100-150 km^3 of ice every year, a volume greater than all the melt water from the Alps. It is retreating further and faster than many experts had predicted.

Research by **glaciologists**, has shown that, since 1992, on Greenland's ice caps:

- winter temperatures are 5 °C higher
- spring and autumn temperatures are 3 °C higher
- ice is melting faster each year
- ice sheets are moving from mountains to the ocean more quickly. One has doubled its speed to about 15 km a year since 1995, while its tip (the glacier snout) has retreated 10 km.

Ice is, therefore, moving faster towards the sea, and melting more quickly as it does so.

Warming island

On top of Greenland's Suicide Cliff, from which old Inuit women used to hurl themselves when they felt they had become a burden to their community, a crack and a thud-like thunder pierce the air. 'We don't have thunder here. But I know it from movies,' says Ilulissat nurse Vilhelmina Nathanielsen, while walking through the melting snow. 'It's the ice cracking inside the icebergs. If we're lucky we might see one break apart.'

It's too early in the year to see icebergs crumple regularly, but the sound is a reminder.

As politicians squabble over how to act on climate change, Greenland's icecap is melting faster than scientists had thought possible. A new island in East Greenland is a clear sign of how the place is changing. It was dubbed Warming Island by US explorer Dennis Schmitt when he found in 2005 that it had emerged from beneath the ice.

Adapted from *The Age*, June 2007

▼ (Left) The extent of the annual ice melt in Greenland in 1992 and 2005

▼ (Right) The extent of the annual ice melt in Jakobshavn, Greenland, 1850-2003

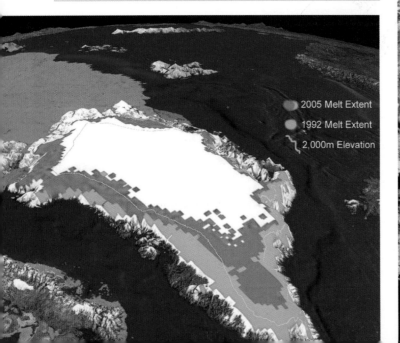

2005 Melt Extent
1992 Melt Extent
2,000m Elevation

So – is it global warming?

Global warming is a fact. Temperatures are rising globally and have increased by 0.75 °C over a period of about 150 years, as you can see in the graph. The rise is not continuous; it varies considerably, but the trend is upwards. The evidence from Greenland is that the rate of increase is getting faster, and is becoming **exponential**. Many predictions are being made about how much temperatures will increase during the 21st century. The **Intergovernmental Panel on Climate Change (IPCC)** has created a number of computer predictions that suggest global increases of between 1.1 °C and 6 °C by 2100.

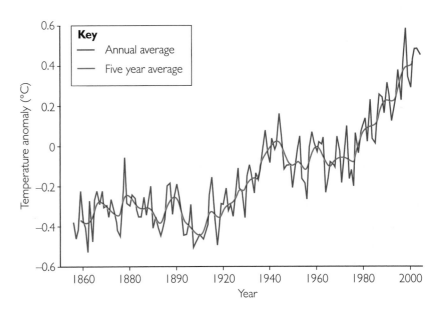

▲ Trends in global temperature since the 1850s

Temperatures are not increasing at the same rate everywhere. Some parts of the world are warming more quickly than others, as you can see from the map. Researchers in Tasmania, Australia, found that average winter temperatures had risen by 1.5 °C and summer temperatures by 2 °C. Its sea temperatures were different, increasing by about 0.8 °C. The map shows how, globally, the variations are even greater; it compares the world between 1995 and 2004 with the 40-year average between 1940 and 1980.

As you saw in Unit 1.8, there is a great debate about the extent of global warming, and especially its causes. In the last 40 years, temperatures have certainly increased. But if we take a longer time period, e.g. 110 years, temperatures have risen and fallen in cycles over that time. The increase in temperatures between 1900 and 1940 was just as great as it is now.

> Why do 'politicians squabble over how to act on climate change'?
>
> **What do you think?**

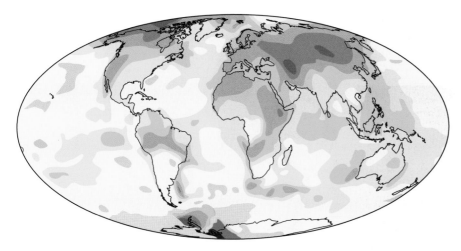

Key
Temperature anomaly (°C)

−2   −1.5   −1   −0.5   0   0.5   1   1.5   2

◄ Variations in global temperature change, 1995–2004 – when compared with the average between 1940 and 1980

The effects of climate change - 2

The impact on the world's oceans

The Arctic currently shows the greatest effects of climate change. These changes could affect the global climate – and especially that of Northern Europe.

Changing salinity

One of the reasons why the Earth is habitable is that oceans and winds help to distribute heat between its warmest and coldest parts. They transport heat from the equator toward the poles. Winds blowing across the sea transport heat through the atmosphere and drive ocean currents towards the poles. The current of warm water in the Atlantic is known as the Gulf Stream; its northern offshoot is the **North Atlantic Drift**, which flows north past the west coast of the UK and so makes the UK's climate warmer than it would otherwise be for its latitude. As it flows, the ocean releases heat into the atmosphere, allowing sub-tropical plants to grow in the Scilly Isles off the coast of Cornwall, 50° north of the Equator.

As ocean water cools in the far North Atlantic, it becomes denser and sinks to the ocean floor. This helps to form an ocean circulation called the **thermohaline circulation** – sometimes called the **global conveyor belt** (see the diagram below).

> ● The **thermohaline circulation** is the flow of warm and cold water that circulates around the world's oceans.

It works like this:
- In the far North Atlantic, the water is both cold and very **saline** (salty), which makes it denser, heavier, and causes it to sink.
- By sinking, it draws warmer water in from the ocean surface above. This, in turn, draws water across the ocean surface from the Tropics.
- Eventually, this movement from the Tropics draws cold water up from the ocean bottom, ready to be warmed again.

In this way a cycle is created.

Scientists have found that this pattern is now being disturbed. More freshwater is entering the Arctic Ocean as a result of global warming, which melts the ice and increases rainfall. Meltwater lowers the salinity, which decreases the density of the ocean, and slows down the rate at which the ocean sinks. This affects the global conveyor belt. Several researchers say that global warming could turn off the North Atlantic Drift, which supplies Europe with warm water and air. If it stopped, January temperatures in Western Europe would drop by at least 5 °C, creating bitter winters.

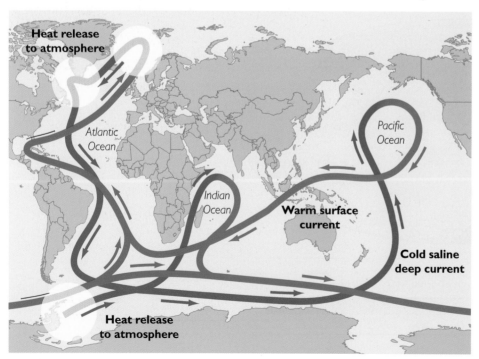

◀ The thermohaline circulation, also known as the global conveyor belt

Increasing river flow

Freshwater flowing into the Arctic Ocean from rivers comes mostly from the six largest Eurasian Arctic rivers – the Yenisey, Lena, Ob', Pechora, Kolyma, and Severnaya Dvina. Together, these drain two-thirds of central and northern Russia and include three of the world's largest rivers. Their average annual flow of fresh water into the Arctic Ocean increased by 7% between 1936 and 1999.

The increase in river flow is caused by warmer surface air temperatures, especially between November and April – when these rivers would normally be frozen. Greater flow creates a significant increase in freshwater flowing into the Arctic, and has the same effect as increasing meltwater – it could slow down or shut off the North Atlantic Drift, affecting the thermohaline circulation and cooling the whole of Northern Europe.

Changes in the polar oceans

The Southern Ocean around Antarctica is one of the Earth's most important natural absorbers – or **carbon sinks** – of CO_2. Cold dense seawater absorbs CO_2, making it an effective way of removing CO_2 from the atmosphere. Natural sinks are vital, because they absorb excess CO_2 and slow down global warming. But now researchers have found that CO_2 sinks have stayed the same since 1981, even though CO_2 emissions have risen by 40%. This might increase CO_2 levels in the atmosphere.

The cause is thought to be increasing windiness. Climate change warms the tropics more than the Southern Ocean, and the temperature difference causes stronger winds. As winds increase, the ocean is stirred up, and CO_2 that would normally stay there is released into the atmosphere.

▲ The catchments of the six Eurasian Arctic rivers that empty into the Arctic Ocean

▲ Storminess in the Southern Ocean – and it could get worse.

● Over to you

1 Study the graph and map on page 35. How could they be used to show **a** that global warming is definitely here to stay, or **b** that it is far from certain?

2 In pairs, draw a sketch map of the Arctic and its surrounding landmasses. Annotate it to show **a** why the Arctic is becoming less saline, **b** why increasing river flow is occurring, and **c** how this might affect the North Atlantic Drift.

3 **a** In pairs, construct a flow diagram or spider diagram to show the possible consequences if the North Atlantic Drift stopped.

b How serious would the consequences be if this happened?

● On your own

4 Define the terms from the text: exponential, IPCC, saline, thermohaline circulation, global conveyor belt, North Atlantic Drift, carbon sinks.

5 Research the IPCC. Find out what its job is, who set it up, and why.

6 Consider, in 500 words 'Does global warming matter?'.

Exam question: Suggest how melting ice in the Arctic might **a** bring advantages and **b** create problems for people there and elsewhere. (10 marks)

THE WORLD AT RISK

In this unit you'll find out about the impacts of global warming on the Arctic and Africa.

The Arctic region lies within the Arctic Circle (the 66.5° line of latitude). It consists of the ice sheet surrounding the North Pole, as well as the northern parts of eight countries – Canada, Greenland, Russia, USA (Alaska), Iceland, Norway, Sweden, and Finland. The area has an incredibly cold climate; January temperatures average -35 °C, and July temperatures -1.5 °C. Much of the Arctic is ice.

However, increasing average temperatures are melting the Arctic ice. In 2006, NASA reported that the amount of permanent sea ice decreased by 14% between 2004 and 2005 – equivalent to an area three times the size of the UK. The rate at which the ice is melting has risen massively. Until recently, 80% of solar radiation was reflected from the polar ice caps. Now the amount of ice has decreased, and the area of open ocean has increased. Oceans are darker than ice and snow, and absorb more energy – converting it to heat. This speeds up the warming effect, which melts more ice and creates a vicious cycle, called the **positive ice albedo feedback**.

At current rates, 50-60% of Arctic ice will be lost by 2100. One theory suggests that it could disappear entirely by 2070.

Impacts on the environment

Increasing temperatures have led to:
- the **tree line** (the edge of the habitat within which trees will grow) moving north, and also to higher altitudes.
- **tundra** ecosystems in Arctic areas (which withstand intense cold) being lost as the climate warms and other plants take over.
- permanently frozen ground (or **permafrost**) thawing out.
- the spread of species such as the spruce bark beetle in Alaska – changing the food chain.
- increases in the number and extent of northern coniferous (or **boreal forest**) fires in Arctic Russia. 10 million hectares burn each year, losing 0.8% of the world's coniferous forest. Boreal ecosystems account for 37% of the world's carbon pool on land, and are effective carbon sinks.

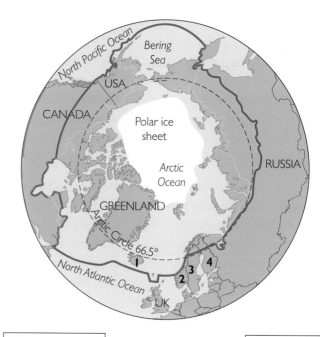

▲ The Arctic region

Key
— Arctic region
— Tree line

Key
1 Iceland
2 Norway
3 Sweden
4 Finland

● **Albedo** is the amount of solar radiation reflected by the Earth's surfaces. Ice and snow (light coloured surfaces) reflect most, and dark rock surfaces reflect least.

▼ The positive ice albedo feedback

There is less ice cover in the Arctic region. → Less solar radiation is reflected away, and more is absorbed by the ocean. → A greater quantity of heat is produced. → The ice melts at a faster rate. →

Fish stocks and polar bears

The marine ecosystem has altered considerably as a result of Arctic warming. It is difficult to assess the true impact of warming because so much commercial fishing of species such as cod has taken place. But the shrinking ice sheets have certainly affected marine species in the Arctic. Warmer water has reduced the quantity of marine plants on which many smaller fish feed. In turn, the reduction in smaller fish species has affected those higher up the food chain, such as cod and halibut, which in turn affects larger marine species such as seals. This has a **negative multiplier effect**. Smaller seal stocks reduce the available food supply for polar bears.

The melting Arctic ice has had a devastating impact on polar bears. They hunt seals on the ice, and the faster annual ice melt has reduced their spring hunting season. Hudson Bay is now ice-free for three weeks longer than it was in 1985, giving polar bears less time in which to hunt the reduced numbers of seals. Female polar bears rely on the spring to build up their body fat to ensure their survival during the summer when the ice they hunt on recedes naturally. Currently, each animal loses 80 kg of fat during the longer summer, making them susceptible to disease, and reducing their ability to reproduce or feed their cubs. Now they face the danger of complete extinction.

Socio-economic impacts

Global warming poses social and economic threats to the 155 000 Inuit living in the Arctic region. Global warming is disrupting their lifestyles, which are adapted to the cold but predictable climate in the Arctic. The impacts include the following:

- Each winter Inuit men take their fishing shacks and equipment on to the ice for three months. Now the weaker and thinner sea ice collapses easily, making it more dangerous.
- The ice used to protect Inuit villages along the coast. However, coasts are now exposed to more ocean waves and storms, causing the destruction of entire villages, and forcing people to move further inland.
- 24 Inuit villages in Alaska are now threatened by flooding.
- 80% of Inuit still hunt caribou, fish, and marine mammals – all of which are declining in numbers. As marine stocks decline, the Inuit rely more on hunting caribou for income, which in turn places greater pressure on caribou stocks. 70% of Inuit income is from paid employment or hunting, so declining stocks hit Inuit incomes hard.

▲ *Polar bears face extinction*

Does it matter if polar bears become extinct?

What do you think ?

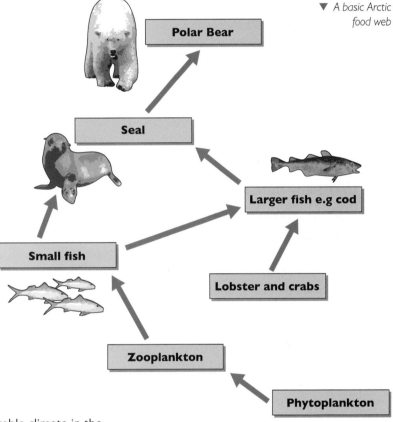

▼ *A basic Arctic food web*

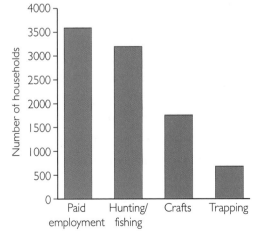

▲ *The distribution of Arctic employment*

The impacts of global warming - the Arctic

Socio-economic impacts continued...

Environmental change creates further problems. Caribou and marine animals provide vital nutrition for the Inuit. Together, caribou, seal, narwhal, fish, and walrus provide over 90% of their food, and reductions in their numbers are dangerous for Inuit lifestyles – especially as a high protein intake is needed to cope with the cold. Imported food is expensive. Clyde River settlement on Baffin Island has 450 residents, who eat 100 tonnes of seal meat annually. To import replacement food would cost US$1 million, and provide less iron, magnesium, and calcium than the natural diet.

Bringing home a seal supper on Baffin Island ▶

Are there any benefits from global warming?

The melting ice creates some commercial advantages for the Arctic region. The Northern Sea Route, north of Canada, is the quickest way of travelling from Europe to the Pacific and Asia, but, until now, the ice has only allowed ships to use it for about 6 weeks between August and October. Now, tourist ships are able to visit northern Canada, and 30% of Inuit now earn income from sculpture or print-making for tourists. In 2007, the North–West Passage between Canada and the Arctic melted sufficiently to allow shipping through for the first time.

However, this brings problems too. Oil tankers have negotiated Arctic waters for nearly 40 years, bringing oil from the shores of northern Alaska. Greater frequency and use of the passage north of Canada increases the risk of pollution and of oil spills in the Arctic. Now, Russia has started to allow nuclear waste disposal in its Arctic waters off the coast of its North Western territories, posing a further threat to the marine ecosystem.

● Over to you

1 Research and design a short presentation on the issues facing the Inuit, e.g. those living in Nunavut, a Canadian Inuit territory.
2 In pairs, classify the effects of global warming on Inuit communities.

	Economic	Social	Environmental
Short-term			
Medium-term			
Long-term			

3 Draw a spider diagram to show the future of Inuit communities if: **a** seal and fish stocks continue to fall, **b** tourist numbers continue to rise, and **c** sea routes around the Arctic open up further.
4 Discuss which you see as the bigger threat: declining fish stocks or opening up the Arctic to increased shipping.

● On your own

5 Define the terms from the text: positive ice albedo feedback, tree line, tundra, permafrost, boreal forest, negative multiplier effect.
6 Research the Arctic climate, using www.worldclimate.com or similar. Compare northern Canada with Greenland and northern Russia. What challenges does the climate present for people?

The impacts of global warming - Africa

There are 55 countries in Africa. In 2007, two-thirds of them were among the world's fifty poorest countries. Their average per capita income was US$0.72 a day. 34% of the population aged 15-24 was illiterate. Many African economies are on the edge of collapse through international debt. Now, on top of everything else, they face the impacts of global warming.

The evidence for global warming

Africa is steadily warming. The continent as a whole is 0.5 °C warmer than it was in 1900. However, temperatures have increased more in the interior – some inland areas of Africa have experienced double the global increase. In Kericho, Kenya, maximum temperatures have increased by 3.5 °C since 1985. This has caused difficult conditions across Africa:

- Droughts have become more common, as areas which are already arid or semi-arid become even drier.
- Rainy seasons are now more unreliable, and overall rainfall is decreasing.
- Rains are more localised. The rains that ended East Africa's drought in 2005-6 were not widespread, and many areas had very little rainfall.

Why is Africa vulnerable?

The clue to Africa's vulnerability to global warming is its economy. Most African economies are heavily dependent on agriculture. Farming is rain-fed and vulnerable to rainfall shortages. Water is already scarce. The United Nations Environment Programme (UNEP) reported in 2007 that 14 countries in Africa were suffering from water scarcity, and that 11 more will join them by 2032. Rainfall is declining in the areas that need it most – areas with high annual totals are likely to maintain or increase them, but the edge of the Sahara (the Sahel region) is likely to see reductions.

▼ *Trends in the African climate during the twentieth century*

> Too much or too little rain can be a matter of life or death in Africa.
> **Oxfam, 2006**

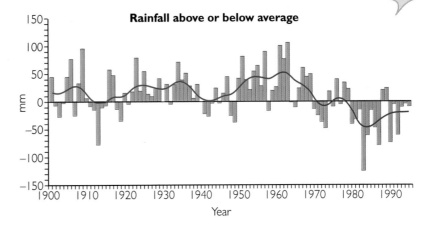

Rainfall above or below average

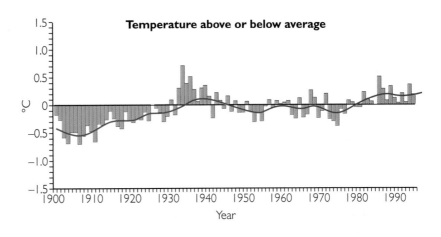

Temperature above or below average

▼ *The northern Kenya drought of 2005-6: walking through a field of dead animals – the image of the future?*

The impacts of global warming – Africa

There are two problems:

- As rainfall becomes more variable, competition for water is likely to arise between countries, particularly when Africa's largest rivers, such as the Nile, cross national borders.
- As scarcity increases, so does dependence on poor-quality sources. This leads to increased water-borne diseases, such as cholera, which puts pressure on health care systems, and therefore on government budgets.

Physical and human impacts

Many of the impacts on Africa are physical. Africa contains about 20% of all known species of plants, mammals, and birds, as well as 17% of amphibians and reptiles. As global warming increases, fragile habitats may not survive and between 20-50% of species in Africa could face extinction.

Many low-lying coastal countries in Africa are vulnerable to sea level rise – with cities, roads and infrastructure at risk from flooding and increased erosion. Already, the low-lying sandy coastlines of Ghana, Togo and Benin are suffering greater erosion from increased storminess and human interference, such as port development and coastal defences. Other parts of Africa are also affected.

However, the main impacts are human.

- The bottom map on the right shows that changes to rainfall – even increases – can reduce the period in which crops tgrow. In some parts of Africa, global warming may make rainfall more seasonal, with a dry season that is just as dry as now, but a wet season where rain may fall in heavier bursts. Only small areas show an increase in crops – the rest of Africa suffers a reduction in food supply.
- Malaria is also on the increase, as increased humidity and rainfall causes the mosquito (the cause of the disease) to spread over wider areas. In South Africa, it is estimated that the area affected by malaria will double, and that 7.2 million people will be at risk – an increase of 5.2 million. The health system will have to deal with many new cases every year. As the infection increases, so does the number of people unable to work.

Should the countries which cause the most global warming pay to reduce its impacts on Africa

What do you think?

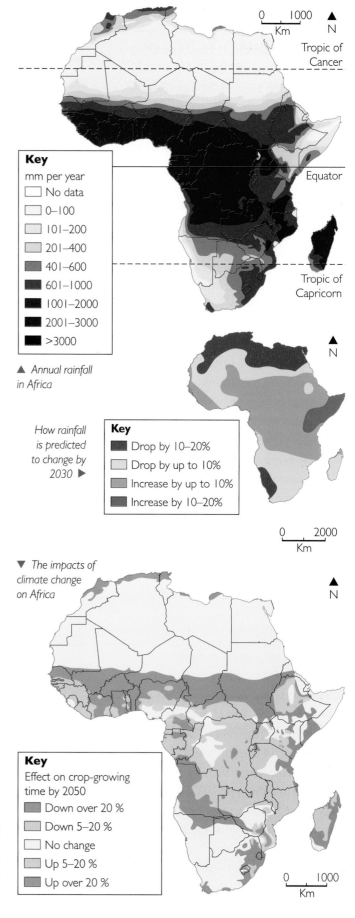

Key

mm per year
- No data
- 0–100
- 101–200
- 201–400
- 401–600
- 601–1000
- 1001–2000
- 2001–3000
- >3000

▲ *Annual rainfall in Africa*

How rainfall is predicted to change by 2030 ▶

Key
- Drop by 10–20%
- Drop by up to 10%
- Increase by up to 10%
- Increase by 10–20%

▼ *The impacts of climate change on Africa*

Key

Effect on crop-growing time by 2050
- Down over 20 %
- Down 5–20 %
- No change
- Up 5–20 %
- Up over 20 %

Global warming and debt

One of the main reasons why Africa is vulnerable to global warming is because of its **debt crisis**. For many decades, the debt burden faced by African countries has been severe. Many countries have now had these debts written off – because the world's richest countries (the G8) decided to cancel them in 2005 (by mid-2007, 18 countries had had their debts written off). However, the burden of debt still affects the majority of African countries.

One of the few ways in which African countries can reduce their debt is to increase overseas trade, mainly through the production of **cash crops**. This has led to forest clearance for commercial farming.

Oxfam believes that global warming is likely to result in even greater forest clearance. They argue that harsher weather conditions, such as drought, will expose soils to erosion by wind. This will place more pressure on existing land to grow food. Such pressure (e.g. over-grazing) often leads to an increase in **desertification**. Famine and malnutrition will increase, placing Africa's **food security** under threat.

- **The debt crisis** is where many African countries have become so heavily indebted that repayments sometimes exceed their entire GNP.
- **Cash crops** are those crops sold for income, rather than one's own food supply.
- **Desertification** is the way in which climate change and over-grazing turn previously fertile land into desert.
- **Food security** is the extent to which a country can rely upon food supplies, e.g. upon the weather, or, if unable to grow all its food, the extent to which it can pay for imports to feed itself.

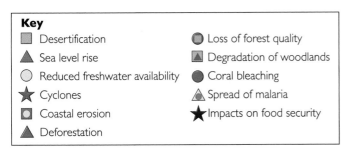

Key
▪ Desertification	⬤ Loss of forest quality
▲ Sea level rise	⬛ Degradation of woodlands
⬤ Reduced freshwater availability	⬤ Coral bleaching
★ Cyclones	◬ Spread of malaria
▪ Coastal erosion	★ Impacts on food security
▲ Deforestation	

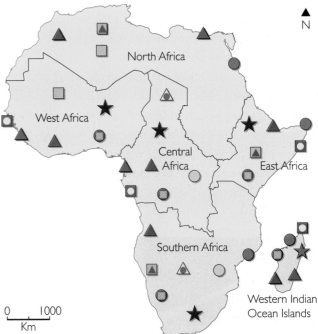

▼ *Africa's vulnerability to climate change*

● Over to you

1 In pairs, draw a spider diagram to show the links between global warming, changing rainfall and impacts on the food supply.
2 Add to this diagram the influence that debt can have.
3 In what ways **a** are African farmers trapped into poverty? **b** does global warming make this poverty trap worse?
4 How might the increased likelihood of malaria impact on this poverty trap, and why?

● On your own

5 Define the terms from the text: desertification, food security, debt crisis, cash crops.
6 Are the consequences of global warming worse for Africa than for the Arctic? Discuss this in 600 words.

Exam question: Why does climate change present potential problems for the African continent?
(10 marks)

▼ *Desertification in the Sahel, on the margins of the Sahara*

Dealing with climate change: mitigation and adaptation

THE WORLD AT RISK

In this unit you'll find out how mitigation and adaptation strategies can be used to cope with climate change.

Global warming – natural or not?

There is a minority – some journalists, US scientists, industrialists and politicians – who believe that global warming is caused by natural processes, not human activity. Most of these individuals or groups do not produce their own data, but simply comment on data produced by others.

However, the Intergovernmental Panel on Climate Change (IPCC) believes that their evidence for climate change as a result of human activity is overwhelming. They believe that their evidence shows that:

- the increase in global temperatures since 1950 coincides with the increase in greenhouse gas emissions since then.
- the changes predicted by scientists if greenhouse gas emissions continued to rise are now beginning to occur – e.g. the melting of the Arctic ice.
- natural processes could not have produced all the changes taking place. Their data show that the speed of change taking place in Greenland is highly unlikely to have been caused by natural processes. Although there have been periods of natural warming before, none have produced ice melt like those occurring at present. The IPCC therefore believes that human causes must be to blame.

What are the choices?

There are two main approaches by which the world can deal with global warming – **mitigation** and **adaptation**.

- Policies which reduce or prevent further global warming are referred to as **mitigation** strategies. These involve reducing carbon emissions and finding ways of dealing with excess greenhouse gases.
- However, even if significant efforts are made to reduce future carbon emissions, policies are needed to deal with the current situation. It will take time to reduce CO_2 to pre-Industrial Revolution levels, because CO_2 has a life of 100 years. Therefore, the world has to cope with the impacts that are already occurring. These policies are known as **adaptation** to global warming.

The table opposite shows the arguments on each side.

> ● **Mitigation of global warming** refers to policies which are meant to delay, reduce or prevent climate changes caused by global warming – such as cutting CO_2 emissions (congestion charging, increasing renewable energy) and increasing carbon sinks (e.g. afforestation).
>
> ● **Adaptation to global warming** refers to policies which are designed to reduce the existing impacts of global warming, such as protection against flooding and coastal erosion.

◀ *Mitigating global warming – increasing renewable energy*

◄ *A summary of the arguments for mitigation and adaptation*

Pro-mitigation	Pro-adaptation
• Emissions have to be reduced immediately if the world is to avoid irreversible climate change. Environmentalists believe that adaptation policies waste time when the world should concentrate on reducing emissions.	• It will take time to get all countries to agree to reduce carbon emissions, let alone cease them.
• Adaptation policies are supported by those who believe in the 'fossil fuel society', so that 'business as usual' continues. Unless this is challenged, change will never occur.	• The time taken to persuade countries to agree to reductions means that climate change will get worse before it gets better.
• Carbon sinks (e.g. afforestation) can be easily achieved, e.g. by government grants to encourage tree planting.	• Climate change is already happening (e.g. ice melting, sea level rising). This needs managing now.
	• Even if all emissions ceased today, greenhouse gases already present in the atmosphere would cause change for the next century. CO_2 has a life of 100 years, so any excess will be around for a long time.
• Cutting emissions in the developed economies has proved politically acceptable (e.g. public support for renewable energies), so more of this should be encouraged.	• The Earth absorbs CO_2 slowly, so even with increased carbon sinks (e.g. afforestation), it will take time to absorb the excess.
• Cutting greenhouse gas emissions in developing countries is essential, as economic growth leads to increases in emissions. This is not the case in developed western economies, whose emissions are now constant or falling.	• Keeping greenhouse gas emissions at current levels is unrealistic, given economic growth in India and China. The world must therefore adapt, because mitigation may not be achieved in time.

Should developed countries tell developing countries to cut carbon emissions

What do you think ?

◄ *Adapting to global warming – protection against further coastal erosion at Barton-on-Sea, Hampshire*

● Over to you

1 In class, divide into pairs and decide what you believe the two strongest arguments are for **a** mitigation, and **b** adaptation. Justify why you believe them to be the strongest.
2 As a class, debate the motion that 'This house believes that mitigation is the only way to manage climate change'.

● On your own

3 Define the terms from the text: mitigation, adaptation.
4 Research the evidence and design a presentation about the anti-global warming debate. A good source is by Christopher Monckton – search 'monckton global warming' at www.telegraph.co.uk. Remember that he is a journalist and not a climate scientist.

In this unit you'll explore options for the future, and consider whether the world has already reached 'tipping point'.

The future?

The Intergovernmental Panel on Climate Change (IPCC), formed in 1988, is a joint organisation of the World Meteorological Organisation (WMO) and the United Nations Environment Programme (UNEP). The IPCC – more than any other organisation – collected evidence during the 1990s which showed that global warming was really happening. To do this, it:

- collected data about the atmosphere
- observed sea levels and the extent of the ice at the poles and at glaciers in mountain regions
- measured greenhouse gas concentrations (e.g. CO_2)
- used super-computers to try to predict what impacts global warming might have.

Using its data, the IPCC has attempted to predict different possibilities – or **scenarios** – regarding greenhouse gas emissions. Its report in 2007 created four scenarios (A1, A2, B1, and B2). Each considers different possibilities about how global warming might work out. Each is uncertain, but helps to predict the future by considering how governments might act.

1 The A1 scenario – a converging world

Situation: There is rapid global economic growth. The global population increases to 9 billion by 2050, after which it decreases. By 2100, global GNP is 26 times that of 2007.

Broad scenario: There is a rapid introduction of new, energy-efficient technology. Global agreements and globalisation enable social and cultural international cooperation. There is a desire to reduce global and regional differences in wealth.

There are three sub-groups, each predicting what different ways of producing energy might do to global warming:

- A1a – technology based on fossil fuels
- A1b – technology based on non-fossil fuel sources
- A1c – technology balanced across all energy resources

▼ *A1a – technology based on fossil fuels*

▼ *The estimated impact of the IPCC's scenarios on climate change*

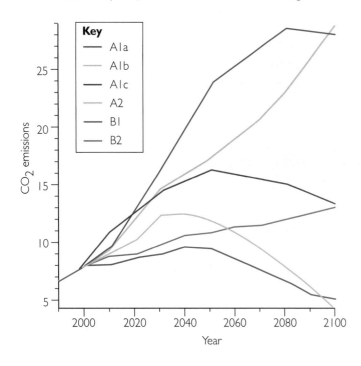

◀ A1b – technology based on non-fossil fuel sources

2   The A2 scenario – a divided world

Situation: There is a divided world, with reduced globalisation, and with countries becoming more independent and keener to preserve their local identities. The global population continues to increase, and fertility rates slowly fall. Economic growth is regional, with little global exchange of ideas or wealth, so that differences in global wealth persist.

Broad scenario: Technological change is very slow.

3   The B1 scenario – a converging world

Situation: This is similar to A1, with reduced global income differences. The global population rises to 9 billion by 2050, followed by a steady decline. The global economy changes to a service and information economy.

Broad scenario: Cleaner, more-efficient technologies are used in an attempt to solve issues of social, economic, and environmental sustainability. The global economy continues to develop, but tries to reduce global warming. By 2100, there is more forest than there was in 1900.

4   The B2 scenario – global social, economic, and environmental sustainability

Situation: A global society is attempting to achieve progress at local levels, by developing in small-scale ways at a local level rather than through large global companies. The global population increases more slowly than in A2.

Broad scenario: There is lower economic growth and, compared with A1 and B1, there are slower but more diverse changes in technology, e.g. those designed to produce energy locally using wind or solar energy. There are attempts to achieve environmental protection and social equality – but at local, not global, levels. As in B1, there is more forest by 2100 than in 1900.

Too late to act?

What the IPCC and others fear is a sequence of events which leads to irreversible change. They fear that the degree of warming could be so great that it would cause shifts in climate belts which would be irreversible – no matter what is done now or in the future to control carbon emissions. The point when change becomes irreversible has been called the **tipping point**.

Some scientists suggest that current rates of warming mean that the world has reached the tipping point. Others believe that the world has passed it. Some believe that we have some time left. They believe that it is possible for the problems of global warming to be prevented, but that just a tiny increase in global temperature could have the following catastrophic and irreversible consequences:

1 **Rising sea level.** Greenpeace estimates that melting of the Arctic ice caps caused a rise in sea level of 10-25 cm during the twentieth century. In the twenty-first century, they predict further rises of 15-95 cm. 95 cm would result in large-scale coral bleaching, killing off hundreds of coral reefs. This would not only threaten the world's most diverse ecosystem, but a 95 cm rise would flood huge areas of inhabited land. If the Greenland and West Antarctic ice sheets (which hold 20% of the world's freshwater) melt, sea level would rise by **6 metres**, flooding, for example, a third of Florida, much of Manhattan, most of eastern England, the entire country of Bangladesh, and almost all of the Netherlands.

2 **Shutting down the Atlantic thermohaline circulation** (see page 36 for thermohaline circulation). Already slower than in the 1970s, there is a 50% chance that this will shut down within 200 years. Bizarrely, this would actually make the **land areas** of Western Europe colder, even though the process of global warming would continue to warm the **oceans** sufficiently to allow further melting of Greenland's ice sheet. The current melting of Greenland's ice sheet is causing annual sea level increases of 0.02 cm. If global temperatures rise by 2 °C more than pre-1750, Greenland's ice sheet will melt irreversibly.

3 **Falling agriculture yields and water shortages** in Africa, Europe, the USA, and Russia, putting the poorest 200 million people in those areas at risk of starvation. 2.8 billion people will suffer water shortages for drinking and irrigation.

● The **tipping point** refers to a point beyond which the Earth cannot recover from the effects of carbon emissions, even with drastic action.

▼ Flooding in Florida in 2006

What is the point in trying to combat global warming if no-one else does anything?

What do you think?

▼ The rainforests of northern Australia contain 255 000 species. Would 40% become extinct if increased global warming leads to reductions in rainfall? According to the Stern Review (see opposite) they would.

The Stern Review

In 2006, the UK government published a review of global warming by Sir Nicholas Stern, former Chief Economist of the World Bank. This was a landmark – for the first time, an important economist joined in a debate normally led by environmentalists.

The Stern Review focused on the impacts of global warming, and the actions needed to deal with them. He argued that the problems of global warming could be reduced if greenhouse gas emissions are stabilised by 2025, and then fall by 1-3% annually after that. To do this would cost 1% of global GDP, because of investment in new clean technology.

Key features of the Stern Review

Environmental impacts of global warming
- Increasing global flood risk.
- Declining crop yields, particularly in Africa, as climate belts shift.
- Rising sea levels, leaving 200 million people permanently homeless in, for example, Bangladesh.
- The extinction of 40% of the Earth's species, e.g. through the destruction of coral reefs.

Economic impacts of global warming
- More extreme weather could reduce global GDP by 1%, e.g. from the costs of hurricane damage.
- A 2-3 °C rise in temperatures could reduce global economic output by 3%, e.g. by reduced crop yields.
- If temperatures rise by 5 °C, 10% of global output could be lost. The poorest countries would lose >10% of their output because they would lack the income to pay for coping strategies, e.g. by storing water in large dams.

Options for change
- Reduce consumer demand for heavily polluting goods and services.
- Make global energy supply more efficient.
- Prevent further deforestation to balance carbon emissions.
- Promote cleaner energy technology.

UK government response
- Set targets to reduce carbon emissions by 30% by 2020 and 60% by 2050.
- Pass laws to set carbon reduction targets, and monitor them.
- Invest in green technology, creating 100 000 new jobs.
- Create a $20 billion World Bank fund to help poor countries adjust to climate change.

Over to you

1 In pairs, decide with reasons which IPCC scenario:
   a is most pessimistic
   b produces the *highest* levels of greenhouse gas emissions
   c produces the most environmentally-friendly society
   d produces the *lowest* levels of greenhouse gas emissions
   e is the least polluting and uses non-fossil fuels
   f is the one you think the world should adopt
   g is the one you think the world is most likely to adopt
2 In pairs, consider the things that are most likely to make a the best scenario happen, b the worst. Present your ideas to the class.

On your own

3 Define the terms from the text: tipping point, scenario.
4 Research the Stern Review and its findings; use the BBC's news website (news.bbc.co.uk) for a summary. Prepare a statement which outlines a its findings, b its best and worst scenarios, c the implications for the UK, and d the implications for the world if nothing is done.

Exam question: Write 500 words to explain why international agreements are essential if anything effective is to be done about global warming. (15 marks)

Strategies for dealing with climate change – 1

In this unit you'll learn how individuals, organisations, governments and international agreements can help to reduce the impacts of climate change.

▼ *Greenpeace offer individuals the chance to take the train instead of flying*

Can individuals make a difference?

Can you or I make a difference to global warming? Opinion polls show that, although most British people support action to control climate change, they are not always willing to pay for it. In 2006, British Airways asked customers to **offset** (see page 52) the carbon emissions that their flight would cause (£5 London-Madrid, £13.50 London-Johannesburg); less than 1% of passengers paid. Similarly, in 2007 the pressure group Greenpeace offered train tickets to domestic BA passengers, saying that their flight would do ten times more environmental damage than the train. Few people took the train.

In the music industry, Coldplay joined up with the company Future Forests to make their second and third albums carbon neutral by **carbon offsetting**. To offset carbon emissions created during production, manufacture and distribution of 'A Rush Of Blood To The Head', Coldplay financed the planting of 10 000 mango trees in Karnataka, India. The trees provide fruit for trade and local consumption, and over their lifetime will soak up CO_2. For their third album, 'X&Y', Coldplay supported a forest in Chiapas, Mexico (see right).

If not individuals ... then who?

Who, then, **should** pay for the impacts of global warming? Private companies? Individuals? Governments? The people who **produce** the fuels that pollute (e.g. fuel companies)? Or the people who **consume** the fuels that pollute (e.g. the motorist)?

The 2006 Stern Review argued that global warming is resulting from the failure of private companies to act, and also of governments to force action. Greenhouse gases, it concluded, are an economic cost that polluters currently do not have to pay. Stern said that polluters impose 'costs on the ... world without facing the consequences'. Therefore, paying for carbon is essential – if people have to pay for it, they may realise its importance.

Coldplay's carbon neutral third album, 'X&Y', funded forest development in Mexico ▶

Can reductions be achieved?

The graph shows measures which could reduce greenhouse gas emissions – but at a cost. Those below the 0 line have a negative cost and are economically worthwhile, such as insulating buildings and using energy-efficient lighting. Low-energy lightbulbs last longer and use less energy, cutting both greenhouse gas emissions **and** costs. Above the 0 line, reducing emissions costs more (e.g. switching from coal to gas in power stations), and results in reduced emissions – but lower profits.

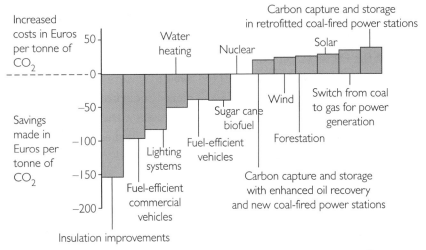

▲ *The costs of cutting carbon emissions in different ways.*

◀ *Two ways of reducing emissions: energy-efficient lighting, which saves costs, and solar energy, which is more expensive. This is a solar thermal power station in the Mojave Desert in California, USA, which generates electricity by using mirrored parabolic troughs instead of photovoltaics*

Reducing pollution costs money. Either companies have to pay more, which is then passed on to their customers, or governments have to tax carbon emissions, e.g. on petrol. Neither approach is popular. Where companies and individuals are not keen to **mitigate** emissions, or to change their behaviour, should they be made to by law? The rest of this unit looks at how well:

- different national and local governments are attempting to reduce greenhouse gas emissions
- global agreements about emissions can be made to work.

Strategies for dealing with climate change - 2

The European Emissions Trading Scheme (ETS)

In 2007, the EU set targets for 2020 to reduce its greenhouse gas emissions by 20% of the levels they were at in 1990. To do this, it set up the ETS as a means of **carbon offsetting**. It set emissions targets for every country, and gave 14 000 factories and power plants in Europe's 'dirtiest' industries (electricity, oil, metals, building materials and paper) permits – or **credits** – to emit certain amounts of carbon.

The ETS aims to:

- **cut emissions** by placing a limit on the total amount emitted.
- get **polluters to pay** for damage they cause by introducing credits for the greenhouse gases that they emit. If their credit is more than they need, companies or countries can sell it; if it is less, they can buy credits from others to allow them to pollute above their limit.
- **create incentives** for companies to invest in cleaner technology.

Over time, the EU plans to reduce the number of available credits to below the level of demand for them. This will create a shortage which will raise the price at which the credits are traded and make carbon emission more expensive. This will then encourage companies to sell credits they no longer need, and pay for cleaner technology. This is the theory, but …

… how effective is the ETS?

The ETS has so far failed in its aims.

- Manufacturing companies have been moving out of Europe, thus reducing the demand for carbon credits and causing the price to fall. It is cheaper for 'dirty' companies to carry on polluting by buying credits than to change to cleaner technology.
- Polluters are not absorbing the price of the credits – quite the reverse. To entice industry to agree to the ETS in the first place, carbon credits were given free to companies. Yet, in electricity generation (the EU's most-polluting industry), companies immediately passed the price of the credits on to their customers! According to one report, UK power-generators alone made £800m ($1.5 billion) profit from the scheme in its first year.
- The low cost of carbon credits is not leading to investment in green technology. Recent high gas prices have led to a shift to coal-fired power stations – the dirtiest sort. Carbon credits are so cheap that it costs less to change to coal and buy credits than to stick with gas.

Background

Carbon offsetting

Carbon offsetting is the name given to a credit system, called **carbon credits**, which aims to reduce greenhouse gas emissions. Carbon credits allow companies to pollute, but at a cost. Each credit costs money which polluters have to pay, and is in proportion to the pollution produced. The cost encourages companies to look for other ways of production by polluting less or not at all.

> ● **Carbon sequestration** is where natural processes, such as plant respiration, are used to offset carbon emissions. Trees absorb CO_2 and give out oxygen, thus removing greenhouse gases from the atmosphere.

Carbon credits come in two forms:

- **Certified** – These are international exchanges, which aim to cut overall emissions, like the EU's Emissions Trading Scheme. Companies and/or countries are given targets, allowing them to pollute a certain amount. Trading is allowed between those with higher or lower levels than they need.
- **Voluntary** – These are payments or projects which offset emissions with equivalent CO_2 savings. They are used where people or organisations volunteer to offset the pollution they create, such as Coldplay whose tree-planting programmes work by removing emissions through **carbon sequestration**.

Carbon offsetting in action

Example 1: Shell

Shell pumps waste CO_2 from an oil refinery in Botlek (the Netherlands) into 500 greenhouses growing fruit and vegetables, which:

- avoids annual emissions of 170 000 tonnes of CO_2
- saves greenhouse owners from burning 95m³ of gas to get the CO_2 they need for heat retention.

Example 2: Alternative energy in Brazil

One sow and her piglets can produce 9 tonnes of CO_2 equivalent annually from the methane emissions from their waste. The effluent collects in lagoons which smell awful and sometimes overflow into water supplies. Now Bunge, a US company operating in Brazil, builds lined, enclosed pools to collect the effluent and capture the methane, which farmers then use to generate electricity. By preventing the methane from escaping, the company gets a carbon credit to sell on the carbon market; the farmer keeps 25%.

Example 3: London's congestion charge

What is it?

- Since 2003, drivers have been charged £8 per day to drive in the Central London congestion zone (during the scheme's operating hours). A new proposal is to charge £25 for larger vehicles, including four-wheel drives (emitting >225g CO_2/km).

Why?

- London has the UK's worst congestion. Drivers in Central London spend half their time in queues – costing London's economy £2-4 million per week.
- Every weekday, the equivalent of 25 motorway lanes of traffic enters Central London. In theory, the congestion charge encourages people to consider public transport.
- By law, the income from the charge is added to London's transport budget to improve public transport. As a result, since 2003, every London bus has been renewed and the old heavily polluting buses have been removed from service.

How effective has it been?

- Traffic levels are down by 15%, and congestion by 30%. Average traffic speeds are up.
- Road traffic accident rates are down by up to 5%.
- There has been a 12% reduction in emissions of NOx and CO_2 within the zone.
- There is 60% less disruption to bus services.
- Retail areas inside the zone have increased sales every year since the scheme began.
- There has been no effect on the numbers of businesses starting up or closing down within the zone, compared to elsewhere in London.
- There has been no effect on property prices.
- Income of £170 million from the charge was invested in London's transport in the first 2 years.

Strategies for dealing with climate change - 3

Example 4: Congestion charging for New York?

In 2007, New York City mayor, Michael Bloomberg, outlined plans to reduce emissions, including:

- cutting greenhouse gas emissions by 30% by **improving power plants**. To pay for this, $2.50 per month will be added to electricity bills. Spending $30 a year until 2015 will save every household $240 a year after that.
- **planting** 1 million new trees by 2017.
- **congestion charging** in southern Manhattan to tackle poor air quality, raising $380 million in its first year to fund public transport.
- forming a **new transport authority** to fund public transport projects with an annual income of $400 million from congestion charging and $400 million from New York City and State budgets.

▲ Traffic congestion in Manhattan, New York

However, these plans need approval from state government, and face opposition from some political parties and many motorists.

● Over to you

1 In groups, assess the successes and failures of schemes from this unit using the table at the bottom.
2 In pairs, reach a rank order of which type of group you think has been most successful in cutting carbon emissions, and explain why.
3 How far do these organisations prove that, in order to cut emissions, we need to 'think global, act local'?
4 In groups, debate whether Sir Nicholas Stern was right to say that 'pollution was a failure of companies to act and governments to force action'?

● On your own

5 Define the terms from the text: carbon offsetting, carbon credits, carbon sequestration.
6 Research carbon offsetting – **a** its meaning, **b** how it can be achieved. Use www.carbonfootprint.com.
7 Research **a** the benefits and **b** the problems associated with building new nuclear power stations. Why are the public so against nuclear power?
8 In 600 words, decide whether you think new nuclear power stations should be built in the UK.

Organisation	Successes	Failures
Private companies (BA, Shell, Bunge)		
Pressure groups (Greenpeace)		
Individuals (Coldplay)		
Local government (London Congestion Charge)		
The EU Emissions Trading Scheme		

Reaching global agreements; the 1997 Kyoto Protocol

What is the Kyoto Protocol?

- A global agreement setting targets for reducing greenhouse gas emissions. 175 countries have signed up.

What are its aims?

- Industrialised countries are to cut their overall greenhouse gas emissions to 5% below their 1990 levels by 2008-12.
- The targets vary, e.g. EU countries are to reduce their overall emissions by 8% (though the actual amount varies between member states) and Japan by 5% (though highly industrialised, Japan is a low polluter because it opted for nuclear power stations a long time ago). A few with low emissions (e.g. Iceland) are allowed to increase them.

Is it succeeding?

- Some countries delayed signing up to Kyoto, e.g. Russia only agreed to back the Protocol in 2004.
- Some countries never signed up, e.g. Australia, the world's second biggest per capita polluter.
- The USA, which produces 25% of global emissions, initially signed up to it but then withdrew in 2001, following the election of President GW Bush.
- Overall, industrialised countries such as the UK did cut emissions to 3% below their 1990 levels by 2000 – achieved mostly by the collapsing economies of the former USSR countries. Emissions from other high-emitting countries actually rose by 8% in this period, and the UN says they are off-target and will emit 10% **above** 1990 levels by 2010.

- Many developing countries signed up, but did not have to commit to actual figures – instead they report emissions levels annually and develop programmes to cut emissions.
- The EU will not meet its target of 8% reduction. France, Sweden and the UK have achieved it, but Spain, Portugal and Ireland have made no progress.
- Many climate scientists believe that the Kyoto targets are much too low. Instead of a 5% cut in emissions from industrialised nations, they believe that 60% is needed.

The implications of Kyoto for the UK

- A change from coal to cleaner gas-fired power stations has reduced emissions.
- A renewable energy policy to produce 10% of electricity by, for example, wind power is near to target, although the UK is unlikely to reach its overall target for renewable energy as a whole. It would be difficult to achieve more than this.
- The government has taxed petrol more highly to try to cut demand. However, there were large-scale protests about this in 2000 and the threat of others is often present.
- Big reductions in emissions can only be achieved with energy conservation and a shift to low emissions using nuclear power. However, opinion polls show that the public are against a programme of new nuclear power stations.

Who should pay taxes on pollution - producers or consumers?

What do you think ?

Energy use, efficiency and conservation - 1

In this unit you'll consider strategies to combat climate change through energy efficiency and conservation.

Stern Review – right or wrong?

It is easy to get caught up in the debate about whether global warming is caused by human activities or not. However, that debate does not get us very far, and the 2006 Stern Review (see page 49) says that what is most important is to reduce greenhouse gas emissions **now**. If it turns out that climate change IS the result of human-related carbon emissions, then reducing emissions is the correct thing to do. If time proves that it IS NOT, then reducing pollution is also good. Reducing emissions *will* cost money, but Stern argues that doing nothing will cost even more – and could even reduce global GDP by 10%.

Up the chimney

Huge amounts of energy are wasted in the UK:

● In even the most efficient electricity power stations – the biggest polluters – 65% of heat generated goes up the chimney. Some is also lost when hot water is released into rivers.

● Energy is also lost in electricity transmission over distance.

● At home, huge quantities of heat are lost through roofs, windows, doors and walls.

The following examples show whether energy conservation and reductions in carbon emissions can be achieved, both in industry and at home.

Combined heat and power in Copenhagen

Copenhagen's CHP system supplies 97% of the city with clean, reliable and affordable heating, and 15% of Denmark's total heating needs. Set up in 1984 as a partnership between local councils and energy companies, it uses a combination of:

● waste heat from electricity production (normally released into the sea or rivers but now taken through pipes into people's homes)

● surplus heat from waste incineration

● geothermal energy

● bio-fuels (wood pellets and straw)

● small amounts of natural gas, oil and coal.

By 2005, annual household heating bills in Copenhagen were 1400 Euros less than if oil had been used for heating. Between 1995 and 2000, the city's annual CO_2 emissions dropped from 3.5 million tonnes to 2.5 million. SO_2 emissions have also been cut by 33%.

▲ *CHP in Copenhagen*

◀ *Losing heat up the chimney?*

Is Stern right – cutting emissions is a win-win situation?

What do you think?

The CHP system is successful because:

1 it is clean
- There are low CO_2 emissions from natural gas and bio fuels.
- It includes a DeNOx system, which removes nitrogen oxide from gas.

2 it is cheap
- CHP is cheaper when compared with other forms of energy. Costs have barely risen in 20 years, and annual costs for consumers are 50% less than with oil heating.

3 it is efficient
- It wastes only 6% of the energy generated.
- The system is computer-controlled so that excess heat in one area can be redirected to another area which needs it more.

4 tax incentives are given to producers
- The Danish government gave tax incentives to the energy companies involved, so they pay less tax if they use CHP.

The UK's nuclear question

Will the lights go out in 2023? That is the question facing UK electricity supply companies.

- Cheap domestic gas supplies from the North Sea are running low, and overseas oil and gas prices have risen dramatically since 2004.
- The UK government has to cut greenhouse gas emissions by 20% of the 1990 level by 2010. Relying on fossil fuels has to change if those targets are to be met.
- In 2006, nuclear energy generated 20% of the UK's electricity. However, most of the UK's nuclear power stations are too old to be efficient and safe. Most were built in the 1970s, and some in the 1950s. All but one will close by 2023, and no new ones are planned. By 2023, only 4% of Britain's electricity will come from nuclear power – from one station in Suffolk (Sizewell B on the right).

Nuclear energy is the most controversial of all energy sources. People are concerned about radioactivity from uranium (the raw material used in nuclear energy), as well as nuclear accidents and how to deal with nuclear waste. The public in the UK is deeply suspicious about nuclear energy following nuclear accidents during the 1950s in the UK, the 1970s in the USA, in 1986 at Chernobyl in the Ukraine, and in 2007 in Japan. The cost of dismantling nuclear power stations after decommissioning is also immense – the UK government admits that it could cost £30 billion.

▲ *Overall emission of CO_2 in Copenhagen, 1995-2000*

▲ *Sizewell B nuclear power station in Suffolk*

▼ *The advantages and disadvantages of nuclear energy*

The advantages of nuclear power	The disadvantages
• Well-constructed nuclear power plants are very clean. • Radioactivity is found in most rock types, including coal. Because of the vast amounts of coal used globally, coal-fired power plants can actually release more radioactivity than nuclear stations. • Nuclear fuel produces vastly more energy than equivalent amounts of fossil fuels. One 2 cm nuclear fuel pellet of uranium or plutonium produces the same amount of electricity as 1.5 tonnes of coal. • Nuclear fuel produces far lower greenhouse gas emissions than fossil fuels.	• Mining uranium is dirty, with added danger from radioactivity. • Nuclear waste is radioactive for many thousands of years. • There is no known safe way to store or dispose of nuclear waste. • Transporting nuclear fuel can be risky, particularly in times of terrorism. • Almost all nuclear accidents can be traced to human error. No technology is ever mistake-free.

Energy use, efficiency and conservation - 2

The future

Can renewable energy sources replace the UK's nuclear power stations and cut our reliance on fossil fuels? UK government advisors believe that:

- renewable energy can only supply 10-15% of the UK's energy requirements
- new nuclear power stations are needed to replace those using fossil fuels if greenhouse gas emission targets are to be met.

In 2004, the government started research into the design and development of new nuclear power stations. Building new nuclear power stations will take time.

How useful is carbon offsetting?

Forests and woodlands cover 10% of the UK. Forest ecosystems absorb CO_2 from the atmosphere and store it in vegetation and soil. In theory, forests and woodlands stabilise atmospheric CO_2 by **sequestering** and storing carbon in plants and soils. Several UK and EU schemes allow individuals and businesses to offset their carbon emissions by planting trees. An estimate is made of the carbon sequestered over the lifetime of a forest, which is then used as a 'carbon credit' against emissions.

However, there is a limit to the amount of carbon that woodlands can sequester, and their contribution to solving the greenhouse gas problem is small as a result:

- The UK's woodlands remove 4 million tonnes of carbon from the atmosphere each year. This is quite high because most UK woodlands are young and still growing. As forests grow older, the rate of CO_2 removal drops.
- However, the UK emits 150 million tonnes of carbon each year, mainly from fossil fuels.
- Therefore, less than 3% of the UK's annual CO_2 emissions are being offset by the UK's woodlands.

With 10% of the UK under forestry, even an increase to 20% would offset only about 5-6% of emissions. Offsetting is, therefore, only useful if combined with emissions reductions.

● Over to you

1 Copy the table at the bottom and show the advantages and disadvantages of the schemes in this unit.
2 Which schemes do you think would work well **a** in the UK as a whole, **b** in your local region? Explain your reasons.
3 In groups, devise a case for a £250 million research programme into ONE of the following:
   - energy conservation and building insulation
   - A new generation of nuclear power stations
   - A CHP scheme for every town with over 100 000 people.

● On your own

4 Define the term sequestration from the text.
5 Can renewable energy sources fully replace fossil fuel power stations? Research Horns Rev wind farm in Denmark, using Google, to find out how far this is possible.

	Advantages	Disadvantages
Use energy more efficiently, e.g. Copenhagen's CHP scheme		
Replace the UK's existing nuclear power stations with new ones		
Conserve and reduce energy consumption, e.g. BedZED housing		
Expand the UK's forests to help with carbon offsetting		

BedZED – an energy conservation project

Households emit 27% of the UK's greenhouse gases, and use 33% of the energy consumed. How can this be reduced? Government research shows that it would cost less to fully insulate every UK home than to build two new nuclear power stations. The answer to the UK's energy problem is simple – use less.

BedZED is a new housing development in Sutton, south London, which attempts to be carbon-neutral. The homes, built in 2004, use heat-efficient natural, recycled or reclaimed materials, which absorb heat during warm spells and release it when cooler. It has its own CHP plant, fed by waste wood from tree surgery that would otherwise become landfill. The CHP system provides hot water, distributed via insulated pipes.

▼ *BedZED, alternative housing in south London*

BedZED – *pluses and minuses*

The diagram below shows methods used to reduce domestic energy consumption. The total cost is £14 000, but central heating systems cost over £5000. Other measures help, such as:

- double-glazing, which cuts heat loss through windows by 50%
- loft and cavity wall insulation, which cut heat loss from walls and roofs by 33%
- low-energy lighting and energy-efficient appliances, which also cut back on household carbon emissions (although initially costing more, the money is repaid by the energy saved).

▼ *How energy-conserving houses work*

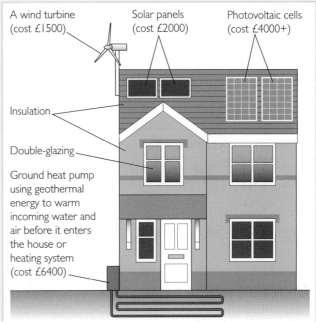

A wind turbine (cost £1500)
Solar panels (cost £2000)
Photovoltaic cells (cost £4000+)
Insulation
Double-glazing
Ground heat pump using geothermal energy to warm incoming water and air before it enters the house or heating system (cost £6400)

Unfortunately ...

- the BedZED CHP system failed in 2005 after months of unreliability
- the reed-beds filtering waste water for use in toilets and gardens were out of operation for seven months
- houses are not cheap and new technology is expensive
- carbon neutrality is difficult to achieve – most people drive their own cars, although there is a car-sharing scheme.

But ...

- the houses are so well insulated and ventilated, that there is little need for heating
- the photovoltaic cells have cut electricity bills
- BedZED uses 33% less water than other developments of its size
- the BedZED houses are in demand – valued at 15% above local house prices
- the residents emit 40% less carbon than average UK households.

1.15 Tackling risk and vulnerability in Bangladesh

In this unit you'll find out how Bangladesh can adapt to the increasing risk of flooding caused by global warming.

● A **delta** is the area of sediment deposited where a river enters the sea or a lake.

Background

Bangladesh

Bangladesh, the world's most densely populated country (population 142 million in 2005), lies on the flood plains of three major rivers, the Brahmaputra, Meghna and the Ganges. These rivers converge in Bangladesh and, together with 54 others, empty into the Bay of Bengal via the largest system of **mega-deltas** in the world. Almost every year, huge areas of the country flood as Himalayan snowmelt adds to monsoon rains and high tides in the Bay of Bengal. 60% of the country lies one metre or less above sea level, and some of the world's poorest and most vulnerable people live here.

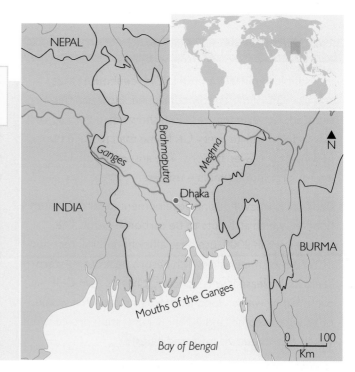

The flood risk

In addition to the Himalayan snowmelt swelling the rivers to bursting point, most of Bangladesh receives 2000 mm of rainfall a year – mostly falling between April and September. Coastal flooding is also a major problem – between March and May, violent thunderstorms produce strong southerly winds of 160 km per hour, which bring in six-metre-high waves from the Bay of Bengal to swamp coastal areas. Also, during tropical cyclones, low pressure reduces pressure on seawater in the Bay of Bengal, which causes it to rise. High tides and storms then surge inland.

Past floods have been devastating. In 1998, two-thirds of Bangladesh experienced flood damage, eroding riverbanks and flooding vast areas.

Bangladesh and global warming

The future looks bleak for Bangladesh. If scientists' predictions about global warming are correct, many millions of people in Bangladesh will be threatened. The very low-lying areas of the country (60%) are where the poorest people live. Their ability to respond to hazards is low, yet they face increasing threats in the future:

- The IPCC predicts that glacier melt in the Himalayas will increase flooding in Bangladesh by 2030. After that, river flows and supplies of freshwater will decrease as the glaciers recede.
- Increased water temperatures will lead to increasing numbers of bacteria and water-borne diseases such as cholera.
- Coastal areas will be at risk from flooding and seawater inundation. This, in turn, will destroy crops and increase the risk of hunger. Indeed, if sea level continues to rise, much of Bangladesh might be permanently

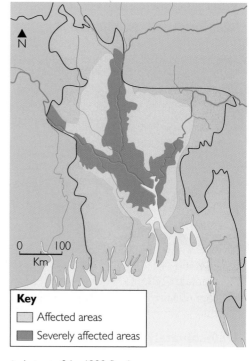

Key
- ☐ Affected areas
- ■ Severely affected areas

▲ Impact of the 1998 floods

lost to the sea, like Tebua in Unit 1.1. This presents the prospect of tens of millions of displaced Bangladeshis seeking new places to live – a significantly larger number of environmental refugees than the potential numbers in Kiribati, and requiring a global response to meet the challenge.

Trend	Farming, forestry and ecosystems	Water resources	Human health	Industry, settlement, and society
Increase in numbers of intense tropical cyclones	Damage to crops; damage to coral reefs	Power failure in storms disrupting water supply	Increased risk of deaths, injuries, water- and food-borne diseases; increased malnutrition, with implications for child growth and development	Disruption by flood and storms; withdrawal of insurance cover (for those able to afford it); loss of property; increased population migration
Increased incidence of extreme high sea level	Damage to crops; likely inundation of freshwater supplies by seawater, affecting irrigation and freshwater supplies		Increased risk of deaths by drowning in floods; increased malnutrition	High costs of coastal and river protection; increased population migration

◀ Impacts of global warming on Bangladesh, according to the IPCC

Adapting to global warming

While the rest of the world considers reducing carbon emissions, Bangladesh has to adapt to global warming now. What are the options?

- **Technological solutions** (hard engineering strategies, e.g. sea walls, river embankments, drainage systems). In the 1990s, the capital, Dhaka (population 12.6 million in 2005), cleared 102 km of drains, constructed 144 km of new drains, and opened up 633 channels to improve drainage. However, this kind of solution involves high costs.
- **Early warning / flood prediction**, which enables people to be evacuated quickly to safer ground, such as flood shelters on natural embankments or raised ground, which have proved effective in the recent past in reducing flood casualties. This option can be targeted at where the most vulnerable live.

- **Behavioural** such as changing land uses, and therefore food choices, e.g. to fish farming. Also banning urban growth in areas of greatest flood risk.
- **Managerial** e.g., improve sanitation, waste management and slum improvement in the poorest areas, so that water infections from disease during floods are reduced. These have always been the biggest causes of death.

Whatever choices Bangladesh makes for the future, there are huge barriers to **adaptation** ahead. What about funding? With such a vulnerable population, who should be protected? Bangladesh has some of the world's most sophisticated flood monitoring, but global warming surely presents the country with huge challenges. Not the least of these is whether the country can actually survive.

> Mass migration is the best way to avoid the flood risk.

● Over to you

1 Copy and complete the table below to show **a** the advantages of some solutions which Bangladesh could adopt to adapt to global warming, **b** barriers in the way of achieving these, **c** whether each solution would help the most vulnerable.
2 In pairs, consider how far the following might be **a** able and **b** willing to pay for protection against flood hazards in Bangladesh:
   a Bangladesh itself

● On your own

   b Aid from the world's wealthier countries e.g. the UK
   c Global organisations such as the United Nations.
3 Define the term mega-delta.
4 Research a datafile on Bangladesh from the World Bank website (worldbank.org) of economic (e.g. GNP) and social (e.g. infant mortality, literacy %) data.
5 Research one flood in Bangladesh, e.g. the one in 1998 or 2007, and its economic, social and environmental impacts.

What do you think?

Type of solution	Advantages	Possible barriers	Will it help those most at risk?
Technological solutions			
Early warning / flood prediction			
Behavioural			
Managerial			

Tackling risk and vulnerability in Indonesia

In this unit you'll explore how Indonesia struggles to cope with the increasing risk of natural hazards and other problems.

Background

Indonesia

Indonesia is located in South-East Asia, between the Pacific and Indian Oceans. It consists of over 17 500 islands, of which 6000 are inhabited. At 1.9 million km², it is nearly eight times the size of the UK. The population of Indonesia is 225 million, with the largest number living on the island of Java. Indonesia is the most populous Muslim country in the world. It is also one of the world's most hazardous countries, situated as it is on the Pacific Ring of Fire – an area of frequent earthquakes and volcanic eruptions.

Devastation in Banda Aceh

Because its large island of Sumatra was situated closest to the epicentre of the December 2004 earthquake which caused the Boxing Day tsunami, Indonesia was by far the worst-affected country – 168 000 people died, and 500 000 were made homeless. Banda Aceh, capital of Aceh province in Sumatra, was closest to the epicentre and was hit by 15-metre-high waves just 15 minutes after the main earthquake. The devastation was made worse because the earthquake caused the Earth's crust there to sink, flooding the city – in some parts permanently (see below).

Political factors played a part too. Indonesian government control of Aceh province was very limited at the time of the tsunami, due to rebel separatist groups there. This affected the response to the tsunami in the short and medium term. Now, Aceh rebels and the government have struck a peace deal, after 30 years, which will make reconstruction and future hazard management there easier.

No early warning system could have prevented all of the deaths from the 2004 earthquake and tsunami. However, the new tsunami early warning system, that started to be installed in the Indian Ocean in 2006 (funded by international aid), should at least help to reduce the effects of a similar natural disaster in the future. Before 2006, no tsunami early warning system existed in the Indian Ocean at all, although such a system did exist in the Pacific Ocean.

▼ *The impact of the tsunami on Banda Aceh – before and after – showing the destruction and flooding caused by the waves*

Some countries cannot afford hazard management. Should there be a global fund to help them?

What do you think?

Managing hazards

Hazard management in Indonesia is made difficult by the number and range of natural hazards that it faces. In addition, Indonesia's struggling economy and political instability make dealing with hazard events more difficult.

- **Low GDP**. In 2006, Indonesia's per capita income was just US$1550 (about £775). In 2005, its GDP was ranked 114th out of 179 countries.
- **The Asian economic crisis**. After rapid growth in the 1990s, Indonesia's economy suffered in the 1997 Asian crisis. Its currency collapsed in value and, in 1998, the economy shrank by 13.7%. Recovery has been slow.
- **Government corruption** has made the problem worse. In a Corruption Index devised in 2006, Indonesia was ranked the 33rd most corrupt country in the world.
- Its economy is also being drained by **debt**, with no prospect of debt relief. Currently, it owes US$140 billion; annual debt repayments cost US$20 billion. 18% of the Indonesian population is classified as being below the poverty line, and 49% live on less than US$2 a day. However, the World Bank considers Indonesia to be a 'middle-income country' – because its *average* daily income exceeds $2.36 per person, it is not eligible for any debt relief schemes for the world's poorest countries.
- **Political instability**, caused by tension between the Muslim majority and Christian minority, has also hindered economic recovery.
- **Terrorist attacks** allegedly carried out by Al-Qaeda have had a short- and medium-term impact on tourism and economic recovery. Few overseas investors are willing to place their money in a country which faces terrorist threats.

What of the future?

Indonesia struggles to cope with existing natural hazards, and now faces a further challenge from global warming:

- Rising sea levels will increase the wave energy in tsunami.
- Much of Indonesia's population lives in low-lying areas; many islands lie just above sea level and are, therefore, at risk from rising sea levels.
- Greater climatic extremes may worsen the impacts of El Niño, resulting in more and worse droughts and forest fires.
- Indonesia's tourist and fishing industries will deteriorate sharply. Coral reefs are vulnerable to temperature change, and increases in sea surface temperature of 1-3 °C are likely to result in coral bleaching and death.

In addition, Krakatau is emerging from the sea again after its 1883 eruption and destruction (see above), and it shows evidence that it may erupt again in the next 20-30 years. An eruption even half the size of that in 1883 would pose huge problems with evacuation and economic damage, due to the much larger population now living in the at-risk area.

Krakatau volcano today

How hazardous is Indonesia?

- Located at the edges of the Pacific, Eurasian, and Australian tectonic plates, Indonesia often experiences earthquakes – and has 129 active volcanoes.
- Of the world's 20 deadliest volcanic eruptions, six occurred in Indonesia, including the top two – Tambora (in 1815), and Krakatau (in 1883), which killed 130 000 people between them.
- Indonesia has two monsoon seasons annually, causing flash floods. In 2007, storms across Jakarta flooded 106 000 acres of rice paddies, and destroyed 100 000 homes.
- During El Niño periods (see p. 29), the country suffers drought.

● Over to you

1 Using the BBC News website (bbc.co.uk/news) research data and photographs to help you put together a presentation on the impacts of the 2004 tsunami in Indonesia.
2 From your research, how far do you think the damage to Indonesia was unavoidable?
3 Draw a Venn diagram with two overlapping circles. Label one 'Economic reasons' and the other 'Political reasons'. Classify the reasons why Indonesia finds it difficult to manage its response to hazards.
4 Explain how well you think Indonesia will be able to face any future hazards.

Unit summary

What key words do I have to know?

There is no set list of words in the specification that you must know. However, examiners will use some or all of the following words in the examinations, and would expect you to know them, and use them in your answers. These words and phrases are explained either in the glossary on pages 297–301, or in key word boxes throughout this chapter.

adaptation
boreal forest
carbon credits
carbon offsetting
carbon sequestration
carbon sinks
deep sea core samples
destructive boundary
disaster
enhanced greenhouse
    effect
environmental refugees
exponential
geomorphological
global conveyor belt
greenhouse effect

greenhouse gases
hazard vulnerability
IPCC
lahars
magma chamber
mega-delta
Milankovitch cycles
mitigation
multiple hazards
natural hazard
North Atlantic Drift
permafrost
positive ice-Albedo
    feedback

pyroclastic flows
radiation
saline
scenario
subducted
tectonic
thermohaline
    circulation
tipping point
tree line
tundra
vent
volcanic emissions
vulnerable populations

Try these questions

1　Using examples, explain why hazard events sometimes bring more problems for some people and societies than others. (10 marks)

2　Explain how evidence can support arguments for both **a** natural and **b** human causes of global warming. (10 marks)

3　Explain how people are attempting to deal with the effects of global warming at **either** a local **or** a global scale. (15 marks)

What do I have to know?

This topic is about globalisation. It has three parts:

● What globalisation means (headings 1-3 below)
● How globalisation affects population in the UK and Europe (headings 4-5 below)
● Its effects on urban growth globally, and on people and the environment (headings 6-7 below).

1 Globalisation: What is globalisation and how is it changing people's lives?

What you need to learn
• Define globalisation, 'connections'
• The factors which have accelerated globalisation
• Effects of globalisation on migration

2 Global groupings: What are the main groupings of nations? What differences in levels of power and wealth exist?

What you need to learn
• Global disparities in wealth and poverty, e.g. MEDC/LEDC
• Why countries group together, e.g. EU
• TNCs and global business and trade

3 Global networks: Why, as places and societies become more interconnected, do some places show extreme wealth and poverty?

What you need to learn
• Global trade and investment switches some places 'on' and others 'off'; some places 'win' and others 'lose'
• Technology in a shrinking world

This section uses examples (e.g. Disney) to show how globalisation occurs, how it has arisen, and its impacts. You do not need in-depth case studies for questions 1-2, but detailed case study information will help you for question 3.

4 Roots: How does evidence from personal, local and national sources help us understand the pattern of population change in the UK?

What you need to learn
• Changing populations (population structure, migration)
• The impacts of social and economic factors on UK population, migration
• The challenge of ageing populations

5 On the move: How is migration changing the face of the EU?

What you need to learn
• International migrations into Europe
• Case studies of migrations within Europe, and its consequences:
  • labour from Eastern Europe
  • Retirement to the Mediterranean

6 World cities: What is driving the 'new' urbanisation? What are its consequences?

What you need to learn
• Rural-urban migration, growth of million / megacities and impacts
• Contrasting megacities – case studies of newcomers (shanty towns) and movers (suburbanisation)

This section is more focused on examples and needs specific case studies. You need to know the UK's population and examples from the EU. For urban case studies, you will need examples of a major city in the developed and developing world.

7 Global challenges for the future: What are the social and environmental consequences of globalisation? Can these be managed for a better world?

What you need to learn
• The impacts of globalisation – a two-speed world, and moral and social consequences, e.g. exploitation
• Reducing the environmental costs of globalisation

A general understanding is needed of the social and environmental impacts of globalisation; you need to know examples of these, and of different ways of reducing or improving each.

2.1 Taking the Mickey to the world?

In this unit you'll look at how Disney has become a major player in the global economy.

It's a small world

These days Disney is not just about theme parks and films; it is a globally recognised brand. It ranks third in the global brand league (after Coca Cola and McDonald's), and competes with AOL/Time Warner for the top slot in media and entertainment. In 2006, its income was $32 billion. It employs 130 000 employees of its own, as well as 40 000 suppliers in 50 countries. From a small animation studio in California in the 1950s, Disney expanded in the 1980s by taking advantage of the satellite and communications revolution. Its programmes and films can be seen almost anywhere – proving that it really is 'a small world'.

The Disney Corporation is involved in many global activities, such as:
- 230 linked satellite and cable TV companies
- 6 film/TV production and distribution companies
- 12 publishing companies, and 15 magazines and newspapers
- 728 shops worldwide, plus galleries and toy companies
- 5 record labels and music publishing
- 2 theatre production companies
- 5 theme parks and resorts, and a cruise line
- sports franchises and teams
- multimedia – producing CD-ROMs and e-games
- property and human resources agencies

How companies grow
- By **expansion** – where a company grows because of the popularity of its product, e.g. Apple, Coca-Cola
- By **merger or takeover** – where a company takes over rivals who are making the same product in order to increase its market share, e.g. BP and Amoco
- By **diversifying** – where companies take over others to broaden their product range, such as MHLV Group, which owns Moet and Veuve Cliquot champagne, Hennessy brandy, Louis Vuitton, Givenchy, Dior, Loewe and other up-market brands

New economy – new heroes?

Disney is typical of the **new economy**, where creating ideas has become as important as producing goods. Disney's creative ideas originate in the USA, but the merchandise (toys, T-shirts, books) is produced overseas. Merchandise alone earned Disney US$23 billion in 2006.

Disney operates a **just-in-time** system in merchandise production. This leads to both benefits and problems.
- For Disney, the company can wait to judge the success of a new film before investing in merchandise. It uses overseas manufacturers – known as **outsourcing** – and demands quick delivery times. This avoids having to operate expensive production lines of its own.
- However, workers overseas, e.g. in China or Vietnam, may receive low wages, be paid late or less than they were promised, and may use toxic substances banned in the USA. In 2007, toys manufactured in China were recalled from shops because of dangerous levels of toxic lead in their paint.

- The **new economy** is where companies and countries are based more on creativity, in finance, media and management, rather than on the production of goods.
- **Just-in-time** is where companies demand goods from suppliers on short timescales, rather than producing and holding large volumes of stock themselves.

However, companies such as Disney are aware of the poor publicity that can result from using production methods of this type, and increasingly they monitor overseas suppliers to ensure higher standards. Disney cancels contracts with companies who abuse workers. However, workers may still be paid low wages.

The other side of Disney's toys

● Before it closed, workers in a Bangladesh textile factory were paid US$0.15 for every US$17.99 Disney shirt they sewed.

● A Chinese toy supplier for Tokyo Disneyland closed in 2006 after a campaign against working conditions in the factory. 800 workers lost their jobs with no compensation, after working 12-16 hours a day. If they failed to meet production quotas, they had to work unpaid overtime and go without lunch breaks. Those assembling stuffed toys suffered skin allergies and sore throats from inhaling fine particles from the stuffing.

*The 'Disneyfication' process
– Disney's approach is seeping into
our everyday lives* ▶

> *Is it right that those who produce clothes for Disney do not earn enough to buy clothes themselves?*
>
> **What do you think?**

Disney and cultural globalisation

Is the world becoming 'Disneyfied'? Disney owns Spanish-speaking radio stations, foreign language TV channels and a Chinese-language radio station in Hong Kong. Several Disney films target specific markets:

● '*Mulan*' marked Disney's entry into China.
● '*Hunchback of Notre Dame*' was launched to re-brand Disneyland in Paris.
● '*The Lion King*' was aimed at Africa, '*Aladdin*' at the Middle East, while '*Rescuers Down Under*' and '*Finding Nemo*' targeted Australia.

Disney aims for global markets, although its characters remain Americanised. Its influence spreads wider – urban planners have imitated Disney's ways of managing theme parks and people movement, as the panel below shows.

Influences on urban planning
● Shopping malls, like Disneyland, often on suburban edge-of-town developments
● Disney-themed fast food outlets
● Crowd monitoring with CCTV cameras
● Resort tourism with everything on site

Media influences
● Disney TV Channels broadcast 24 hours a day in North Africa, the Middle East, Europe, Australia, Malaysia and even Cuba, and Disney also owns shares in commercial TV channels in Europe and Brazil.
● Chinese state TV uses Disney's ESPN for sports coverage.

Influencing governments
● The US government enforces copyright protection for US companies such as Disney
● The French government paid $2 billion towards EuroDisney, providing 30 000 jobs.

● Over to you

1 In pairs, complete a table showing the benefits and problems of the ways in which goods are produced for:
   a Disney
   b their workers
2 On a world map, locate and show, using different symbols, where:
   a Disney's company decisions are made
   b goods are produced
   c products are consumed
   d governments are influenced
3 How far is Disney a truly global company?

● On your own

4 Explain what the 'new economy' means for those in a wealthy countries and b poorer countries.
5 Explain the meaning of economic and cultural globalisation.
6 Research a short report on 'The pains and pleasures of working for Disney', using these websites for information: http://www.americansc.org.uk/Online/disney.htm#Parks

http://www.hartford-hwp.com/archives/43a/index-da.html

2.2 What is globalisation?

In this unit you'll find out what globalisation means.

Global webs!

Imagine a British tourist wearing a T-shirt made in Guatemala, scrolling through the menu of a Japanese-designed MP3 player (made in Taiwan), selecting the music of an American band, while relaxing in a hardwood chair imported from Thailand, having enjoyed an Indian meal on the veranda of a South African hotel managed by a Spanish leisure chain.

▲ *Familiar names in a Hong Kong street*

On her first visit to remote rural Africa she was invited to a local home for an evening's entertainment. She expected to find out about the traditional pastimes of this isolated community. Instead, the evening turned out to be a viewing of the film '*Basic instinct*' on video. The film at that point had not even reached cinemas in London.

Professor Giddens describing a European's first visit to Africa in 1992

▲ *Familiarity in a shrinking world*

While physical distances between parts of the world remain the same, changing communications have reduced the time it takes to trade or pass information globally from months or weeks to seconds, as the diagram on the right shows. What impacts does this have?

What is globalisation?

Globalisation, as we know it, has only really taken off since the 1970s. In some ways, it is like the world trade that has happened for centuries – between the wealthy who invest and who manufacture goods, and the poor who supply them with cheap raw materials. But in other ways it is very different. Manufacturing has now shifted to poorer countries, and globalisation has widened in scope – as the panel opposite shows.

FALLING COSTS OF COMMUNICATIONS

1500–1840

Best average speed of horse-drawn coaches and sailing ships was 10 mph

1850–1930

Time/space convergence affects some places more than others. It depends on the connections

Steam trains averaged 65 mph
Steam ships averaged 36 mph

1950s

Propeller aircraft 300–400 mph

1960s

Jet passenger aircraft 500–700 mph

1990s
Cyberspace information in seconds

Morse code
to telephone
to satellite
to fibre optics/ internet
and broadband

REDUCING TIME LAPSE OF INFORMATION TRANSMISSION

● **Globalisation** means the processes by which people, their cultures, money, goods and information can be transferred between countries with few or no barriers.

Is the creation of a global village a good thing?

What do you think?

Impacts of globalisation

On finance

- Global capitalism is now spread by large transnational corporations (TNCs) with greater turnovers than the GDP of many individual countries – using cheaper labour in developing economies to supply consumers in America or Europe.
- Trillions of dollars are exchanged electronically each day around the world in payments, loans, share purchases and debt.
- There is a worldwide reduction in consumer prices – 'a race to the bottom price'.

On politics

- Companies such as News International (which owns Sky TV in the UK, and the *Sun* and *Times* newspapers) influence how people vote and think about issues. Politicians seek their support because voters' opinions are affected by what they see and read.
- There has been an expansion of international political organisations, such as the EU.
- This has led to a loss of national identity, as TNCs and political organisations gain power over national governments. Governments can lose the ability to control events in their own countries.

- Some global trade barriers, like **tariffs** and **quotas**, have been removed.

On people

- Those with skills in management, finance, and IT are moving around the world to where they are most in demand.
- Migrant labour is flowing to areas of higher wages and better standards of living, e.g. Eastern European migrants coming to the UK for work.

On culture

- Global tourism is transporting people to far-flung overseas locations.
- Increasingly inexpensive global phone and Internet connections (between those who have them) are bringing information by e-mail or global media (BBC World, CNN).
- A 'global village' is emerging, where people share common interests, such as sport, music and films.
- Music, TV and media encourage a loss of individuality by spreading European or American values and attitudes.

Like Disney, those who decide to manufacture overseas, who determine and influence consumer tastes, and who provide investment, come mainly from the world's wealthy countries in North America, Japan and Europe, plus wealthy investors from areas such as the Middle East. 'New players' such as China, India and South-East Asia have become the manufacturers of the world, joining others such as Brazil who expanded their manufacturing in the 1970s. However, the 'have-nots' – mainly in Africa and the world's poorest countries – become more detached and isolated, with little economic influence.

The term 'globalisation' gets strong reactions. Those FOR globalisation think that it brings people closer together and that trade spreads wealth more equally across the world. Those AGAINST globalisation think that existing inequalities will be made worse and, as a result, there will be more winners and losers.

● Over to you

1 Survey what globalisation means to your class, by finding out:
   a where class members and their parents were born
   b where the clothes they wear, the music they listen to and the films they watch come from.
   c where parents are working and for whom
   d where family and friends are for gap years.
3 In pairs, draw spider diagrams to show what you think the gains and losses are for **a** you, **b** your local area, and **c** the UK as globalisation progresses.
4 Explain who you think the main winners and losers of globalisation are, and why.

● On your own

5 Does it matter if national boundaries lose their importance? Draw up a list of the benefits and problems of this.

Exam question: In 600 words, explain the benefits and problems associated with globalisation, using examples. (10 marks)

In this unit you'll explore the origins of global connections and how globalisation began.

Early connections

1492 was a major turning point in the world's history. In that year, Columbus first landed in America and established European control. He began a pattern of trade and development that shaped the global economy. Thousands of people had their way of life changed as these early **connections** were made. Some gained, while others lost.

> ● **Connections** are how places and people are linked together, e.g. by trade, by transport and information links, and by political control.

The story of Tenochtitlan

Spanish soldier Bernal Diaz wrote the passage on the right about Tenochtitlan, the Aztec capital and site of the future Mexico City, when the Spanish first conquered it in 1519.

In 1519, Tenochtitlan had 300 000 residents, five times more than London. It was an agricultural city divided by canals, and was self-sufficient for food. Using mud from the canals as fertiliser, three harvests were possible each year.

Within a generation, Tenochtitlan (Mexico City) had changed beyond recognition as European colonial systems were imposed by the Spanish.

- The old farm systems fell into ruins. By 1538, the first food shortages occurred.
- By the eighteenth century, large Spanish-owned estates had increased agricultural production and the valley was self-sufficient again. But most of the native population had lost their land and were pushed into areas of poorer soil.
- Clearing forests for large estates caused environmental destruction.

By 1984, the Mexican government reported to the United Nations that: 'land use problems are linked to the lack of soil nutrients. Overgrazing and deforestation have led to resource depletion in the valley. There are serious pollution problems, due to waste from homes and industrial zones. Water shortages mean the 17 million people depend on water from the surrounding mountains.'

> We arrived at a broad causeway and saw so many cities and villages built in the water … we were amazed by the diversity of trees and gardens, planted in order and with tanks of fresh water with streams flowing in at one end and out the other. All the roads have shelters made of reeds or straw or grass so that the people can retire when they wish to and purge their bowels unseen, and also in order that their excrement shall not be lost.

▲ *Mexico City today – one of the most polluted cities in the world*

Early economic connections

Like the fate of Tenochtitlan, today's patterns of wealth, trade and development have been shaped by the past. European countries became wealthy by trading with their colonies, which supplied raw materials, food and labour to their colonisers. Once these supply lines were established, European nations could then become manufacturers.

The impacts on the colonies were considerable:

- By 1560, Spain had overthrown the powerful Aztec and Inca empires of **Central and South America**, gaining land and precious metals.
- In the **Caribbean**, profitable sugar cane plantations replaced subsistence food crops grown for local people. Britain and France controlled most of the Caribbean islands, using slaves from Africa to work the land.
- In **Latin America**, 7-8 million slaves were brought to Brazil alone between 1550 and 1850. The impact on **Africa** of removing millions of young men and women was to impoverish formerly wealthy and powerful African empires.

A wealthy **core** and **periphery** emerged. These trade patterns – see below – largely survive even now.

▲ *Crates of tea from India being unloaded at the British East India Company Docks in London in the 1860s*

▼ *Traditional flows of commodities and finished products*

The Fisher-Clark model of employment change over time ▼

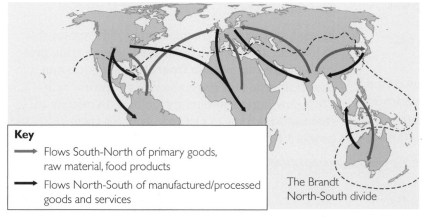

Key

→ Flows South-North of primary goods, raw material, food products

→ Flows North-South of manufactured/processed goods and services

The Brandt North-South divide

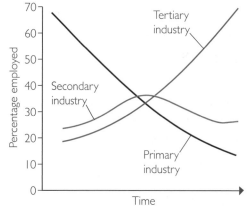

Background

The theory of core and periphery

Geographers use core and periphery theory to explain the process by which some countries become wealthy and others poor. The core is where most wealth is produced. Global core areas include North America, Europe and Japan. This core:

- owns and consumes 80% of global goods and services
- earns the highest incomes
- makes most decisions about the global economy, e.g. what goods are produced
- provides most global investment.

Meanwhile, the poorer periphery is usually distant from core markets. The global periphery of poorer countries:

- owns and consumes 20% of global goods and services, despite having 75% of global population

- earns low incomes – 2.5 billion people live on under US$2 per day
- makes few decisions about the global economy
- provides little global investment.

Recent global shifts of industry mean that:

- manufacturing has fallen in the old core and risen in 'new' peripheral areas because of cheaper labour. However, core countries profit because they dictate to the new production lines.
- now, new global connections add flows of finished and semi-finished goods from peripheral countries to the traditional flows of commodities and raw materials. But investment and decision-making remain in the core.

In the graph above, economists Fisher and Clark show how employment patterns shift as countries change and develop over time.

Getting disconnected

The flow of wealth from the colonies prevented them from developing as much as their colonisers, making them economically **disconnected**. Core nations depended on their investments in the periphery for increased wealth. The periphery nations depended on the core nations for their development, and grew export crops like cotton and tobacco instead of subsistence food crops. Each depended on the other – known as dependency theory.

A few side effects of colonialism have proved valuable, such as India's extensive railway system. But, for most, colonial ties have lasted beyond independence. Most African states still depend on Europe for their trade, and many Caribbean plantations survive today with the descendants of the original European owners still in charge. Some people believe that colonialism lives on, as in the case of Guatemala below.

> ● **Disconnected** in this context means that some countries become less influential and less involved in economic decision-making.
>
> ● **Modernisation theory** is the means by which the world would become more economically uniform, using Western investment aimed at reducing poverty.

Getting disconnected – Guatemala's cotton

In the 1980s, 75% of Guatemala's cotton crop was exported. The money earned was used to buy pesticides, machines and equipment for the next year's crop.

However, if Guatemala had processed the raw cotton into finished clothes, and then exported these instead, the country would have earned far more. More significantly, only 1% of the land devoted to growing cotton would have been needed to produce the same income, and Guatemalans could have grown other crops, opened new markets and grown enough food to feed themselves.

Instead, trade arrangements made it difficult for this to happen, and Guatemala was tied to exporting raw cotton until competition from other suppliers and materials killed off its production altogether by 2005. TNCs now import foreign cotton into Guatemala so that the workforce in clothing factories there can produce T-shirts for overseas markets.

Getting more connected

Economic modernisation – keep the communists away!

The end of the Second World War in 1945 led to the spread of communism in Eastern Europe, China and North Korea. To counter this, the Americans produced an economic model written by Rostow, subtitled 'A Non-Communist Manifesto' which aimed to prevent the further spread of communism by reducing poverty and increasing development, using **modernisation theory**.

- American overseas aid, investment and loans were granted to countries 'at risk' from communism. Investment was poured into Japan, Singapore, South Korea, Taiwan, Thailand, the Philippines, and Mexico – many of whom were ex-colonies. They became dependent on aid and investment, but their economies grew rapidly.

- The result was a growth in wealth from exports – four of the five new entries in the export table on the right for 2000 were countries which had received American aid.

- India introduced the Green Revolution during the 1940s-1970s, with high-yielding cereals and modern farming techniques. Although tying farmers into buying machinery, fertiliser, and seeds from the USA, it increased output and incomes, and probably helped to prevent communist revolution.

	Global exporters	
	Rank in 2000	Rank in 1963
USA	1	1
Germany	2	2
Japan	3	5
France	4	4
UK	5	3
Canada	6	12
China	7	N/A
Italy	8	6
Netherlands	9	9
Hong Kong	10	15
Belgium	11	7
South Korea *	12	N/A
Mexico *	13	N/A
Taiwan *	14	N/A
Singapore *	15	N/A

* recipients of US aid

▲ The world league table for export trade in 2000, with 1963 as a comparison

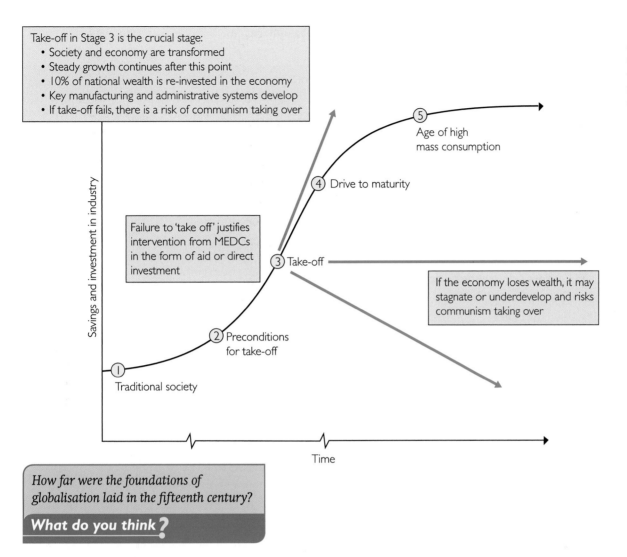

Take-off in Stage 3 is the crucial stage:
• Society and economy are transformed
• Steady growth continues after this point
• 10% of national wealth is re-invested in the economy
• Key manufacturing and administrative systems develop
• If take-off fails, there is a risk of communism taking over

Failure to 'take off' justifies intervention from MEDCs in the form of aid or direct investment

If the economy loses wealth, it may stagnate or underdevelop and risks communism taking over

⑤ Age of high mass consumption

④ Drive to maturity

③ Take-off

② Preconditions for take-off

① Traditional society

Savings and investment in industry

Time

◄ Rostow's model of development – cutting the risk of communism.

How far were the foundations of globalisation laid in the fifteenth century?

What do you think?

Population and the global labour force

The second half of the twentieth century saw dramatic changes in population in many developing countries, brought largely by aid. These included:

• a rapid decline in death rates and increased life expectancy
• inoculations, vaccinations, improved water supplies and diets, reducing infant mortality
• birth rates falling more slowly than death rates
• growing youthful populations, which provided a large labour force, as China and India discovered to their advantage
• children being seen as economic assets; the International Labour Organisation estimates that 120 million children are in the workforce in developing countries.

● Over to you

1 Prepare a case for/against a debate motion that 'The colonies gained benefits from their colonisers'.
2 In a table, summarise the impacts of colonialism in two columns: **a** for the 'connected' nations and **b** for the 'disconnected' nations.
3 How might Guatemala's fate have been different had the country remained 'connected'? How might this have happened?

● On your own

4 Define the following terms: connections, disconnected, core and periphery, modernisation theory.
5 In 250 words, say whether you think it is justifiable for one country to try to defeat the spread of communism in another.
6 Research the exports and imports of one developed and one developing country. How far do they match the trade patterns described in this unit? What problems are there with this kind of trade?

Global differences

In this unit you'll think about whether globalisation reduces global differences.

Recognising differences

Three worlds .... ?

The term **third world** originated during the Cold War. The advanced, industrialised, free-market economies became known as the **first world** (West), and the communist (also industrialised) nations became the **second world** (East). However, these two groups only accounted for 25% of the world's total population, so the remaining 75% became known as the third world.

Differences within the third world were enormous, but images of poverty and low levels of well-being dominated the media and people's perceptions of the Third World – and have been difficult to shift since.

.... or two worlds ....?

In 1980, The Brandt Report acknowledged that human well-being depended on more than economic growth, and it established a wider range of criteria for judging levels of development around the world. It advocated global reforms and urged the **developed** North to share its wealth with the **undeveloped** or **developing** South.

... or one world?

If globalisation – together with the loosening of trade barriers and decades of international aid – has been successful, the gaps between rich and poor areas should have declined. Countries are now classified according to their levels of income: **High**, **Middle** and **Low**.

NORTH – developed core

- 25% of global population
- 80% of global income
- life expectancy of over 75 years now
- most people well fed
- high levels of education
- 90% of global manufacturing output
- 95% of global spending on research and development
- 99% of global ownership of patents
- controls global finance and trade

SOUTH – developing periphery

- 75% of global population
- 20% of global income
- life expectancy of 50 years in 1980, but mostly 65 or more now
- 20% suffer from malnutrition or hunger
- fewer than 50% attend school
- 10% of global manufacturing output
- limited spending on research and development
- dependent on the North for medical supplies
- dependent on loans and aid for development

The North-South divide ▲

The percentage of world average wealth per person in 2006 ▼

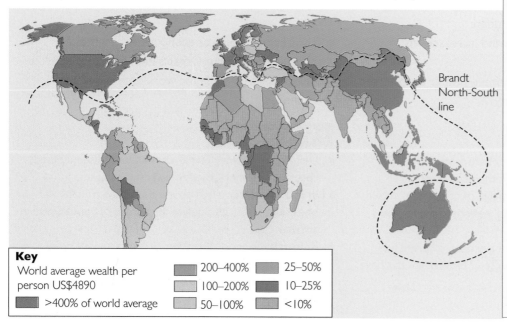

Brandt North-South line

Key
World average wealth per person US$4890

- [ ] >400% of world average
- 200–400%
- 100–200%
- 50–100%
- 25–50%
- 10–25%
- <10%

- The poorest 20% of the world's population received 1.5% of global income in 2006 (down from 2.3% in 1998).
- The richest 20% increased from 70% to 85% of global income in the same period.
- Sub-Saharan countries have seen their incomes fall to below 1980s levels.
- The developing world spent $13 on debt repayment for every $1 it received in aid in 1998; very little has changed since then.

Measuring development

1 Economic indicators

Economic activity is measured in US dollars, using key indicators:

- **GNP (Gross National Product)** is the value of all goods and services earned by a country, including companies working abroad.
- **GDP (Gross Domestic Product)** is the same as GNP but excludes foreign earnings.
- **Per capita** statistics provide an average 'per person' figure, e.g. GDP per capita means average income per person.
- **Purchasing power parity (PPP)** relates average earnings to prices and what it will buy, because a dollar buys more in some countries than in others.

A map of the Human Development Index in 2006 ▼

2 Human development indicators

To balance the emphasis on economic data, the UN uses three indicators about human well-being as a way of measuring progress:

- The **PQLI (Physical Quality of Life Index)** includes life expectation, literacy rates and infant mortality rates. The higher the score, the better the quality of life. Money spent on health care and education implies a higher level of development.
- The **HDI (Human Development Index)** is the same as for PQLI, plus school enrolment and PPP to help judge standards of living.
- The **HSI (Human Suffering Index)** adds daily calorie intake, access to clean water, inflation rate, access to communications, political freedom and civil rights. In this way it indicates quality of life.

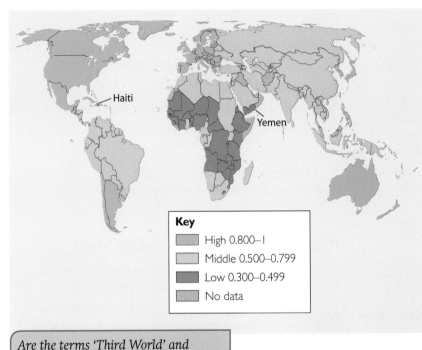

Key

▨	High 0.800–1
▨	Middle 0.500–0.799
▨	Low 0.300–0.499
▨	No data

Are the terms 'Third World' and 'North/South Divide' useful ones now?

What do you think?

Human development trends – the HDI and beyond

- In the last 30 years, almost all world regions increased their HDI score.
- East Asia and South Asia have accelerated since 1990.
- Central and Eastern Europe, and the Commonwealth of Independent States (CIS), declined in the early 1990s after the fall of communism, but have recovered.
- The major exception is Sub-Saharan Africa. Since 1990 it has stagnated, mainly because of the catastrophic effect of HIV/AIDS on life expectancy.
- Eighteen countries have a worse HDI score today than in 1990 – most in Sub-Saharan Africa. Today, 29 of the 31 low human development countries are in Sub-Saharan Africa.

● Over to you

1 Choose 15 countries from the map above – 5 low, 5 middle and 5 high. Complete a table showing their life expectancy, infant mortality, PPP, HDI index. Use the web sites www.worldbank.org and http://hdr.undp.org/hdr2006/statistics/ to research the data.

2 Are these data more useful than GNP or GDP per capita? Explain your thinking.

3 In pairs, think of four other indicators that would help to measure development and research them.

4 Does your evidence suggest that development is bringing a better life for all countries?

● On your own

5 Using Google, research the UN Millennium Development Goals, and the chances of these being met.

Staying connected: global groupings

In this unit you'll learn how different global groupings control economic development.

Mamadou Niang's story

Mamadou Niang is one of thousands of rice farmers in Senegal who are facing an uncertain future because of unfair trade rules. 'Each year the cost of fertiliser, which is imported, increases, but the price I'm paid for my rice decreases.' he says. Rice farmers in Senegal used to be given fertiliser but this was stopped after pressure from the World Bank and International Monetary Fund (IMF).

Last year Mamadou made enough money to last five months, and then he was forced to sell some of his goats to get by. 'My herd of goats is now much smaller than it used to be. This is my bank account, my savings.'

As with most poor countries, the Senegalese government has had to promise the World Bank and IMF that it will open up its markets to foreign competition as a condition for receiving much-needed loans. But the effect of this on farmers like Mamadou is devastating. They cannot compete with the cheaper rice flooding in from Asia and the USA, and much of their rice ends up being unsold.

You're brilliant at stopping me from selling my rice. You're brilliant at destroying my market with cheap imports. You're brilliant at making my children's future uncertain. You're brilliant at keeping me dependent on foreign handouts. You're brilliant whoever made up these trade rules.

A Christian Aid campaign poster from 2003 ▲

Global connections

Mamadou is a tiny player in a huge web of world trade. World trade is dominated by three regions – North America, Europe and East Asia – which produce most goods, control most trade and absorb most investment. In 2001, these three generated 85% of the world's manufacturing output and 81% of exports (an increase from 76% and 71% in 1980). East Asia has increased its share the most since 1980, from 16% to 29% for manufacturing output and 17% to 26% for exports.

Future connections – 'Chindia'

India designs and innovates, and China manufactures. The two economies collaborate. Constant updates in technology mean that many TNCs are having their products designed in India and made in China. Every year, a million engineers and scientists graduate from college in the two countries. Fast broadband links make the distance between India's low-cost laboratories and China's massive low-waged factories irrelevant.

> ● **Trading blocs** are countries which group together to improve their economic interests and trade patterns.

Regional connections – trading blocs

There are a number of trading blocs around the world, including the EU, NAFTA, ASEAN, MERCOSUR and OPEC. They group together countries in geographical areas, and encourage trade within the bloc by removing duties or tariffs on goods for members (e.g. for UK residents buying alcohol in the EU), and creating barriers to outsiders by placing tariffs on goods from outside. For some, economic growth has been huge – e.g. Asia's NICs (Newly Industrialised Countries) such as Thailand have grown dramatically in trade, and the USA–Mexico Border Zone.

Some people argue that blocs protect their members using subsidies, and only encourage trade between members rather than with outsiders. By promoting some goods and preventing others, they prevent the development of a global economy.

Global trade – who's in charge?

As well as trade blocs, several other organisations exist to promote trade of different types and from different countries. These include:

- **WTO** (World Trade Organisation). Formed in 1993, it aims to cut trade barriers (subsidies, tariffs and quotas) that stop countries trading freely, so that goods can flow more easily. It sets conditions under which trading can occur. It has been accused of acting in the interests of wealthy countries, especially the USA, through its fixation with reducing trade barriers.
- **OECD** (Organisation for Economic Co-operation and Development). It is a global 'think tank' for 30 of the world's wealthiest nations.
- **OPEC** (Organisation of Petroleum Exporting Countries). Established to regulate the global oil market, stabilise prices and ensure a fair return for the 11 member states who between them supply 40% of the world's oil.
- **G8, +5**. The Group of 8 (G8) consists of Russia, the USA, the UK, France, Canada, Germany, Italy and Japan. The G8 represents 65% of the world's trade and meets annually to discuss economic development. In 2005, G8+5 included China, India, Brazil, Mexico and South Africa to create deeper international co-operation and an understanding of climate change and international trade.
- **G20**. This actually has 23 members from the developing world, and was formed in 2004 with a focus on agricultural trade.
- **World Bank**. Exists to promote investment globally and provide loans for countries who agree to conditions (like the IMF).
- **IMF** (International Monetary Fund). It forces countries to privatise (or sell off) government assets, which are then bought by large TNCs, and open up trade in return for re-financing debt. Many believe that this has forced poorer countries to sell their assets to wealthy TNCs.

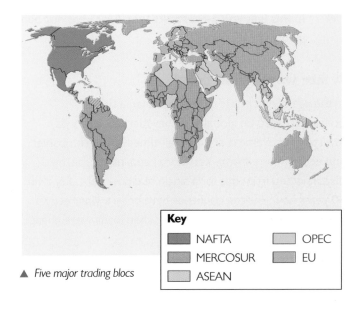

▲ Five major trading blocs

Key
- NAFTA
- MERCOSUR
- ASEAN
- OPEC
- EU

Can free trade ever be fair trade?

What do you think?

Over to you

1 In pairs, draw a diagram to show the different organisations and groups who affect Mamadou's economic fate, and how they do this.
2 In pairs, research one of the trading blocs shown in the map, so that your class covers all of them. Design a leaflet explaining the purpose of one of these trading blocs.
3 Compare your findings. Which blocs have **a** similar and **b** different aims?
4 Using examples, suggest why the membership and influence of blocs might change over time.
5 How and why might trading blocs affect Mamadou?

On your own

6 Study the trade maps on http://ucatlas.ucsc.edu/economic.php and explain the differences in imports and exports between trading blocs.
7 Use the following web sites to show how effective **a** the G20, and **b** OPEC nations are:
www.g-20.mre.gov.br/members.asp
www.opec.org/home
www.ces.ncsu.edu/depts/agecon/trade/eight

GOING GLOBAL

In this unit you'll investigate why Africa has not benefited from global connections.

A fate worse than debt?

In the age of electronic cash, more than a trillion dollars is traded on global currency markets every day. Investments overseas are at an all-time high as companies and economies expand. Yet Africa remains stubbornly disconnected from this for a single reason – debt. For over 30 years now, African countries have been trying recover from debts dating from the 1970s, when loans were cheap.

One of the biggest lenders was the World Bank, which draws together investment money from banks in the developed world and lends it to developing countries. One of its former chief officers, Susan George, exposed how African countries got into a debt crisis and became unable to repay their loans. In her book *A fate worse than debt?* she explained this in a series of eight stages, shown in the panel on the right.

Africa's crisis – the debt burden

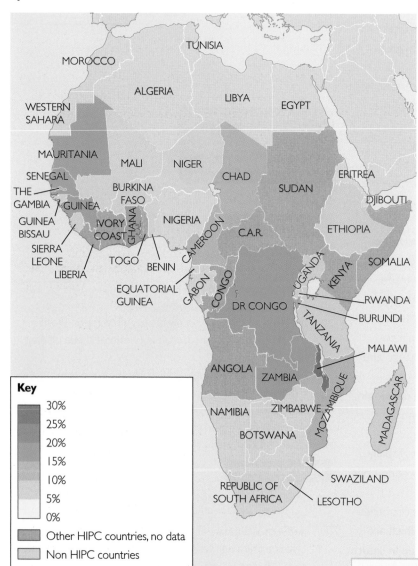

Key

	30%
	25%
	20%
	15%
	10%
	5%
	0%
	Other HIPC countries, no data
	Non HIPC countries

▲ *The cost of servicing debts in some African countries in 2003, as a percentage of government revenue. The **Heavily Indebted Poor Countries initiative (HIPC)** referred to in the key was set up in 1996 by the IMF and World Bank to reduce the debt of the poorest countries.*

> *Africa's debt burdens are the new economy's chains of slavery.*
> *Jesse Jackson, 1998*

> *Debt, of course, is not Africa's only problem. Nor are all countries in the same situation. But 33 of the 41 countries identified by the World Bank as heavily in debt are in Africa. North African countries are using almost a quarter of their export earnings to pay off debts.*
> *Africa Action, 2005*

> *Must we starve our children to pay our debts?*
> *Former Tanzanian President Julius Nyerere, 2003*

> *Development aid, which has been in steep decline in recent years, does not make up the gap. In 1996, sub-Saharan African countries were paying out $1.30 on debt repayment for every $1 received in grant aid from donors.*
> *Jubilee 2000*

- *UNICEF estimates that debt has caused 5 million children and vulnerable adults to die in sub-Saharan Africa since the 1980s.*
- *3 million more children are expected to die in the poorest countries of sub-Saharan Africa by 2015.*

● **The International Monetary Fund (IMF)** is a global banking organisation based in Washington (and largely controlled by the USA), which exists to ensure global banking stability by assisting countries with investments and debt.

How the world's poorest countries got into debt

1 In the 1970s, **OPEC** raised the price of petroleum twice – massively increasing the earnings of its members.

2 The OPEC countries banked their earnings in Western banks.

3 The banks then lent this money to developing countries for huge infrastructure projects, such as dams and power stations.

4 By the 1980s, global interest rates had more than doubled – increasing the repayments on the loans made in the 1970s.

5 Developing nations became crippled by debt. Some could not meet their repayments.

6 To prevent a collapse of the world's banking system because of the developing countries that could not repay their loans, the **International Monetary Fund (IMF)** constructed a solution called **Structural Adjustment Packages (SAP)**. It re-scheduled the loans to make them more affordable – but in return for cuts which the IMF imposed on government spending in those countries.

7 Without IMF approval, no debtor country in trouble would get further credit. SAPs therefore became compulsory.

8 The biggest government spending by developing economies is almost always on health and education. Compulsory IMF cutbacks on government spending therefore affected these – and had the greatest impacts on the poorest people.

Susan George claims that the debt in these countries has caused malnutrition, child mortality, and a halving of living standards for ordinary people. Only recently has debt been reduced. By the early 21st century, the estimated total debt of developing countries was over $3 trillion. Many African countries now use over 40% of their annual income to repay loans (see the map opposite). Is it any wonder that economic development seems to pass them by?

Background

Africa's global disconnections

In order to understand Africa's lack of economic progress, it helps to see the last 50 years in four stages:

Part 1: The 1950s and 1960s – decades of 'progress'?

- Between 1945 and 1965, Western technology boosted farming in Africa. Exports boosted economic development.
- Between 1965 and 1975, American grain surpluses flooded markets in developing countries. It became cheaper to import grain than to grow it.
- Cheap grain caused the collapse of local markets. African governments encouraged farmers to grow cash crops for export instead, e.g. coffee, nuts, beans, and cotton – using taxes on the farmers' earnings to fund development projects.

Part 2: The 1970s – the dangerous decade?

- The oil price increase made imports more expensive, because of increased transport and energy costs.
- The increased costs threatened African development projects unless these countries could earn more by exporting. Therefore they grew even more cash crops to help pay for imports.

Part 3: The 1980s – a decade of squeeze?

- International interest rates doubled, causing a global recession.
- Demand for products from overseas fell in the richer countries, so the value of African cash crops fell.
- Falling incomes in Africa meant that more cash crops were needed just to make the same income as before. Countries found themselves in debt.
- Poorer farmers were forced to use marginal land. Soils became exhausted, dry, and eroded. A series of natural droughts caused famine in East Africa.

Part 4: The 1990s and 2000s – decades of hope?

- The UN set its **Millennium Development Goals** to encourage wealthier nations to reschedule debt repayments and to make debt either affordable or to cancel it altogether.
- The 2005 G8 Gleneagles Agreement cancelled $40 billion of debt owed by 18 of Africa's poorest countries. In return, these countries are expected to liberalise their economies and privatise key industries. Yet these 18 countries still owe $523 billion according to the World Bank. They had already paid back $550 billion yet they only borrowed $540 billion in the first place!

Strained connections – 2

A marginalised continent?

In 1945, Africa could feed itself. But from the 1980s onwards, many Africans have starved and have made little economic progress. What changed?

The following case studies investigate the situation in Zambia, Tanzania and Kenya, in order to explain what changed. They ask three questions about Africa's inability to 'connect' with the rest of the world:

Africa's plight 1: a lack of resources?

Zambia is rich in copper deposits. Copper once shaped Zambia's economy, and paid for its schools, health care and prosperity. For 27 years, it provided 90% of Zambia's foreign earnings and 25% of its GDP. However, in the 1990s, its value fell as it was replaced by fibre-optic cables in modern telecommunications technology. Zambia's exports fell, its mines closed and people were put out of work. GDP fell as a result, and Zambia's debt reached two and a quarter times its GDP. The government was forced to cut spending on health and education, and Zambia's Human Development Indicators (including life expectancy) have fallen as a result.

Source: World Bank, 2007 ▼

Zambia data file	
External debt (% of Gross National Income - GNI)	83.3
GDP (US$ billion)	7.3
GNI per capita (US$)	500
Life expectancy at birth (years)	38
Total population (millions)	11.7
Population growth (annual %)	1.6
Primary school enrolment (%)	88.9
Surface area (km²)	752 600

An abandoned copper smelter in Zambia ▶

Africa's plight 2: poor-quality farmland?

Tanzania used to pride itself on excellent health care and education, in spite of its low GDP. It has some excellent farmland on fertile volcanic soils, and its wealth came from cotton, the price of which was high in the 1970s. Farmers were encouraged to grow cotton, and did well. However, in the early 1980s, global cotton prices fell by half due to global over-production. The price was determined in London, Tokyo and New York. At the same time, prices of manufactured goods were rising – which Tanzania was unable to pay for. Farm incomes fell, and the government received less tax as a result – forcing cutbacks in spending on education and health. Cotton prices have remained low ever since, because global production remains high.

Source: World Bank, 2007 ▼

Tanzania data file	
External debt (% of GNI)	64.4
GDP (US$ billion)	12.1
GNI per capita (US$)	340
Life expectancy at birth (years)	46
Total population (millions)	38.3
Population growth (annual %)	1.8
Primary school enrolment (%)	91.4
Surface area (km²)	945 100

Picking cotton in Tanzania ▶

Africa's plight 3: too little food?

During the 1990s, Kenya had Africa's fastest growing population, at 3.5% per year. It was also suffering from debt and needed to develop new ways of earning income. Now, land has been devoted to cash crops and flowers. In 2001, the UN reported that sub-Saharan Africa supplied 43% of the cut flowers imported into the EU. The value of these flowers was over €250 million.

In 2001, 23 of the countries involved in the flower trade faced severe food emergencies as drought struck, and over 60% of their people suffered a 20% decline in food per head. In the same year that water shortages occurred, Kenyan farmers diverted water supplies in order to produce 52 million tonnes of flowers for the EU. Water from the slopes of Kilimanjaro is piped to the outskirts of Nairobi to support flower farms.

The area around Lake Naivasha is the main commercial flower-growing district, but water supplies there are now dangerously low, the ecosystem is suffering and people are suffering food and water shortages. Meanwhile their land and water were used to provide decorative flowers for a market 4250 miles away.

▲ *Roses being grown in Kenya for export to the EU*

Kenya data file	
External debt (% of GNI)	33.1
GDP (US$ billion)	18.7
GNI per capita (US$)	540
Life expectancy at birth (years)	49
Total population (millions)	34.3
Population growth (annual %)	2.3
Primary school enrolment (%)	79.9
Surface area (km²)	580 400

▲ *Source: World Bank, 2007*

Should Africa's debts to the rest of the world be cancelled?

What do you think **?**

● Over to you

1 Use the World Bank website (www.worldbank.org) to research development indicators (e.g. GNP, infant mortality) for five African nations which are heavily in debt and five which are not. What do you notice?

2 In groups of three, use the examples of Zambia, Tanzania and Kenya to decide whether a lack of resources, poor-quality farmland, and a lack of food help to explain Africa's slow rate of development.

3 Divide a selection of sub-Saharan countries among the class, e.g. Niger and Mali. Research them, report back, and produce a display map to include:
- each country's debt
- its resources and exports
- items that class members have bought which come from there

- the numbers of international companies operating there.

4 Comment on the patterns that emerge.

● On your own

5 Use Susan George's findings, and the background box on Africa's global disconnections, to complete a table identifying the economic, social, and environmental impacts of events over the past 50 years.

6 For each event, identify who is responsible and how. What do you notice?

7 Using the BBC News website, search for 'Africa economy', and explain why Africa cannot feed its people.

Exam question: Using examples, explain how some countries have been affected by debt, and the impacts that this has had on them. (10 marks)

Corporate connections – 1

In this unit you'll find out about the role played by TNCs in the globalisation process.

Corporate colonialism

We think of TNCs as recent developments, but from the seventeenth century parts of India were run by the East India Company. It controlled trade routes and ruled 20% of the world's population. This is probably one of the world's first TNCs. Today there are over 60 000 TNCs, and the top 200:

- account for 25% of the world's economic activity
- employ 1% of the world's total workforce.

In addition, six of the top ten TNCs originated in the USA – and some make more money in a year than some countries produce in GDP.

The Top 10 TNCs, 2006 ▼

Rank	Company	Country of origin	Revenues ($ billion)	Profits ($ billion)	Compared to country GDP
1	Exxon Mobil	USA	339.9	36.1	More than Saudi Arabia
2	Wal-Mart Stores	USA	315.6	11.2	More than Saudi Arabia
3	Royal Dutch Shell	The Netherlands/UK	306.7	25.3	More than Saudi Arabia
4	BP	UK	267.6	22.3	More than Denmark
5	General Motors	USA	192.6	-10.6	Equal to Finland
6	Chevron	USA	189.5	14.1	Equal to Finland
7	DaimlerChrysler	Germany	186.1	3.5	Equal to Portugal
8	Toyota Motor	Japan	185.8	12.1	Equal to Argentina
9	Ford Motor	USA	177.2	2.0	Equal to Hong Kong
10	ConocoPhillips	USA	166.7	13.5	More than Venezuela

How do TNCs grow?

How and why do companies become so large? Three factors are important: motive, means and mobility.

Motive

Under a capitalist system, there is one motive – profit. Therefore, those companies which become dominant do so by controlling costs – which means mostly the price of raw materials and the costs of producing goods and services. This involves mergers and take-overs, which occur in three ways:

- *Horizontal integration* – by buying up the competition. Ford moved from mass-market sales to up-market sales by acquiring Jaguar, Volvo and Land Rover. In 2000, Italian and Deutsche Telecom merged in a US$81.2 billion deal; the new company now reaches 25% of Europe's population, has 20% of online traffic, and 33 million mobile subscribers.

- *Vertical integration* – by controlling and owning every stage of production from exploration and research through to sales, e.g. Exxon owns oil wells, oil tankers, oil refineries and petrol stations.

- *Economies of scale* – companies expand production (e.g. Apple computers) to increase efficiency and reduce unit production costs.

2006 set a record of $3.4 trillion worth of company mergers and acquisitions, the same value as all of the debts owed by the world's poorest countries.

Means

The banks provide the means for companies to grow by providing finance. HSBC supports businesses throughout the world. Companies invest overseas to boost their market presence or to take advantage of labour and environment laws, e.g. where environmental regulations are less strict than at home. Flows of money around the world connect businesses and countries in complex global webs.

The top five countries investing overseas have for a long time been: the USA (21%), the UK (15%), France (8%), Germany (7%) and Japan (5%). In 2005, half of all global investment came through London's banks.

But a growing trend this century has been for TNCs from developing countries to invest overseas. A kind of **reverse colonialism** is happening with Hong Kong, Singapore, Taiwan, China, South Korea, Malaysia, India and Brazil now accounting for more than 15% of overseas investment. Examples of this, from the first part of 2007, include

- Indian firms, led by Hindalco and Tata Steel, who bought 34 overseas companies for a combined US$10.7 billion.
- Big Blue sold its personal-computer business to a Chinese multinational, Lenovo.
- Russia overtook India, investing US$11.4 billion abroad, and was in the running to buy Alitalia, Italy's state airline.

> While globalisation has opened new markets to rich-world companies, it has also given birth to a pack of fast-moving, sharp-toothed new multinationals that is emerging from the poor world.
>
> The Economist, April 2007

Mobility

Other factors which have helped companies to grow relate to mobility, transport and communications, and include:

- accelerated and cheaper transport – including the use of ever-larger container ships.
- accelerated communications systems – using fibre-optics, satellite and digital technology
- production technology – 'just in time' (see Unit 2.1) and production systems that can provide cheap and fast turn-around, enabling companies to be faster than their competitors.
- A reduction in transport and communications costs – even as oil prices have risen, companies have responded with fuel-efficient technology. Travel between the UK and Australia costs the same in 2007 as it did in 1985.

Supermarkets as superpowers

The new global supermarket companies are huge and control massive volumes of food and non-food goods. Wal-Mart is the world's biggest, but in the UK (and increasingly in Europe) Tesco is the giant. In 2007, £1 in every £8 spent in the UK was spent in Tesco. In 2007, Tesco operated in 12 countries around the world, had over 2300 shops and had expanded from groceries into non-food products, finance (everything from car to pet insurance) and petrol. In April 2007, its profits exceeded £2 billion.

In the late 1990s, Tesco expanded into Central and Eastern Europe, and South-East Asia. 57% of Tesco's floor space is now outside the UK. These new host countries lacked supermarkets, and planning controls were less restrictive than in the UK.

The Tesco philosophy is 'to be everywhere, sell everything and sell to everyone.' ▼

Key
- Countries in which Tesco currently operates
- Countries with planned or speculated future Tesco operations
- Countries from which Tesco has withdrawn

Corporate connections – 2

Supermarkets are contentious. In the developed world, where working people are increasingly cash-rich but time-poor, the opportunity to do all the shopping under one roof, with free parking, ease of access and – increasingly – at any time of day or night, is appealing. Like many major companies, the big supermarkets are subjected to criticism – but they have their own arguments too, as the table shows.

Searching globally, selling locally

Criticisms of supermarkets are more to do with the way that most food in the UK is sourced and sold. Large supermarkets search the world in order to deliver the best-priced or guaranteed quality goods to their customers. The bigger the company, the stronger its purchasing power is, and its ability to control suppliers and prices.

- In the case of fresh food, this means that produce is shipped or flown from one side of the world to the other to supply out-of-season fruit or vegetables.
- In the case of non-food items, the companies search for the cheapest. Often, this raises the same issues as have been shown in Unit 2.1 on Disney.

Critics would say …	Leading supermarkets would say …
The supermarkets' profits do not benefit the communities or countries in which they operate.	We are commercial companies that provide both a service that people want and also local employment.
Large supermarket chains cause local shops to close.	A number of supermarket chains have now opened 'local' or 'express' shops which are open for 16 hours a day or more, and which provide goods of known quality. Opening a store brings more shoppers into the area than there were before.
Supermarkets monopolise the market	We provide a range of services all under one roof – and people can shop elsewhere if they want to.
Supermarkets have a major impact on food suppliers – often undercutting prices.	Without the large supermarkets, many food suppliers would grow produce without necessarily having a guaranteed market.
Local suppliers are undercut by cheaper goods from elsewhere, e.g. Polish milk is sold throughout Central Europe	Consumers want low prices and guaranteed supplies – so these goods have to be sourced where both are possible.

Women in Bangladesh making clothes for UK supermarkets work 14 hours a day and earn as little as 5p an hour – below what they need to feed their families. ▶

▼ *Supermarkets exposed – messages from around the world.*

In South Africa

Casual women farm workers:

'I get 378 Rand [£32.50] every two weeks, which is not the minimum wage. I can't afford school fees for my daughter or go to school functions or buy school uniforms.' *Tawana Fraser*

'A (supermarket) buyer … picks up the phone and says x is offering me apples for £1 a carton cheaper; meet him or I take you out of the programme,' says a farmer and supplier from the Western Cape,

'They also change the prices – £1.49 is the price, then suddenly they put it on sale and make it 99p. Then they sell it in bulk.' *Apple grower*

'Supermarkets … have all the power in the world, and we have to cut costs as far as we can. We're really at their mercy.' *Wine producer*

Elsewhere in the world

Workers in Costa Rica producing bananas for export to all major UK supermarkets are earning 33p an hour – a wage so low that they cannot afford to take an hour off when dangerous pesticides are being sprayed.

Women in India processing cashew nuts for supermarkets are scarred by the corrosive oil produced by the nuts as they are shelled.

'One supermarket wanted us to change their grape packaging from open to sealed bags. The new bags were three times as expensive. And productivity in the pack-house went through the floor, because it took workers 20 to 30% longer to seal those bags. But the price stayed exactly the same – That's the way it goes.'

Growing locally, selling globally

Almeria province was one of Spain's poorest regions until it discovered the polytunnel. The UK's demand for year-round fruit and vegetables has transformed the plain of Dalías and the Alpujarra hills, putting a National Park at risk. 350 km² of market gardens grow produce for the UK. Crops grow in bags beneath hectares of plastic, with serious environmental impacts as the panel below shows. Cheap migrant labour (often women) from Mali, Colombia, Ukraine, Romania, Poland and Morocco help to maintain farmers' incomes as supermarkets force prices down.

▲ Long hours of sunshine and mild sea breezes make Almeria ideal for growing, but WWF claim that supermarkets are causing an environmental catastrophe.

Growing under plastic

How it is done:

- The grow bags contain oven-puffed grains of perlite stone.
- Chemical fertilisers are drip fed to each plant by computer.
- 40 kg of pesticides are used per hectare.
- Strawberries are grown on soils which are sterilised using ozone-depleting methyl bromide.
- There is continuous cropping from Oct-July for tomatoes, and from January-July for strawberries.

Is globalisation just another name for twenty-first century colonialism?

What do you think?

The system has serious impacts:

- Trees are cleared and terraces are constructed on hillsides.
- There is a risk of serious flooding after heavy rain.
- Riverbeds are dried up by water demands from farmers and ground water levels have fallen by 50%.
- The natural wetlands of the Doñana National Park/World Heritage Site are threatened, together with the habitats of the rare species e.g. Iberian lynx.
- There is illegal use of water wells.
- Streams are polluted, plastic litter and pesticide drums are abandoned.
- There is an increased risk of breast cancer from pesticide usage.
- Small villages become dependent on single crops.

● Over to you

1 a TNCs can be categorised into resource-based or producing, manufacturing and retail and services. List five examples of each.
   b Give examples of companies outside these categories, or which cut across more than one.
2 In threes, research one of each category, using these headings: locations of its headquarters; where it operates; workforce size; products; how it supplies products; impacts on environment and societies.
3 Complete the table below about each of the three types of growth.

● On your own

4 Define the following term: reverse colonialism.
5 Why is Tesco expanding its number of overseas stores? In a table, list the positive and negative consequences of supermarkets expanding overseas for **a** the company, **b** the UK, and **c** the country into which they expand.

Exam question: Using examples, assess how far TNCs bring benefits and problems for their suppliers and consumers. (10 marks)

Type of growth	What it means	Examples of companies	Its benefits (and for whom)	Its problems (and for whom)
Horizontal integration				
Vertical integration				
Economies of scale				

In this unit you'll learn about the emergence of China and India as global players, and consider whether everyone wins.

Emma's here for Christmas!

On 4 November 2006, the giant Danish container ship *Emma Maersk* docked at Felixstowe, Suffolk, after a journey that took in Gothenburg in Sweden, Yantian in China, Hong Kong, and Tanjung Pelepas in Malaysia. It was carrying 11 000 containers – mostly for the UK Christmas market – holding:

- Christmas crackers, poker tables, bingo sets, drum kits, electronic toys and gadgets (including 12 800 MP3 players), 40 000 rechargeable batteries, children's building blocks and 1 886 000 Christmas decorations
- 22 280 kg of Vietnamese tea, thousands of frozen chickens, 150 tonnes of New Zealand lamb, pumpkins, 10 tonnes of mussels, along with unspecified quantities of swordfish, tuna, noodles, biscuits, jams
- other assorted items included potato mashers, slotted spoons and graters, toothpicks, leather sofas, spectacles, televisions, Disney pyjamas, and 138 000 tins of cat and dog food.

▲ The **Emma Maersk** *is the world's largest container ship and, at 397 metres long, is the longest ship of any type currently in use*

China – the world's fastest growing economy

The *Emma Maersk* delivered 45 000 tons of consumer goods from China, and is the latest stage in the shift in manufacturing from Europe and the USA to China, which has already taken in Japan, Taiwan and South-East Asia. In 2006, the port of Felixstowe reported an increase of 16% in Chinese imports. Since 1980, the UK has experienced a thirty-fold increase in trade with China – exporting goods worth £2.8 billion and importing goods worth £30 billion in 2005 alone. The *Emma Maersk* is a true symbol of globalisation.

China's growth as a manufacturing nation means that European consumers can purchase cheaper goods, and therefore inflation remains low. Ironically, Maersk Shipping explained that the *Emma Maersk* would be used to ship one of the UK's biggest exports to China – waste plastic – which, in turn, would be turned into new soft toys and decorations ready for Christmas 2007!

> The Emma Maersk brought the goods that Europe used to make. Whole sectors of global trade are now being dominated by companies operating in China. The real cost of the goods that the Emma Maersk is bringing in should include the environment, the markets destroyed in developing countries and the millions of jobs lost.
>
> *Caroline Lucas, Green MEP for South-East England*

▼ *The distribution of key energy resources in China*

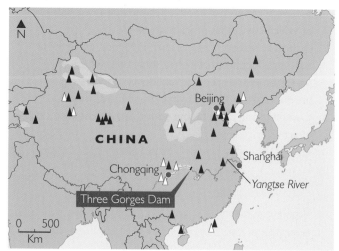

Key

▨	Coal field
△	Gas field
▲	Oilfield

How has China grown so quickly?

China has undergone massive change. Since the 1970s, China's economy has doubled every eight years, while America's economy has doubled just once. China has the largest sustained GDP growth ever seen and is no longer a low-income country. Its largest city, Shanghai, has been transformed in the last twenty years, probably more than any other city on Earth. Its shiny business towers offer a glimpse of China's future, when manufacturing will have moved elsewhere and the 'new' knowledge and information sector take over. How has this change happened?

Local reasons

- From 1949 to 1976, Mao Zedong's People's Republic kept communist China deliberately separate and disconnected from the rest of the world. The economy was planned centrally, goods produced for the consumption of China's own people, and little individual wealth or business was permitted.
- Following his death in 1976, new rulers developed an 'Open-Door Policy' to investment from overseas from 1986 onwards.
- In the 1990s, China transformed to a more capitalist economy, allowing individuals to accumulate wealth by producing goods and services, without State interference.
- China's massive population (see right) and huge natural resources (see left) provide workers and raw materials for industrialisation.
- Prolonged spending on health and education over 50 years has provided a healthy, literate and skilled workforce.
- The creation of industrial **Export Processing Zones** has stimulated cheap mass manufacturing.

Global reasons

In 2001, China joined the World Trade Organisation (WTO). TNCs now invest in China to take advantage of the low labour costs and the designation of manufacturing Export Processing Zones. Since 2000, China has been the largest recipient of overseas investment, and 53% of its exports are produced by foreign-owned companies or those in partnership with Chinese companies.

Shanghai today ▲

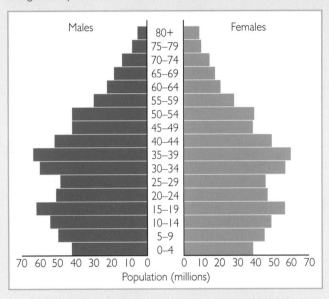

▲ *A population pyramid for China in 2005. By the middle of 2007, the total population of China had reached over 1.32 billion, with huge reserves of people of working age – giving China a massive available labour force.*

● **Export Processing Zones** are zones in which businesses are free to import raw materials, process and manufacture them, and re-export without paying duties or tariffs – helping to keep costs down.

60% of the increase in world trade since 2004 has been as a result of China's industrialisation. China has now overtaken the USA as the world's largest consumer:

- 67 million tonnes of meat (USA 39 million tonnes).
- 258 million tonnes of steel (USA 104 million tonnes).
- 382 million tonnes of wheat, in 2004 (USA 278 million tonnes).
- Before 2050, China will be consuming more oil and paper than the world now produces.
- China already consumes 50% of the world's concrete, 40% of its coal, 25% of its iron and nickel, and 19% of its aluminium.

Global remix: winners and losers – 2

Switching on new connections

The last twenty years have seen huge economic shifts, as countries that suffered from weak connections in the past 'switch on' to providing goods and services for the wealthier nations, and, in turn, grow themselves.

- Increasing economic activity in Pacific countries (Japan, Malaysia, Taiwan) is now being followed by China and India.

The impacts of China's industrialisation

Chongqing, on the Yangtze River, is the world's fastest growing city. At the heart of developments around the Three Gorges Dam, this port is in the centre of Sichuan province. It has a population of 10 million – one third of the population of Sichuan live here – and is expected to double in size by 2020. The city represents modern China, with giant factories and skyscrapers. 50% of the population are migrants, and half a million new workers arrive each year. Since 2000, over US$250 billion has been spent on roads, bridges, dams and power supplies. Living conditions are often crowded and squalid for workers like Yu Lebo. But incomes there are three times higher than on farms.

Chongqing is also one of the dirtiest cities on Earth – China now has 16 of the top 20 – and its filthy air causes thousands of premature deaths. The air quality was so bad in 2005 that it did not even reach the government's own safety standard for 25% of the time.

- The countries of the former communist Soviet Union and Eastern Europe – which, like China, were deliberately isolated and 'switched off' from the rest of the world – have changed into a new capitalist 'bloc', whose energy resources are sure to influence Western Europe in coming decades.

▲ Chongqing, China – the world's fastest growing city

Yu Lebo shares a three-room apartment in Chongqing with his wife and three other couples, all of whom are porters, cleaners or odd-job men. There are two double beds in one room, separated by a thin sheet, a third in a tiny room next door and another in the kitchen.

'I used to be a farmer, but I could not afford to raise my two children. So we left them behind with relatives. I earn about 20 yuan (£1.50) for 12 hours work. Most of this, and the money my wife earns as a cleaner, goes on rent and food, but as long as we stay healthy we can send money home to buy clothes and books for our children.'

Education and health care – once free – are now the biggest burden on peasants.

Adapted from C4/Guardian Films, March 2006 'Invisible City'

The shift to an urban economy

- The movement to cities of people in China is the largest rural-to-urban migration ever recorded – 8.5 million people per year.
- Over 140 million Chinese have left the countryside since the 1980s.
- China now has 90 millionaire cities, i.e. of over 1 million people
- 45 million Chinese are expected to move by 2012. Most head for factory towns by the coast.
- The interior of China is being transformed by industrial development.

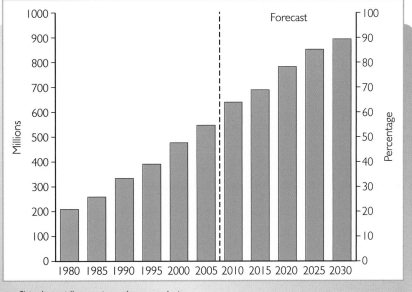

▲ China's rapidly growing urban population

Small commodities – big markets

On the south-eastern coast of Zhejiang province is the rapidly industrialising city of Wenzhou (7.7 million). This poor rural area has become a dynamic Export Processing Zone, and export values rose from $5 million to $29.4 billion between 1978 and 2002. Wenzhou's family-run workshops have turned into factories where the motto is 'small commodity – big market'. Over 70% of the world's cigarette lighters are made here, 30% of the world's socks, millions of door handles, light switches, bra strap fixings, plugs and adapters. The government's '1000 project' for this region developed 1000 km of high-quality roads, 1000 kw power stations, 1000 km of coastal defences and improved 1000 farms. Incomes have risen, education is compulsory for 9 years, and literacy is 100%.

Moving mountains

Inland to the south-west is Lishui, where industrialisation has literally moved mountains. A factory zone was established in 2002. By 2006, over 200 factories were in full swing and 30 000 migrants had arrived. The city's population has risen from 1.6 million in 2000 to 2.5 million in 2005. $8.8 billion has been invested in the area's infrastructure:

- The new migrants work for very low incomes – often below the local minimum wage of 40 cents an hour.
- 108 hills have been flattened to accommodate new businesses. Another 400 are to be removed to permit further expansion by 2020. Thousands of rural peasants have been relocated.

The consequences of China's growth

China's growth has been remarkable – but it has been at a cost.

- 20% of China's population live on less than US$1 a day.
- 70% of China's rivers and lakes are polluted.
- 100 cities suffer from extreme water shortages.
- 360 million people do not have access to safe drinking water.
- Water quality in 207 of the Yangtze River's tributaries is not fit for spraying on farmland, yet 1 in 12 of the planet's population depend on this river for drinking water.
- Tap water in Chongqing contains 80 of the 101 substances forbidden under Chinese law.
- The Huang He River (one of China's biggest) dried up because of over extraction.
- 85% of trees along the Yangtze have been cleared, causing erosion and dust storms
- 30% of China suffers from acid rain caused by emissions from coal-fired power stations. China overtook the USA as the largest emitter of CO_2 in 2007 – over 6 billion tonnes.

Global remix: winners and losers – 3

India today

Motorola, Hewlett-Packard, Cisco Systems and other high-tech giants rely on teams in India to devise software platforms and multimedia features for next-generation devices. Google principal scientist Krishna Bharat is setting up a Bangalore lab to work on core search-engine technology. Indian engineering houses use 3-D computer simulations to tweak the designs of everything from car engines and forklifts to aircraft wings for such clients as General Motors and Boeing. Financial and market-research experts at outfits like B2K, OfficeTiger, and Iris analyse the published accounts of blue-chip companies for Wall Street. By 2010 such **outsourcing** work is expected to quadruple to $56 billion a year.

Adapted from Business Weekly, an American magazine, August 2006

● **Outsourcing** means employing overseas labour in their home country to do a job under contract to provide a service for a company in a developed country. Usually associated with IT software development, banks, and service companies (e.g. call centres).

Bangalore's brain gain

India, together with China, is the world's next emerging superpower. Since the 1990s, economic growth has averaged 6% a year. Language and technical skills are being used to establish an economy that services the world. Outsourcing of computer- and telecommunication-based activities has been substantial. TNCs take advantage of the lower Indian wages and a large well-educated workforce (2 million English-speaking graduates every year).

The Internet has been at the root of India's rapid rise since the government opened up the economy. Services earn the country US$25 billion a year. High-tech companies producing quality goods and services have taken over parts of Bangalore. Indian companies are now producing state-of-the-art equipment and services at 20% of the European prices. Large numbers of Indians have been returning from North America and Europe to work. Bangalore is India's fastest growing city (from 1.6 million in 1970 to 6.5 million in 2006), and has the highest-paid workers. But the gap between rich and poor remains. Roadside tents, overcrowding and squalor sit next to the bright lights of a flamboyant city centre.

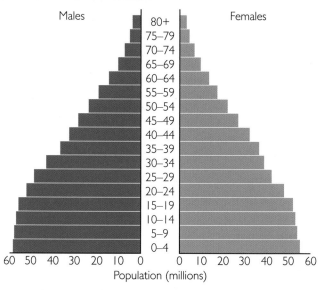

▲ *India population pyramid*

◀ *'I find Bangalore one of the most exciting places in the world,' says Dan Scheinman of Cisco Systems. 'It's like Silicon Valley in 1999.'*

Are the social and environmental costs of China's and India's economic growth worth it?

What do you think?

Switching off rural areas

India's growing economy does not spell good news for everyone. Low international commodity prices, reduced subsidies, and free trade, have each sucked in cheaper food imports from overseas, thereby hurting millions of farmers who previously supported themselves. India's farmers still make up 70% of the population.

Visions of the future – a world of BRICs

In 2007, China overtook the UK as the world's fourth largest economy – and by 2008–9 will overtake Germany as the third. Only Japan (second) and the USA (first) lie ahead, as China grows towards being the world's biggest economic superpower.

By 2050, Brazil, Russia, India and China (the so-called BRICs) will be in the G7 – and only the USA and Japan will remain in the G7 from the current list. Germany, Italy and France will join the UK as former members.

- Brazil and Russia are two of the world's largest suppliers of raw materials.
- Russia controls massive reserves of oil and natural gas, which it supplies to Europe.
- 50% of Chinese foreign investment is in Latin America and the Caribbean.
- Chinese, Indian and Russian companies are acquiring European businesses and opening plants in the USA.

The balance of trade and power is shifting. Where next?

India's winners and losers

No more agriculture for us, it does not feed us, it does not feed our children. We will move to the city and work as tea bearers, and live in slums if we have to – it is better than starving. We see the glittering city on TV and look around us at the shocking levels of deprivation in parts of India that were once prosperous.
Dharampul Jarundhe (rural farmer)

Software professionals are earning money their parents never dreamed of, driving Mercedes and partying all night. I am making more money than my parents could imagine, and for an Indian woman that is totally liberating. I don't need to depend on my parents. I don't need a husband. I can choose when to marry. I might not even need to get married.
Devika (urban call centre worker in Mumbai)

◄ ▼ *Who will be the next superpower?*

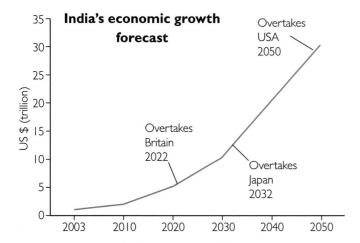

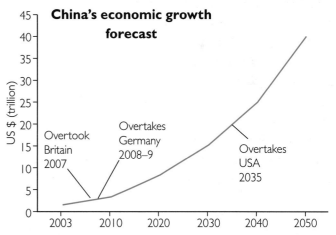

Over to you

1. In pairs, use the World Bank and UN Human Development Report websites to research economic, population and quality of life data for Brazil, Russia, India and China. Report your findings back to the class.
2. Rank the four countries by their levels of development in 1950, 1975, and 2005 – and predicted data for 2020. What do you notice?
3. Is any country likely to catch up and overtake China in the next 25-50 years? Justify your decision.

On your own

4. Define this term: outsourcing.
5. Explain why most economic growth in both China and India has taken place in urban and not rural areas.
6. In 400 words, compare India's development to that of China.
7. What might make Russia a superpower in decades ahead?
8. In what ways might the BRICs pose **a** threats and **b** opportunities for the 'old' economies (e.g. the USA, the UK, France)?

World on the move

In this unit you'll explore the link between globalisation and international movements of people and information.

Migration

International migration is at an all-time high. In 2005, there were 191 million international migrants. In 2006, 3% of the world's population (192 million people) were on the move. Recently, more and more migrants have become concentrated in a smaller number of developed countries. However, there are also areas where people never move.

The total number of migrants has more than doubled since the 1960s, and distinct geographical variations exist ▼

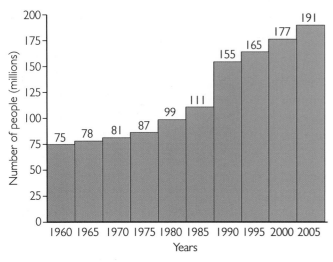

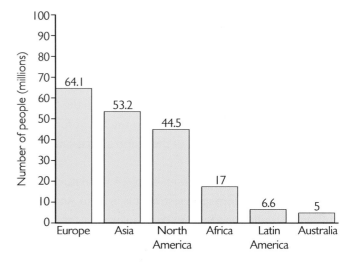

In the last 30 years, the number of migrants has increased due to economic globalisation. For many, such as teachers or doctors, the decision to move is one of necessity rather than choice, because they have so few opportunities in their own countries.

> ● **Source nations** are those from which migrants have come.
> ● **Host nations** are those where migrants decide to go, or to which they are admitted.

Who is moving?

We often assume that it is poor people in developing countries who are keen to move to wealthier nations for higher wages and a better lifestyle. But the reality is very different. Statistics show that the poorest people would prefer to stay in their own countries, and that it is the educated and professional groups already on higher incomes, such as doctors, who are more likely to move. Skilled and educated members of the population in the **source nations** are more aware of the opportunities overseas, and are able to move to fill job shortages for skilled workers in **host nations**.

The average number of migrants entering and leaving Britain in a typical day in 2006, and the main countries involved ▼

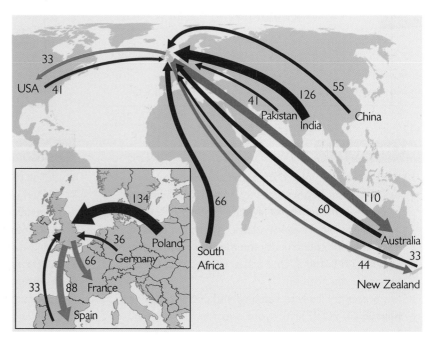

The impact on host nations

Voluntary international migrations (as opposed to forced migrations due to reasons such as civil war) are basically flows of labour. The host nations gain because they fill gaps in their labour markets – with the new workers often prepared to work for lower wages than the local workers. In 2002, the UK government expanded its Highly Skilled Migrant Programme, which is intended to:

- attract the brightest and best foreign workers
- fill skills shortages
- attract lower- or intermediate-skilled migrants where it proves difficult to fill job vacancies
- balance the UK's own ageing population.

Australia, New Zealand, the USA, Canada and some European countries compete with each other for highly skilled workers. Expanding global trade means that leading nations require ICT-skilled, and business-minded workers, in order to compete in the world economy.

The impact on source nations

On the plus side, huge sums of money (called **remittances**) are sent home by international migrants. The World Bank estimates that the total value of remittances exceeds the amount given in aid to individual countries. Up to US$233 billion was sent home in 2006, providing vital income for the families left behind. On the down side, the migrants are usually the most skilled people from the source countries, which creates a **brain drain** and helps to slow down the countries' development.

- Some migrant labour is low-paid, low-skilled and seasonal.
- Overseas Filipina women in domestic jobs sent US$12 billion back to the Philippines in 2006.
- Moroccan textile workers no longer go to France because companies there contract work out to Morocco where labour is cheap.

Rapid communications systems

The twentieth century saw the opening up of airspace and cyberspace to millions of people around the world. Falling costs and increasing availability mean that networks of communications bring people and places closer together. Well-connected places experience the full impacts of globalisation.

The digital divide

Mobile phones, texts, faxes and e-mail now dominate business and increasingly enhance learning, communication and spending. Rapid movement of data via the Internet brings people even closer together; slow or no movement keeps them apart. This creates a **digital divide**. African nations, such as Uganda (see pages 282-85), remain 'switched off', with many people outside the reach of the Internet. Africa only had 34 million Internet users in June 2007, compared with 437 million in Asia and 322 million in Europe. Isolation increases the gulf between the haves and the have-nots, but it can also protect areas from westernised culture and influence, and preserve the uniqueness of places.

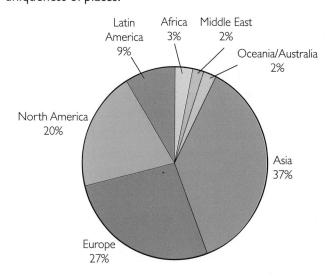

Proportion of Internet users by region ▲

2.10 Migration into Spain

In this unit you'll learn about migration into Spain, and the reasons for it.

In they come!

Spain's population is growing at a rate of around 2% each year – equivalent to an increase of 910 000 people. Only around 220 000 of these additional people are due to **natural increase** – the majority of the 2% growth is due to **immigration**. Foreign residents now make up 9.3% of Spain's total population (there were 4.1 million foreigners living in Spain at the beginning of January 2006).

What are the benefits for Spain?

The Spanish government believes that immigration creates many benefits for the Spanish economy. Spain, like many other Western European countries, has an **ageing population**. Spain's **fertility rate** is low, and the government is concerned that soon Spain's indigenous population will be too small to create the strong and educated workforce necessary for a successful economy. It therefore encourages migration of both skilled and unskilled workers. Nearly 65% of all immigrants to Spain are aged between 16 and 44, so the majority of them are young adults capable of contributing to the Spanish economy.

Where have they come from?

The largest group of migrants to Spain is from Morocco in North Africa, followed by those from Ecuador and other parts of South America (see below). This reflects Spain's historic, social and cultural ties with South America. Spain colonised many countries in Central and South America during the sixteenth century. Spanish is the national language in these countries, and many families have personal links with Spain, so it is natural that migrants should move to and from there.

> ● **Natural increase** is the difference between birth rate and death rate in a population.
> ● **Immigration** is the inward flow of people into a country.

> ● An **ageing population** means that the average age of the population is increasing, usually as a result of increasing life expectancy and a falling birth rate.
> ● **Fertility rate** is the average number of children born per woman.

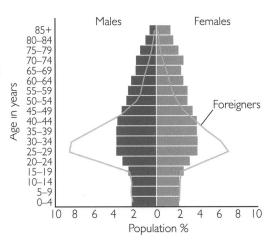

▲ *A population pyramid for Spain in 2005, showing the population breakdown for native Spaniards and for foreign residents of Spain.*

Ecuador	461 310
Colombia	265 141
Argentina	150 252
Bolivia	139 802
Peru	95 903
Brazil	72 441
Dominican Republic	61 071
Venezuela	51 261

◀ ▲ *Migration to Spain in 2005*

UK: 274,722
Poland: 45,797
Germany: 150,490
Ukraine: 69,893
France: 90,021
Romania: 407,159
Italy: 115,791
Bulgaria: 101,617
Portugal: 80,635
Algeria: 47,079
Morocco: 563,012

0   500
Km

N

Many of Spain's migrants are also from other countries of the European Union (EU), including significant numbers of skilled workers, business people and retired people from the UK. Residents of the EU can get work permits to work in other member states, but Spain cannot recruit enough people for jobs which are less well-paid – especially the **three Ds** – jobs that are dirty, difficult or dangerous, such as farming or construction. It therefore accepts migrants from non-EU countries. Some are **seasonal workers** – people with permission to work for up to six months in Spain. They come as members of the **Seasonal Agricultural Workers Scheme (SAWS)**, an EU scheme allowing people (mostly students) from countries such as Ukraine, Moldova and Belarus to work in EU countries for limited periods.

The largest concentrations of overseas-born people in Spain are along the Mediterranean coast and in Madrid. Within Valencia, there is a high proportion of retired people from other EU countries (particularly the UK). Within some parts of Alicante, on the south-east Mediterranean coast, over 40% of the population is from overseas. However, the concentration of retired immigrants in some coastal regions has created social and environmental concerns:

- In some areas, the overseas retired community has not integrated with the local population. They live in isolated **enclaves** with little social contact with local people.
- The increasing density of the coastal population is putting enormous strain on the region's limited water resources. Water is under increasing demand for leisure use, e.g. watering golf courses or topping up swimming pools. Local farmers are concerned that tourist resorts and retirement enclaves are using too much water for non-essential uses.
- There has been a reduction in the amount of fertile farmland near the coast, where the land has been sold for building and lost to farmers.

Should immigration be used to fuel economic growth, even if there are environmental costs?

What do you think?

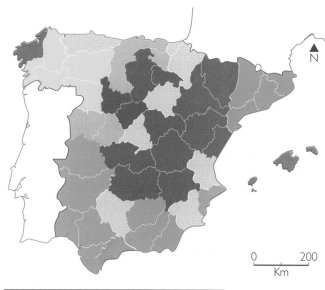

Key

	Portugal		Argentina
	Colombia		Bulgaria
	Ecuador		Morocco
	Germany		Romania
	United Kindom		

▲ *The predominant foreign nationality in each region of Spain in 2005*

- An **enclave** is a group of people living in a community which is isolated or separated from the rest of the population.

● Over to you

1 In pairs, discuss and list the reasons that migrants might have for moving to Spain. Identify different reasons for each of the following people:
   a university or gap year students
   b a young Ukrainian man or woman
   c a recently retired person from the UK
2 Discuss and complete a table about the benefits and problems caused by migration for the source country and the destination or host country.
3 Discuss in groups: 'Is immigration good or bad?'

● On your own

4 Define the following terms: natural increase, immigration, ageing population, fertility rate, enclave.
5 Use the population pyramid to describe the similarities and differences in population structure between Spain's foreign residents and native Spaniards.

GOING GLOBAL

In this unit you'll examine the reasons for, and consequences of, migrant flows within the European Union (EU).

Changing migration patterns

Migration patterns into and out of the UK have traditionally reflected strong historical and cultural ties with members of the Commonwealth. This is a group of 53 countries – including India, Australia, South Africa and Pakistan – with a combined total population of over 1.7 billion people. The Commonwealth countries established strong trading, cultural and demographic links with Britain when they were part of the British Empire, and movements of people between the UK and members of the Commonwealth are still an important part of the UK's immigration and emigration flows. However, the table below shows that migration flows are now strong between the UK and other EU countries, especially those who joined the EU in 2004 – such as Poland.

The changing pattern of immigration into the UK. The numbers of overseas nationals entering the UK and being allocated a National Insurance number, 2002–2006 (in thousands) ▼

2002/2003		2003/2004		2004/2005		2005/2006	
India	25.0	India	31.3	Poland	62.6	Poland	171.4
Australia	18.9	South Africa	18.4	India	32.7	India	46.0
South Africa	18.6	Australia	17.1	Pakistan	20.3	Lithuania	30.5
Pakistan	16.8	Pakistan	16.8	South Africa	19.3	Slovakia	26.5
France	13.8	Portugal	14.0	Australia	16.6	South Africa	24.0
Philippines	11.8	China PR	13.3	Lithuania	15.6	Australia	23.8
Spain	11.7	France	13.1	France	13.3	Pakistan	22.3
Zimbabwe	10.3	Spain	11.9	China PR	12.6	France	17.2
Iraq	10.1	Poland	11.2	Portugal	12.2	Latvia	14.2
Portugal	9.8	Philippines	10.7	Slovakia	10.5	Germany	13.3
Total	**349.3**	**Total**	**370.8**	**Total**	**439.8**	**Total**	**662.4**

EU enlargement and its impacts on migration

The UK also has strong trade and political links with other members of the EU. Before 2004, this was a trading bloc of 15 countries. Movement between member states was easy, and the UK welcomed economic migrants from any of the other 14 members. Then, on 1 May 2004, the EU15 were joined by another 10 (known as the Accession 10) and, on 1 January 2007, by Romania and Bulgaria (known as the Accession 2) – making 27 members in total (for now, because other countries, like Turkey, also want to join the EU). The Accession countries have smaller economies than those of the EU15 (e.g. Germany, France, Italy and the UK). Furthermore, unemployment in some of the new members, such as Poland, is high. The table shows that UK immigration in 2004/2005 followed a different pattern to 2002/2003 as a result of the enlargement of the EU.

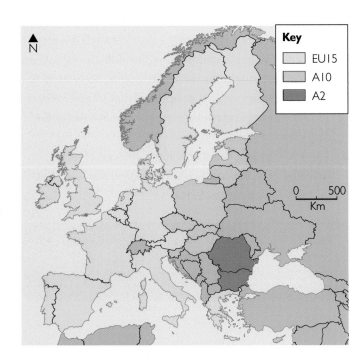

N

Key	
	EU15
	A10
	A2

0   500
Km

Why do people move?

Most people become migrants because they think that their move will be beneficial – providing a better job and quality of life. Such migrants are usually called **economic migrants**. They expect that their move will create economic benefits, such as:

- opportunities to gain seasonal employment
- better chances of full employment than in their own country
- better pay and better training and career opportunities.

These benefits are usually described as **pull factors**, which attract migrants to new locations. The table on the right suggests that people are generally much better off in the EU15 countries than in the A10 or A2. This difference in standard of living is a powerful pull factor for migration. In the table, GDP per capita has been compared to purchasing power (i.e. the cost of living) in each country. The figure is then calculated as an index, where the average for the 27 EU countries is 100. If the index of a country is above 100, people are better off than the EU average; if the index is below 100, people are worse off than the EU average.

Economic migrants also base their decision on **push factors** – problems that are pushing them away from their existing homes. These push factors include:

- unemployment or under-employment. For example, many of the economic migrants who have moved from Poland into the UK since 2004, are well qualified but could not find permanent jobs in Poland, where unemployment has been exceptionally high for several years.
- economic and political change. Of the A10 countries joining the EU in 2004, the eight from Central and Eastern Europe had quite different patterns of employment than the existing EU countries. After the Second World War, these eight countries were forced to become communist satellites of the USSR, and had **centrally planned economies** – the State took close control of nearly all business and industry, and there was little opportunity for private enterprise. Then, between 1989 and 1991, the whole region experienced a massive political change, with the end of the Cold War and the fall of communism. East and West Germany were reunited, the USSR collapsed, and the countries of Central and Eastern Europe gained their independence. This meant that the new governments of the eight countries were able to change radically the ways in which their economies were run. Their governments sold the ownership of many factories to private owners, a process known as **privatisation**. These political changes also caused massive changes to patterns of work and employment. Many old industries had received little new investment under the communists and could not compete on the world market. Consequently, unemployment rose significantly as the new owners tried to cut costs and make their businesses more competitive.

Luxembourg	260.1
Ireland	139.8
Netherlands	125.5
Austria	123.1
Denmark	121.6
United Kingdom	117.5
Belgium	117.1
Sweden	116.5
Finland	112.7
Germany	110.1
France	106.2
Italy	98.5
Spain	97.6
Cyprus	87.6
Greece	85.6
Slovenia	84.7
Czech Republic	78.1
Malta	74.0
Portugal	69.5
Estonia	69.0
Hungary	63.9
Slovakia	63.5
Lithuania	57.5
Latvia	57.2
Poland	53.4
Romania	37.5
Bulgaria	36.3

Key

EU15	The fifteen countries of the European Union in 2003
A10	The ten countries which joined the EU on 1 May 2004
A2	The two countries which joined the EU on 1 January 2007

▲ The Gross Domestic Product (GDP) in Purchasing Power Standards (PPS) for the 27 countries of the European Union (2007).

What opportunities are created by migration?

The declining birth rate, and the increasing proportion of retired people, is causing an acute labour shortage in some European countries. There are labour shortages in skilled occupations, such as the medical profession, right through to semi- and unskilled jobs on construction sites and in agriculture. Many Western European politicians believe that this problem can be solved by **replacement migration** – where migrant workers are encouraged to move from EU member states which have a labour surplus (generally the newer members), to states which have labour or skill shortages. The majority of migrants are young and often well educated, and they clearly benefit countries such as Spain, where the rate of natural increase in the population is low. Migration should also have a positive impact on the wider European economy – reducing unemployment in some regions of the EU increasing productivity in others.

Why are migrants needed?

- Many dirty, difficult and dangerous (so-called 3D) jobs, which migrants are prepared to do, are increasingly being avoided by local workers in industrialised countries.
- The current lifestyle of many Europeans is sustained by a wide variety of service jobs, such as childcare, house cleaning, and pizza delivery, which are undertaken by foreigners who cannot easily be replaced by locals.
- Migrants also respond to another type of labour demand – low-skilled jobs in the informal, underground economy, which in EU countries now engages 10-20 million workers, many of whom are illegal migrants.
- Migrants also traditionally fill jobs in sectors such as farming, road repairs and construction, hotel, restaurant and other tourism-related services, which, although not completely avoided by local workers, often suffer from seasonal shortages of labour.
- There are several skill- and knowledge-intensive industries in industrialised countries, notably in the fast-moving IT sector, with a demand for highly skilled migrants. There is a scramble for additional skills to develop new technologies, enhance competitiveness and create new jobs.

Migration not only benefits the receiving country, but it also creates opportunities for the migrants and their families. Many migrants move from countries with high rates of unemployment to find more opportunity and better-paid work in Spain, Italy, the UK, France, Germany and other Western European countries. These migrants are often able to send home some of their savings (known as **remittances**) to support relatives living in their countries of origin.

The social, cultural and environmental challenges of immigration

The number of migrants moving into and around the EU has risen sharply over the last ten years. Immigration has become a hot topic that is much debated by politicians and by the European press. Some journalists and politicians claim that there are too many migrants – that migrants take the jobs of local people and put a strain on local services, such as schools and housing – while others emphasise the benefits to the economy created by migration.

In the UK, the Institute for Public Policy Research (IPPR) has recently conducted research in England and Scotland into attitudes towards migrants. Some hostility was voiced by interviewees towards migrants, although only a small minority of the views expressed in the IPPR report were racist. The main concerns voiced by interviewees were over:

- *jobs and wages* – however, most analyses of the empirical effects of immigration on labour markets in the UK have demonstrated that the impact of immigration on wages and employment prospects is minimal
- *welfare entitlement, social housing and community support* – the belief that migrants receive preferential treatment from the Welfare State
- *public services* – extra strain on GP services and schools
- *cultural differences and integration* – a fear that English culture is under threat.

- **Natural increase or decrease** is the difference between birth rate and death rate in a population.
- **Net migration** is the figure arrived at when natural increase is subtracted from the total change in population

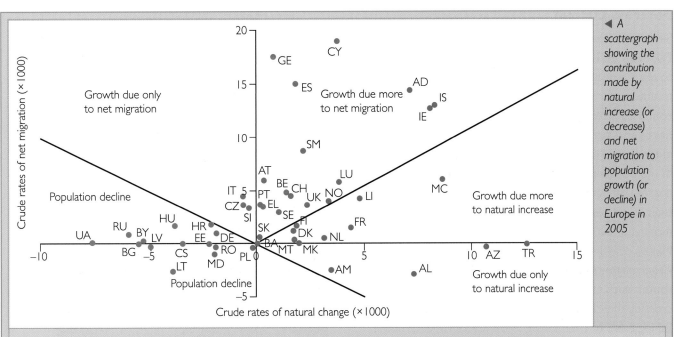

◀ A scattergraph showing the contribution made by natural increase (or decrease) and net migration to population growth (or decline) in Europe in 2005

KEY

European Union (EU) member states: Belgium (BE), Czech Republic (CZ), Denmark (DK), Germany (DE), Estonia (EE), Greece (EL), Spain (ES), France (FR) Ireland (IE), Italy (IT), Cyprus (CY), Latvia (LV), Lithuania (LT), Luxembourg (LU), Hungary (HU), Malta (MT), the Netherlands (NL), Austria (AT), Poland (PL), Portugal (PT), Slovenia (SI), Slovakia (SK), Finland (FI), Sweden (SE), Bulgaria (BG), Romania (RO), and the United Kingdom (UK)

Eurozone: Belgium, Germany, Greece, Spain, France, Ireland, Italy, Luxembourg, the Netherlands, Austria, Portugal and Finland

EU candidate countries: Croatia (HR), the Former Yugoslav Republic of Macedonia (MK) and Turkey (TR)

European Economic Area (EEA): all the European Union Member States, Iceland (IS), Liechtenstein (LI) and Norway (NO).

European Free Trade Association (EFTA): Iceland (IS), Liechtenstein (LI), Norway (NO) and Switzerland (CH).

Council of Europe member states: all the European Union member states, EU candidate countries, EFTA countries, Albania (AL), Andorra (AD), Armenia (AM), Azerbaijan (AZ), Bosnia and Herzegovina (BA), Georgia (GE), Monaco (MC), Moldova (MD), Russian Federation (RU), San Marino (SM), Serbia and Montenegro (CS) and Ukraine (UA).

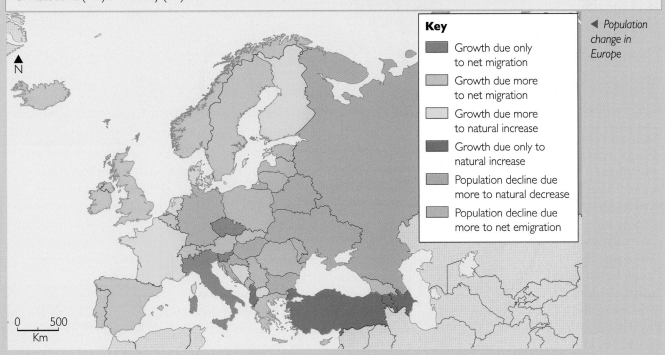

◀ Population change in Europe

Key
- Growth due only to net migration
- Growth due more to net migration
- Growth due more to natural increase
- Growth due only to natural increase
- Population decline due more to natural decrease
- Population decline due more to net emigration

Fortress Europe

Since 2004 and the accession of the A10 countries to the EU, many EU15 governments have considered the need to place a quota on the number of immigrants allowed to enter their country from both A10, A2 and non-EU countries. Prevented from entering the EU legally, there has been a steep rise in the number of people trying to enter the EU illegally. Countries on the southern periphery of the EU, such as Spain, Italy and Malta, have stepped up coastguard patrols to prevent illegal migrants from entering their countries in small fishing boats. During the first six months of 2006, over 1000 illegal migrants landed on Malta's coastline. As patrols have become more effective, illegal migrants have risked longer and more dangerous routes, such as trying to sneak into Fortress Europe by crossing the Mediterranean from North Africa in a small fishing boat, or even sailing to the Canary Islands in the Atlantic Ocean. By doing so, they put themselves at great risk and many drown during the attempt.

Migrants are helped aboard a rescue vessel when their small boat is intercepted by coastguards near Almeria on the Mediterranean coast of Spain ▼

Routes for illegal migrants attempting to enter 'Fortress Europe' ▼

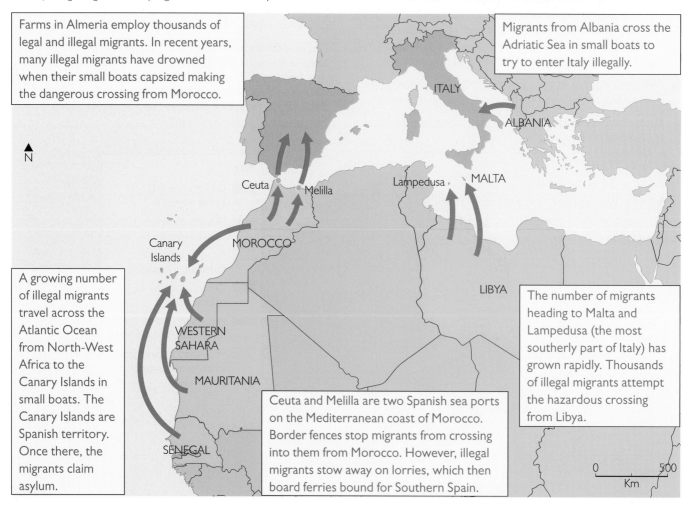

Farms in Almeria employ thousands of legal and illegal migrants. In recent years, many illegal migrants have drowned when their small boats capsized making the dangerous crossing from Morocco.

Migrants from Albania cross the Adriatic Sea in small boats to try to enter Italy illegally.

A growing number of illegal migrants travel across the Atlantic Ocean from North-West Africa to the Canary Islands in small boats. The Canary Islands are Spanish territory. Once there, the migrants claim asylum.

Ceuta and Melilla are two Spanish sea ports on the Mediterranean coast of Morocco. Border fences stop migrants from crossing into them from Morocco. However, illegal migrants stow away on lorries, which then board ferries bound for Southern Spain.

The number of migrants heading to Malta and Lampedusa (the most southerly part of Italy) has grown rapidly. Thousands of illegal migrants attempt the hazardous crossing from Libya.

ITALY

ALBANIA

Lampedusa

MALTA

Ceuta

Melilla

MOROCCO

Canary Islands

WESTERN SAHARA

MAURITANIA

SENEGAL

LIBYA

N

0   500
Km

Exploitation

If the **illegal migrant** manages to enter Europe safely, their problems do not end there. The majority of illegal migrants, and some legal migrants and seasonal workers, end up working in the **informal sector** of the economy. In most cases this means exploitation – with long hours, little health and safety protection, and low pay. Often, with seasonal employment – such as fruit picking and other agricultural work – the employer also provides accommodation. The migrant then has part of their pay taken to pay their rent. In some cases, migration is also associated with organised crime. Criminal gangs in Africa or Central Europe organise the movement of people into the EU15 in exchange for a large payment. In the worst cases, women and children are trafficked by criminal gangs to work in sweatshops or in the sex industry.

The rise of extremism

In a few cases, migrant workers have been the victims of harassment or racial abuse and violence. One of the worst cases was in Almeria, Spain, in 2000. The region has a large concentration of legal and illegal immigrants from North Africa working in greenhouse agriculture. Fierce riots were fought between immigrants and Spaniards chanting racist slogans. More recently, in April 2007, clashes broke out between Chinese immigrants and locals in the Italian city of Prato. This region of Italy is the centre of the textile industry and has attracted a large concentration of legal and illegal Chinese immigrants who have experience of working in the textile industry. The Chinese workers were tolerated while they stayed largely hidden from view, working in the textile factories, but extremist views amongst locals have become more vocal as the Chinese community has expanded into commerce and therefore become more visible in the Italian city's streets. Examples like this prove that EU governments face the major challenge of maintaining **social cohesion** while still encouraging immigration. In other words, immigrants need to be effectively integrated into the wider society if they are not to feel victimised and excluded.

▲ Migrants behind barbed wire at a refugee camp on Lampedusa, a small Italian island that receives thousands of illegal migrants and asylum seekers each year

● Over to you

1 Working in pairs, produce a poster which explores the push and pull factors that explain migration from countries such as Poland, Latvia and Romania into the UK.

2 Study the population change scattergraph and map.
   a Describe the geographical distribution of countries where the population is:
      i increasing due more to net migration
      ii decreasing due more to net migration
   b Offer three suggestions which might explain this distribution pattern.

3 Work in pairs. Use the Internet to research examples of different points of view on the topic of migration in the EU. Try to find at least one site which has a positive view of the opportunities created by migration, and one that takes a negative point of view and focuses on the challenges created by migration. Produce a short presentation which summarises these points of view and which assesses whether or not you think these attitudes have any value.

● On your own

4 Produce a 500-word report on the consequences of migration in Europe. Make sure you comment on both the positive and negative consequences. Include short sections on the economic and social impacts of migration.

Exam question: Explain the benefits and problems that economic migrants can bring for **a** the host country and **b** the source country (15 marks)

In this unit you'll learn about recent changes to the UK's population, which have helped to create the Britain of today.

In 2005, the population of the UK reached its highest total ever – 60 million. In future, this total is predicted to rise slowly – reaching 65 million by 2023 and 67 million by 2031. Why has the UK's population grown so much?

The components of population change

The population of any country or region is constantly changing – geographers describe this as a **dynamic system**. **Inputs** (e.g. births and immigration) and **outputs** (e.g. deaths and emigration) vary over time to create either growth or decline in the population. The factors that influence population change can be broadly divided into social, cultural, political, economic and demographic. For example, the rate of immigration into the UK varies depending on factors such as the number of jobs available in the UK, and the ease with which migrants are able to enter the country. Unit 2.11 examines in more detail the reasons for the recent growth in migration within Europe, while the diagram below summarises the main factors that affect population change.

> ● **Population structure** is the proportion of people in each age group compared to other age groups.

▼ Factors that affect population change

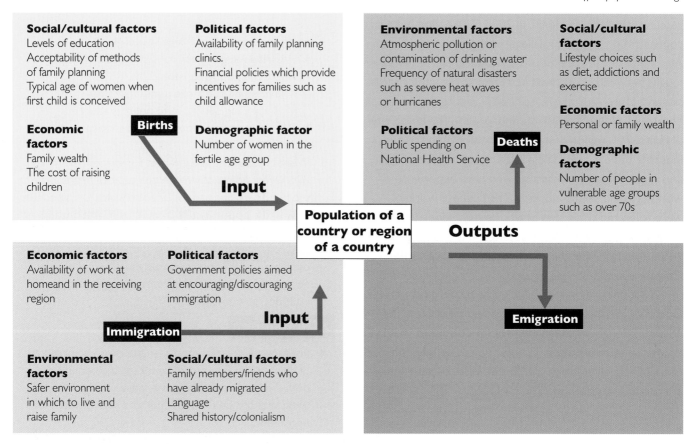

Social/cultural factors
Levels of education
Acceptability of methods of family planning
Typical age of women when first child is conceived

Political factors
Availability of family planning clinics.
Financial policies which provide incentives for families such as child allowance

Economic factors
Family wealth
The cost of raising children

Births

Demographic factor
Number of women in the fertile age group

Input

Economic factors
Availability of work at home and in the receiving region

Political factors
Government policies aimed at encouraging/discouraging immigration

Input

Immigration

Environmental factors
Safer environment in which to live and raise family

Social/cultural factors
Family members/friends who have already migrated
Language
Shared history/colonialism

Population of a country or region of a country

Environmental factors
Atmospheric pollution or contamination of drinking water
Frequency of natural disasters such as severe heat waves or hurricanes

Political factors
Public spending on National Health Service

Deaths

Social/cultural factors
Lifestyle choices such as diet, addictions and exercise

Economic factors
Personal or family wealth

Demographic factors
Number of people in vulnerable age groups such as over 70s

Outputs

Emigration

The reasons for this change are complex. The declining number of children has been caused by one set of social, economic and demographic factors, while the increased longevity of the population is the result of other factors. For example, longer life expectancy can be explained by a combination of:

- economic factors – greater national wealth has led to massively increased funding for the UK National Health Service, while greater personal wealth means that people can afford to take out private health care plans
- political factors – the willingness of both Conservative and Labour governments to invest in the National Health Service and the wider Welfare State (which includes the provision of such things as day care facilities for the elderly)
- cultural factors – an improved understanding of the importance of a healthy diet and levels of personal fitness.

As a result of these factors, not only are more people living into retirement, but many more people are living into old age – there are already 1.2 million people aged over 85 living in the UK.

An ageing population

More people in the UK are living into old age, a situation that has been termed the **greying** of the population. At the same time, the number of children being born is in decline. These two trends have been working together over the last 35 years to change the **population structure** of the UK.

- In 1971, there were almost twice as many under 16s (14.3 million) as there were people aged 65 and over (7.4 million).
- By 2005, the numbers of people in these two age groups were much closer at 11.6 million (under 16s) and 9.6 million (age 65 and over)
- By 2014, it is estimated that the number of people aged 65 or over will exceed those aged under 16 for the first time.

▲ The Who recorded 'My Generation' in 1965, and The Zimmers re-recorded the song in 2007. The Zimmers performed as part of a documentary that highlighted the isolation felt by many elderly people in Britain.

▼ Improving life expectancy in the UK, 1901–2021

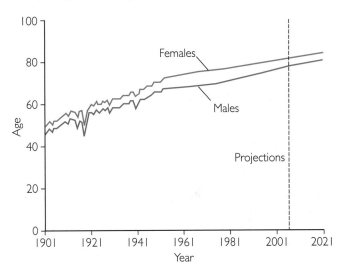

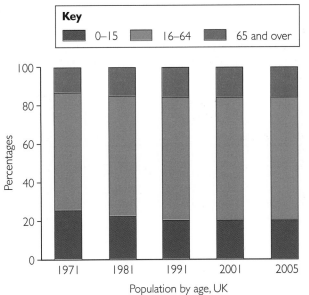

The changing population structure of the UK, 1971–2005 ▶

Analysing population change in the UK – 2

Patterns in the birth rate

The UK experienced four **baby booms** during the twentieth century. The graph below clearly shows that the number of births each year in the UK fluctuated greatly during this period. However, what is significant is that the final years of the twentieth century saw generally low numbers of births, with less than 700 000 births each year in 2001 and 2002. Since then, the number of births has increased slightly, and projections suggest that the number of births will continue at a relatively stable level at between 695 000 to 724 000 each year.

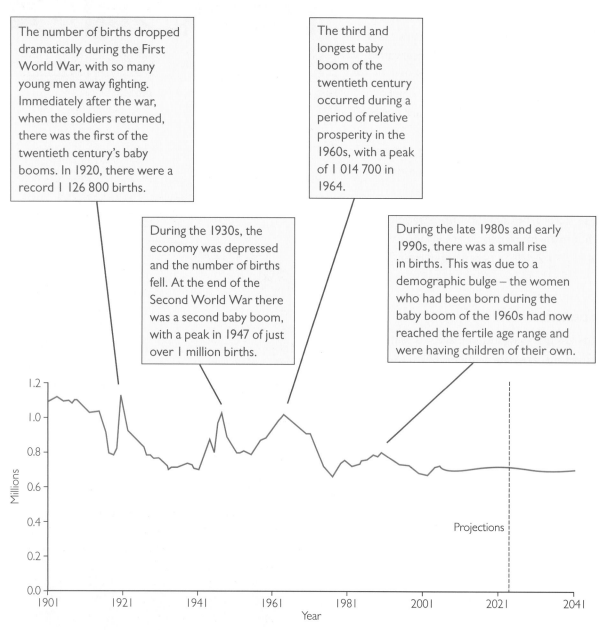

The number of births dropped dramatically during the First World War, with so many young men away fighting. Immediately after the war, when the soldiers returned, there was the first of the twentieth century's baby booms. In 1920, there were a record 1 126 800 births.

During the 1930s, the economy was depressed and the number of births fell. At the end of the Second World War there was a second baby boom, with a peak in 1947 of just over 1 million births.

The third and longest baby boom of the twentieth century occurred during a period of relative prosperity in the 1960s, with a peak of 1 014 700 in 1964.

During the late 1980s and early 1990s, there was a small rise in births. This was due to a demographic bulge – the women who had been born during the baby boom of the 1960s had now reached the fertile age range and were having children of their own.

▲ The number of actual and projected live births in the UK each year, 1901–2041

Changes in fertility

Fertility is another important measure for monitoring population change. **Total Fertility Rate (TFR)** is the average number of children who would be born per woman if she were to live to the end of her childbearing years and follow normal patterns of fertility. In 2005, the TFR in the UK was 1.79. In other words, while some women had no children, and others had one, two, three or more, the average number of children born per woman in the UK in 2005 was 1.79. The **replacement level** needed to maintain population size is 2.1. Fertility rates in most European countries have now dropped below 2.1, which indicates that, over the long term, the populations of many European countries will not replace themselves by natural change. This explains why many European politicians believe that immigration has to be encouraged to maintain a strong workforce and economy.

In the UK, the TFR fell to a record low in 2001 and 2002. Since then, it has gradually climbed back to 1.79, but the Office for National Statistics cannot predict whether or not this is the beginning of a sustained rise. 20% of women in the UK do not have children, either through choice or because of infertility. The age at which women in the UK become pregnant for the first time is also getting later – the birth rate for women in their twenties is falling, while the birth rate for women in their thirties is rising. This is due to a number of social changes, including:

- higher levels of divorce or separation, which may mean that many women do not start a family in their twenties with their first partner
- many women delaying the start of a family into their thirties so that they can pursue a career.

But when women choose to start a family later in life, they automatically face the consequence that they are less fertile. The chances of conception begin to decrease rapidly in the thirties, making it likely that family size for the older mother will be small.

Recent changes in the crude birth rate, fertility and crude death rate in selected EU countries. ▼

	Crude birth rate			Fertility			Crude death rate		
	2000	2004	2005	2000	2004	2005	2000	2004	2005
France	13.3	12.9	12.9	1.89	1.92	1.94	8.9	8.3	8.5
Germany	9.3	8.6	8.3	1.38	1.37	1.34	10.2	9.9	10.1
Italy	9.5	9.7	9.5	1.26	1.33	1.34	9.8	9.4	10.0
Poland	9.8	9.3	9.5	1.37	1.23	1.24	9.6	9.5	9.6
Spain	9.9	10.6	10.7	1.23	1.33	1.34	9.0	8.7	8.9
United Kingdom	11.5	12.0	12.0	1.64	1.77	1.80	10.3	9.7	9.7
Romania	10.7	10.0	10.2	1.31	1.29	1.32	11.7	11.9	12.1

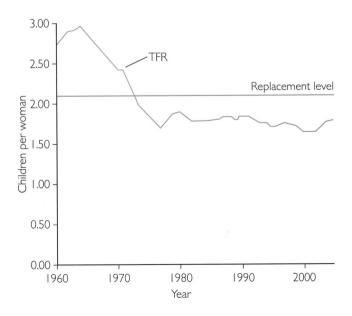

◄ *Changing patterns of UK fertility, 1960–2005*

● Over to you

1 Copy and complete the diagram on page 102, to suggest the factors that control emigration.
2 Summarise five key changes in the UK population since The Who first recorded 'My Generation' in 1965.
3 Study the graph of the number of live births in the UK. Use it to illustrate the way in which the birth rate is influenced by political, social and economic factors.

● On your own

4 Prepare a short report on changing patterns of UK and European fertility. Include a graphical representation of selected data from the table above. Suggest how these patterns may be perceived by politicians. ●

Analysing population change in the UK – 3

The impact of migration on the age structure of the UK

The UK's current multi-racial composition and age structure is largely due to the migration patterns that occurred during the twentieth century. As on page 96, migration patterns in and out of the UK reflect the country's strong historical and cultural ties with members of the Commonwealth, as well as with other – mainly European – nations.

The relative importance of migration to the UK's changing population varied significantly during the twentieth century. For example, during the 1950s and early 1960s, the UK experienced immigration from India, Pakistan and the Caribbean. However, this also coincided with a baby boom (see page 104). This natural increase was responsible for 98% of the UK's population growth at the time, while net immigration (more immigration than emigration) made up the other 2%. Since the 1980s, however, net immigration has steadily contributed a higher percentage of the UK's population growth and, with the UK's lower fertility levels, natural increase has become less significant for population growth.

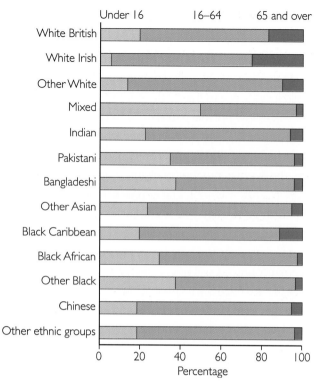

▲ UK population structure by ethnic group in 2001

Economic migrants from Ireland dominated migration to the UK during the early part of the twentieth century.

During the 1950s, the British government encouraged people from the Caribbean, India, Pakistan and Bangladesh (which, at the time, was still part of Pakistan) to emigrate to the UK to fill many job vacancies available at the time.

During the 1960s and 1970s, more people emigrated from the UK (to places like Australia) than immigrated, but natural increase due to higher birth rates more than compensated for the loss.

1900 1910 1920 1930 1940 1950 1960 1970 1980 1990 2000 2010

India gained independence from Britain in 1947, and it was partitioned into modern-day Pakistan and India. There was a massive movement of people between the two new countries. In the ensuing violence, many families from both India and Pakistan emigrated to the UK.

A significant influx of Chinese immigrants arrived in the UK.

A second wave of Chinese immigrants arrived in the UK, as officials relaxed emigration controls in China. This decade also saw increased migration from African nations.

In May 2004, the European Union was enlarged to include 10 countries, mainly from Central and Eastern Europe. This allowed economic migration from countries such as Poland, Lithuania and Slovakia to the UK.

Between 2001 and 2005, immigration outstripped emigration. During this period, net immigration was responsible for an extra 182 000 people living in the UK each year, whereas natural increase only added an extra 92 000 people each year. Between 2005 and 2021, it is expected that natural change and net migration will account for similar amounts of the overall population change.

The graph on the left clearly shows that ethnic population groups in the UK have a different age structure to the White British population. These age structures reflect the history of that ethnic group's migration, as well as differing fertility patterns. For example, large numbers of the Black African population migrated to the UK in the mid 1980s, when they were young adults. By 2001, when the data for the graph was collected, many of these people had young families of their own, which explains the relatively large percentage of under 16s.

A selection of British celebrities, each of whom is the son or daughter of a twentieth century immigrant to the UK ▶

● Over to you

1 Select specific examples of key migrations to the UK and explain them in terms of:
   a social/cultural factors
   b economic factors
   c political factors
2 Study the graph of population structure by ethnic group. In the light of the fact that different ethnic groups migrated to the UK at different points in the twentieth century, explain:
   a the large percentage of White Irish who are over 65, and the small proportion who are under 16
   b the large proportion of mixed race under 16s.

● On your own

3 Prepare an essay plan to: 'Explain how changing social and economic factors in the twentieth century have had an impact on UK population and migration'. In your plan make sure you identify specific examples of:
   a social factors that have influenced birth rates/ fertility and patterns of migration.
   b economic factors that have influenced birth rates, death rates and patterns of migration.
   Complete your plan by weighing up how important these various factors have been compared to each other, and compared to other factors (such as government policies or environmental factors). ●

Sanjeev Bhaskar: comedian, born 1964 in Essex. Sanjeev's parents emigrated from the Punjab after Indian partition in 1947.

Lenny Henry: comedian, born 1958 in Dudley. Lenny's parents emigrated from Jamaica in the 1950s.

Monty Panesar: England cricket player, born 1982 in Luton. Monty's parents emigrated from the Punjab in the late 1970s.

Saria Khan: TV presenter and star of The Apprentice, born 1970 in Long Eaton. Saria's parents emigrated from Kashmir in 1965.

Issues facing society as the UK's population ages

The ageing, or greying, of the population creates a number of issues for British society. These issues can be seen as either challenges (problems to be overcome) or opportunities, depending on your point of view. For example, a large number of business opportunities have emerged as a result of the increased number of people who have retired with generous private pensions. Companies selling leisure products, including DIY, gardening and holidays have benefited from the so-called **grey pound** of the wealthy retired. However, many elderly people do not have surplus cash for these new leisure opportunities. Poverty amongst the elderly is a major challenge for them as individuals and for non-government organizations – such as Age Concern – who campaign on their behalf.

Health care and social housing

The World Health Organisation (WHO) warns that the ageing European population will lead to a big rise in cancers, heart disease, and diabetes. Providing treatment for the diseases associated with ageing is becoming increasingly expensive as the proportion of elderly people increases.

Health problems, such as dementia, diabetes or heart disease are certainly a challenge for the elderly individual and their family members or friends who care for them. They are also a challenge for the National Health Service, which has to prioritize how it spends its limited funds, and for the voluntary sector, which relies on donations of money and time from volunteers. However, the growing private health care sector regards our ageing population as an opportunity. Firms who provide private health care insurance, such as BUPA, or companies such as Boots (who offer an increasing range of health care services) regard the growing elderly population of the UK as a business opportunity.

A further issue created by the ageing population is the need for new housing. Many elderly people have specific housing needs – smaller properties without stairs; wider doorways and lower kitchen units for those who have limited mobility; and **sheltered accommodation** for those who need carers living on site. The current shortage of such housing represents a challenge for elderly people and their carers, but is an opportunity for house builders and developers.

- Increased spending by the wealthy retired
- The provision of housing adapted to the needs of the elderly
- A decline in the size of the workforce

Issues facing British society as a result of population change

- Increased demand for specialised health and social care
- The need to integrate economic migrants into British society
- Smaller class sizes in primary and secondary schools

The pensions crisis

The ageing of the population is creating a pensions crisis that will affect future generations for years to come. State Pensions in the UK are supported by national insurance contributions made by those who are in work. As more people live to a greater age, there are more people claiming a pension for a longer period of time. As a result, the ratio of people in work compared to people over 65 is falling. In the UK, in 2000, there were 3.71 people in work to every person over 65. By 2040 it is forecast that there will only be 2.1 people in work for every person aged over 65. This means that lower tax and national insurance contributions are being collected by the Government at a time when the Welfare State needs to increase its payments to the increasing number of pensioners and elderly people who need NHS treatment. Europeans are being encouraged to take out more private pension schemes while they are young. In future, it is likely that only the least well off will be able to rely on a State Pension when they are old.

	Males		Females	
	1981	2002	1981	2002
Life expectancy	70.9	76.0	76.8	80.5
Healthy life expectancy	64.4	67.2	66.7	69.9
Years spent in poor health	6.4	8.8	10.1	10.6
Disability free life expectancy	58.1	60.9	60.8	63.0
Years spent with disability	12.8	15.0	16.0	17.5

▲ *Recent changes in life expectancy and implications for health*

Investigating your local population

There are a number of useful websites that you can use to find out data about the population in the UK. One of the most useful is www.statistics.gov.uk, which is the official website of the UK National Census. One of the best ways to use the site is to click on the link to 'Neighbourhood' at the top of the home page. You can then search for data using either place names or postcodes. Enter the postcode for the neighbourhood you want to investigate and click on the button for ward.

This will select an area with a population of around 5000 people and (in an urban area) covering an area a little larger than 1 km². The next screen will give you access to a whole range of data for the ward you have chosen. A map in the top right of the screen shows the extent of the ward. The hyperlinks on the left of the screen will take you to useful information about the ward. The best place to start is probably the second link: 2001 Census: Key Statistics (31 datasets).

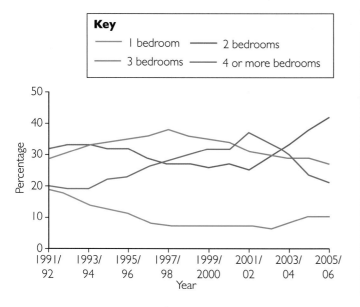

Key

— 1 bedroom — 2 bedrooms
— 3 bedrooms — 4 or more bedrooms

◄ *Completion of new houses in the UK by number of bedrooms, 1991–2006*

● Over to you

1 Use the table above to identify opportunities and challenges created by the UK's ageing population.
2 Analyse the house-building trends in the graph on the left. Does it show whether or not the UK's house-building industry has risen to the challenge of the UK's ageing population?

● On your own

3 Use www.statistics.gov.uk to conduct an investigation into the population structure of your own locality. Prepare a report which focuses on how the local population compares to the national averages in terms of its age structure and ethnic background.

In this unit you'll look at the rise of the 'megacity', and think about whether places as large as Los Angeles can be sustainable.

Los Angeles – an American dream?

Los Angeles, California, conjures up lots of different images. Beverly Hills, Hollywood, the Beach Boys – and sun. While it represents the American dream for some, for others it is not all glitz and glamour, as this unit shows.

> Come to Los Angeles! The sun shines bright, the beaches are wide and inviting, and the orange groves stretch as far as the eye can see. There are jobs a-plenty, and land is cheap. Every working man can have his own house, and inside every house, a happy, all-American family … Life is good in Los Angeles … It's paradise on Earth.
> *Sid Hudgens, tabloid journalist in the 1997 film L.A. Confidential, set in the 1950s*

Growth of the megacities

Urbanisation is increasing. By 2007, more than 3.2 billion people – over half the world's population – were living in towns and cities. In many MEDCs, over 90% of the population were urban inhabitants. Not only that, but there has been an increase in the number of very large cities which have populations of over 5 million people – so-called **super cities**. **Megacities**, which have populations of over 10 million people, are also growing – and tend to dominate countries in terms of their size, location of manufacturing and investment. Being big is not always good news, because it can bring problems for things like housing, employment, water and sanitation, health and education. The graph on the right shows the projected growth of megacities from 2000-2015, and the map shows their location.

▼ *Expected megacity growth between 2000 and 2015*

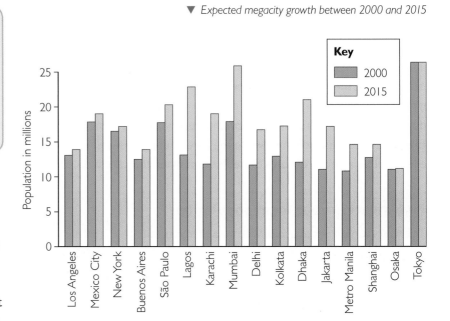

Key
- 2000
- 2015

Population in millions

Los Angeles, Mexico City, New York, Buenos Aires, São Paulo, Lagos, Karachi, Mumbai, Delhi, Kolkata, Dhaka, Jakarta, Metro Manila, Shanghai, Osaka, Tokyo

▼ *The world's megacities*

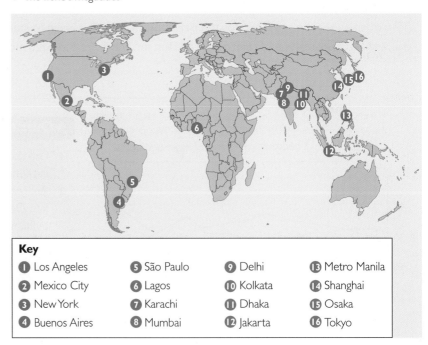

Key
- ① Los Angeles
- ② Mexico City
- ③ New York
- ④ Buenos Aires
- ⑤ São Paulo
- ⑥ Lagos
- ⑦ Karachi
- ⑧ Mumbai
- ⑨ Delhi
- ⑩ Kolkata
- ⑪ Dhaka
- ⑫ Jakarta
- ⑬ Metro Manila
- ⑭ Shanghai
- ⑮ Osaka
- ⑯ Tokyo

World cities are important because of their connections with the rest of the world. There are four main types of world cities, as the map shows:

- Major capitals and commercial cities in developed countries (the core) e.g. London and Tokyo
- Major capitals and commercial cities in developing countries (the semi-periphery) e.g. Rio de Janeiro and Singapore
- Less important capitals and commercial cities e.g. Sydney, Madrid
- Less important developing world cities e.g. Bangkok and Mexico City.

> ● **Urbanisation** is the increase in the percentage of people living in towns and cities.
> ● A **super city** has a population of over 5 million.
> ● A **megacity** has a population of over 10 million.
> ● **World cities** are cities of power based on trade, political strength, innovation and communications.
> ● A **megalopolis** (see page 112) is a very large urban area that contains several metropolitan centres.

World cities are connected to other places ▼

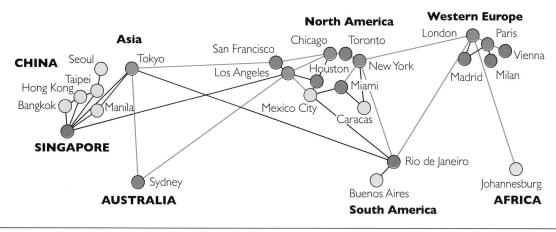

Key

- Core: primary city – developed country – major capital; commercial city
- Semi-periphery: primary city – developing country–major capital; commercial city
- Core: secondary city – developed country – less important capital; commercial city
- Semi-periphery: secondary city – developing country – less important capital; commercial city
- — Linkages between core cities
- — All other linkages

Global connections

Los Angeles is not only a megacity, but a world city too. It does not exist in isolation, but is linked, or connected, to other cities around the world, as the diagram above shows. The connections between cities are what hold the global economy together.

People living outside cities depend on them for certain goods and services and, therefore, are connected to the city. Through their links with the city, they are also connected to other places around the world. Perhaps we should now think of the world as a 'global city' and not just a global village.

Los Angeles: megacity, mega problems?

Why did Los Angeles grow?

Not everyone would agree with Sid Hudgen's view of Los Angeles, but the city has certainly attracted people in their millions to California. Los Angeles forms part of the SanSan urban corridor, connecting San Francisco to San Diego – sometimes called SoCal (Southern California). This **megalopolis** is home to 24 million people.

Los Angeles is the second largest city in the USA – but why did it grow?

- The arrival of the railway in 1876 stimulated rapid growth, with half a million people arriving within 40 years.
- The discovery of oil, the opening of a Ford car plant and numerous manufacturing industries meant continued growth.
- The aircraft industry took advantage of the good weather for civil and military test flights and production sites.
- The development of the film industry in Hollywood in the 1920s and 1930s was a further factor in Los Angeles' growth.
- By the 1970s, Los Angeles was the fastest growing city in the USA, and by 2000 over 14 million people lived in the metropolitan district. It has an average population density of 3000 per km^2, and a growth rate of 1% pa. The fastest-growing areas have been on the eastern extent of the metropolis, in Riverside and San Bernardino counties, where growth is up to 6% pa.

But Los Angeles, like all major urban areas, has problems and its biggest is suburban sprawl and the related problem of pollution.

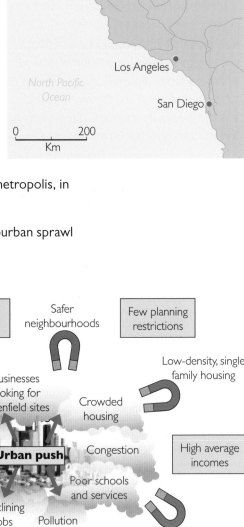

Suburban sprawl

Los Angeles is a sprawling urban mass. The arrival of electric tramways in the 1920s and 1930s and, later, freeways (motorways) meant that people could live further away from work on their own plot of land. The 1980s saw massive urban growth spreading out as far as the mountains and deserts – 2 hours travelling time from Los Angeles. Time spent travelling to and from work creates social problems – there may be no time left for family and friends. Another result of urban sprawl is that many communities now exist only as night time dormitory settlements.

Why have the suburbs sprawled?

As the diagram on the right shows, there are many reasons for the sprawling suburbs – related to factors which are specific to America (background factors) and others that either push people from urban areas, or pull them to the suburbs.

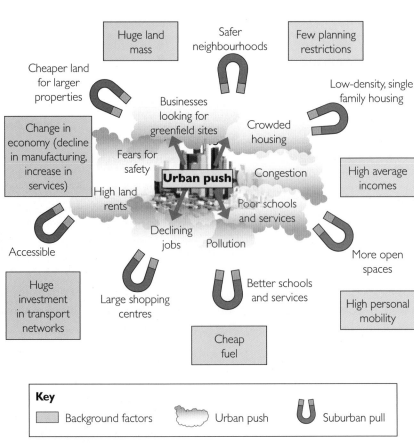

Key

Background factors   Urban push   Suburban pull

What are the impacts of suburban sprawl?

Suburban sprawl may mean nicer places to live for some – but is Los Angeles sustainable? Congested freeways, air pollution, loss of some of the best farmland and the decline of central Los Angeles are just some of the impacts of the sprawl.

▲ The greater Los Angeles metropolitan area extends over 88 000 km² and includes 100 distinct towns. 10 million car owners create an environmental hazard – smog.

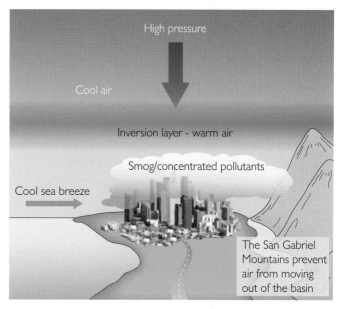

▲ Los Angeles sits in a basin. Cool surface air with warm air above creates an inversion and pollution is trapped above Los Angeles. Exposure to pollution doubles the chances that children will develop cancer later in life.

Other problems facing Los Angeles

Other problems facing Los Angeles include:

- **housing shortages** – caused by high rates of migration
- urban tension – due to ethnic differences. In 1965 and 1992 there were serious race riots in the run-down inner city districts of Watts, South Central and Compton.
- **work** – a changing economy (**deindustrialisation**), has resulted in the loss of thousands of manufacturing jobs. Replacement jobs are often low paid.
- **water** – piped in from 350 km away causes disputes with neighbouring counties and states.
- **waste** – 24 million people produce 50 000 tonnes of waste every day.
- **health and education** – many migrants are excluded because they cannot afford to pay for these services.
- **transport** – there are 10 million vehicles on the road, only 30% of people use public transport.
- **massive energy use** – as the photo on the right shows.

● **Deindustrialisation** is the decline in manufacturing (secondary) industry, and the growth in tertiary and quaternary industry.

Los Angeles – one bright city! The 22 622 megawatts of power delivered at 4 pm on 21 July 2006 overtook the previous record of 21 934 megawatts supplied a year earlier. ▼

Los Angeles today

'Donut' city

Los Angeles has been described as a 'donut city' – a city with a hole in the central (downtown) area. Why?

- Car, tyre, steel and aircraft factories closed due to competition from overseas and changing technology.
- Businesses and people moved out of central areas.
- Modern high-tech electronics, aerospace and light-manufacturing industries wanted large sites with car parks on the edge of the city.
- The central areas declined, leading to dereliction and concentrations of poorer people in segregated areas.
- The Long Beach and Santa Ana freeway areas lost a million jobs, leaving many migrants unemployed. Cudahay, Bell Gardens, Watts and Huntington Park became the poorest districts.

As new industrial sites developed on the periphery of the city limits, Los Angeles was also referred to as 'the city that turned itself inside out'. **Edge cities**, such as Anaheim, Irvine and Ontario, developed along the freeways, with concentrations of new industries next to office developments, giant shopping malls and leisure zones. Today the downtown 'hole' has been filled and is dominated by the headquarters of TNCs (see above).

▲ TNC headquarters now fill LA's 'donut'

Land use in Los Angeles

Land use patterns in Los Angeles have developed like a patchwork – not like the land use models of Burgess and Hoyt (see page 241). Wealthy areas are found in the inner city next to poor areas, and the same happens on the edge of the city. Migrants and others of different ethnic groups often live in **ethnic enclaves**. Some of the ethnic enclaves are rich, others are poor, making the land use pattern more complicated. The map below shows the distribution of different ethnic enclaves across Los Angeles.

- An **ethnic enclave** is part of a city in which the population is mainly from one ethnic group, often with its own religious beliefs and places of worship, together with shops and community centres.

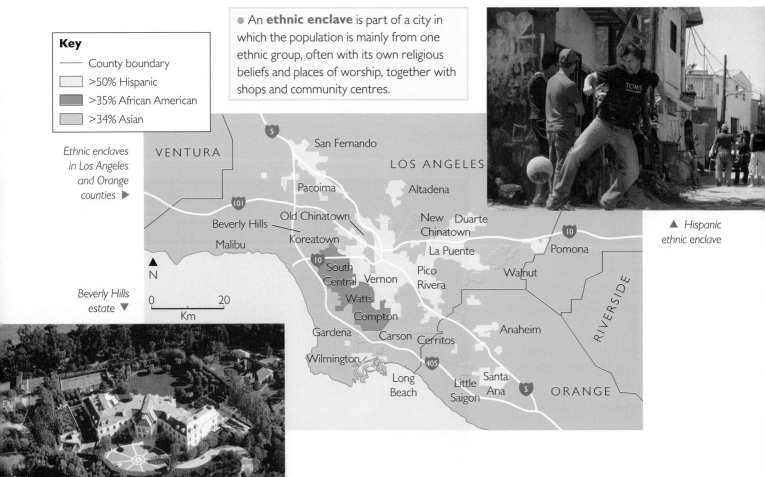

Key

- — County boundary
- ☐ >50% Hispanic
- ☐ >35% African American
- ☐ >34% Asian

Ethnic enclaves in Los Angeles and Orange counties ▶

Beverly Hills estate ▼

▲ Hispanic ethnic enclave

Can Los Angeles be sustainable?

Los Angeles, as you have seen, has its problems – but the Progressive Los Angeles Network (PLAN) proposes an agenda for sustainable living. The emphasis is on issues ranging from transport, the urban environment, food and nutrition, to economic development, housing and workers' rights. It really means giving residents more control over the future of their city.

> Megacities – can they ever really be sustainable?
>
> **What do you think ?**

Some of PLAN's proposals for a sustainable city

- Require employers to pay their workers a decent wage
- Provide community benefits such as childcare and affordable housing
- Increase urban parks and clean up contaminated brownfield sites
- Promote clean fuel vehicles and green energy
- Require developers to build affordable housing in all new residential developments
- Attract food markets, farmers' markets and community gardens
- Improve public transport with 'clean' buses and new rapid bus lines

- Universal low fare card that allows easy transfer between buses and trains
- Promote safe, walkable and bikeable neighbourhoods
- Promote smart growth land use where people can drive less and live nearer to where they work, shop, study, and play
- Ban new 'big box' retail stores which undermine local retail and community activity, and steer future development to locations near existing and planned transit stops

● Over to you

1 Discuss in pairs the reasons why people choose to live in suburbs instead of inner city areas.
2 Complete the table at the bottom to show the impacts of suburbanisation.
3 In groups of four, research the following aspects of Los Angeles: water supplies, waste disposal, traffic management, pollution (www.ci.la.ca.us will help). How well is Los Angeles doing?

● On your own

4 Define the following: megacity, megalopolis, suburban sprawl, 'donut' city, edge city, ethnic enclave.
5 Use the following websites to find out how the city authorities are making Los Angeles more sustainable. Write a 300-word report on what you find out.
www.sustainlane.com/us-city-rankings/losangeles.jsp
www.journalism.berkeley.edu/projects/greenla/sustain.html
www.cicle.org/cicle_content/pivot/entry.php?id=1048
6 What else could improve Los Angeles' sustainability?

	Positive impact	Negative impact
Social, e.g. housing, communities, services		
Economic, e.g. jobs, traffic, energy, time		
Environmental, e.g. land use, quality, management, resources		

GOING GLOBAL

Megacities: Mumbai

This unit looks at the rapid changes taking place in Mumbai, and asks whether India's largest city can also be a global city.

Global Mumbai

Mumbai is the commercial capital of India. It is a lively, busy, cosmopolitan city. The growth of Mumbai and India has brought wealth and power to some people, but not all.

- Mumbai provides 33% of India's entire tax revenue.
- 40% of international flights to India land in Mumbai.
- Mumbai is the home of Bollywood, the centre of India's film industry, and it produces more films than Hollywood.
- Rents in the most exclusive parts of the city are higher than in London and New York.

Globalisation has provided Mumbai with international banks, world-class restaurants, smart cars and the headquarters of new Indian transnational corporations like Tata Steel, Mukesh Ambani Oil and Godrej Retail – businesses that are now taking over their European and American rivals.

Mumbai – a city of contrasts

Mumbai, like many large cities, is a city of contrasts. Not everyone has enjoyed its growth and success. While some live in comfortable upper- and middle-class homes, over a million people live in Dharavi, one of the world's largest shanty towns.

> If you want to catch a glimpse of the new India, with all its dizzying promise and turbocharged ambition, then head to its biggest, messiest, sexiest city – Mumbai.
>
> *Time magazine*, 2006

▲ *Modern Mumbai*

Two cities – dual societies

In Mumbai, I flitted in and out of the two Indias. One is on the streets – the beggars, the pavement dwellers, the slums, the street children, the tiny labourers who pick through the litter for recyclables when they should be laughing on a playground. It's noisy, in your face.

The other India is one of quiet, air-conditioning, service, amenities, middle- and upper-class people living their lives much as the wealthier live their lives anywhere. They are doctors, professors, engineers, computer programmers and other professionals.
They live in beautiful gated homes or modern flats and spend evenings in top restaurants and trendy nightclubs with drinks that cost as much as they would in New York or London.

Mostly, the two Indias exist separately, as if each is unaware of the other's existence.

▲ *The other side of Mumbai – the Dharavi shanty town*

◄ *Adapted from an online article in June 2007, 'How the other half lives' by Shelley Seale on Worldpress.org*

Hyper-urbanisation

Mumbai is at the heart of the growing Indian economy. It is also growing rapidly. Home to 17 million people, it is estimated to receive 600 new migrants a day. By 2020 it is expected to have a population of 26 million and could become the world's largest city.

Mumbai is experiencing **hyper-urbanisation**. Economist Susan George said 'Shanty towns built by desperate people fleeing rural poverty and trying to create a better life for themselves have become a common sight around Asian cities.' They are typical of rapidly changing economies, where the rural population is forced off their land into cities and where governments have not provided the basic requirements of urban life – drainage, clean water, public health and education facilities.

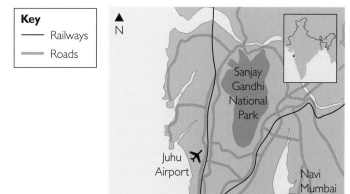

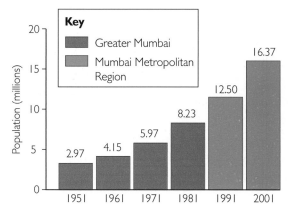

▲ *Population growth in Mumbai, 1951-2001*

● **Hyper-urbanisation** is where the increase in the urban population is happening so rapidly that the city cannot cope with the needs of the people.

Why is Mumbai growing so quickly?

Susan George was describing **rural-urban migration**, and it is not confined to Mumbai. It is typical of many cities in developing countries. The diagram on the right shows that there are factors which 'pull' people into cities, and others that 'push' them from the countryside. There are background factors at work too, e.g. increased overseas investment in cities, creating new jobs which may not have existed a decade ago.

The reality of moving to the city often does not meet people's expectations. Instead of better housing and quality of life they could end up as one of Mumbai's thousands of slum dwellers.

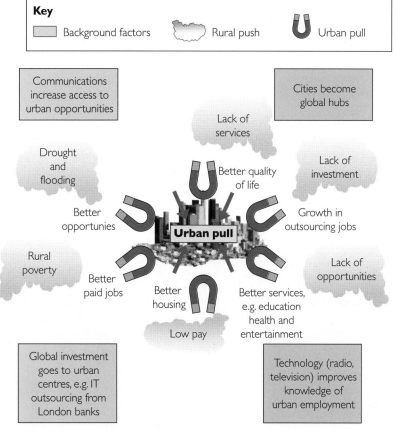

Dharavi

Mumbai or Slumbai?

Mumbai is keen to be the next Shanghai. But what does this mean? Since the 1990s, Shanghai has been a major centre of investment and growth – and Mumbai is keen to follow in its footsteps. Shanghai has epitomised China's rapid growth. Yet 60% of Mumbai's population is living in poverty (in places like Dharavi), and missing out on India's growth. The old manufacturing industries have been shedding jobs and the **informal sector** has been increasing.

Dharavi is situated between two main railway lines. It provides cheap accommodation for the low-skilled workers who arrive in the city in search of a better life. Homes have a solid look and many have electricity. Small-scale workshops produce cheap pottery, plastic toys (the sort we get in Christmas crackers), embroidered clothes and leather handbags for export markets.

Average incomes are low. Rakesh Pol, a leather worker, earns about £40 a month. He can rent a room for about £12 a month. Over time, families acquire building materials to improve their homes – few of them can afford to move out of Dharavi because the rest of Mumbai is too expensive.

Sustainable Dharavi

One million residents live illegally in Dharavi, in closely knit communities where self-help health clinics, food halls and meeting places exist alongside the well-organised cottage industries. Most work is informal and illegal too, but the box below illustrates how Dharavi supports itself and is sustainable.

> ● The **informal sector** means self-employed work that is irregular and with little security, e.g. street trading.

Recycling

- 80% of Mumbai's waste is recycled in Dharavi. Nothing is considered to be rubbish.
- The recycling industry in Dharavi is estimated to be worth nearly $1.5 million a year.
- It employs almost 10 000 people, including children.
- Barefoot child workers, in 15 000 single-room factories, collect and carry plastic, glass, cardboard, wire hangers, pens, batteries, computer parts and even soap.
- The workshops range from those where aluminium smelters recycle drink cans to those where vats of waste soap from hotels are melted and remoulded into new bars of soap (see photo below).
- The aim is to turn everything into something new and useful.

Other small businesses
Clothing

- Families work at sewing machines, making shirts in assembly-line fashion. One does the cutting, another sews the sleeves, another the collar and so on.
- The finished product sells to a retail store buyer for 15 rupees, about 17 pence!

Pottery

- Kumbharwada is the pottery zone on the edge of Dharavi.
- Women mix clay for the men who make pots and bowls of all shapes and sizes.
- The pottery is sold to a trading company across the city that exports them around the world.

Vision Mumbai

Mumbai attempts to become the new Shanghai and sell its soul

The grand vision to turn India's biggest city into a thriving megacity has finally got under way with the demolition of Dharavi and the sale of the land to Indian and foreign investors as part of 'Vision Mumbai'.

The area is to be transformed and the population housed in new homes as part of a $2.3 billion project.

May 2007

Mumbai has two main problems: its economic growth has slowed down and its quality of life has deteriorated. Slums have multiplied, and congestion, pollution and water problems have rocketed. However, Mumbai has a plan – *Vision Mumbai* – to tackle these problems and turn the city into a world-class location by 2013, with state-of-the-art transport systems and higher-quality housing. Work on the plan began in 2004. If economic growth can be sustained at 8-10% per year, it is hoped that 1.1 million low-cost homes will be built – and slum populations will fall to 10-12% of their 2000 level. General safety, air pollution, water, sanitation, education and healthcare will all be improved. Part of *Vision Mumbai* means that Dharavi and other slums will have to go.

The land occupied by Dharavi is worth $10 billion. The plan for Dharavi, and other slums – as the above article says – is to sell the land and redevelop it. Private developers can buy the land for less than it is worth, and for every square foot of new housing they build for the poor, they can have 1.3 square feet for commercial development. High-rise blocks for slum dwellers will be built next to new shopping malls, offices and apartments for the rising middle classes. But not everyone is happy with the plans for the slums, as the following article shows.

Voices from the shanties

'It is cold, we have no food and we cannot go to work because there is nowhere to keep our belongings,' said Sunita Kaude, who came from Marathwada with her husband and two children in 1996 and found a place in a slum in south Mumbai.

'We cleared the land, which was a swamp, and made it liveable. Now they want the land. The politicians promised to legalise our home if we voted for them. What happened to that? The politicians have backtracked on their promise'.

But, Sunita has a weak case, for she and her neighbours cleared mangroves to build huts, which is illegal.

'We don't have a target but our aim is to get rid of every illegal shanty in the city,' said Prakash Patil, an Assistant Municipal Commissioner.

Adapted from *Frontline*, India's national magazine, January 2005

How will Vision Mumbai work?

Vision Mumbai is based on six core targets.

- Increase housing availability to reduce the number of people living in slums and make housing more affordable
- Raise adequate financing and reduce administrative expenditure
- Upgrade other infrastructure (safety, environment, water, sanitation, education and healthcare)

Vision Mumbai

- Improve transport infrastructure, providing more train carriages and buses; increasing freeways and expressways and the amount of parking space
- Boost economic growth to 8–10% pa, e.g. by focussing on services and making Mumbai a 'consumption centre'
- Make governance more efficient and responsive, e.g. reduce the time needed for building approval

In addition, part of *Vision Mumbai* is based on 'quick wins' – things that could go ahead quickly to improve Mumbai, for example:

- promote NGO and corporate sponsorship to clear, restore and maintain 325 open and green spaces
- build an extra 300 public toilets
- widen and 'beautify' main north-south and east-west roads.

By 2007, 200 000 people had been moved and 45 000 homes had been demolished in Mumbai's slums, as 300 hectares of land were cleared for development.

Future Mumbai

Vision Mumbai envisages creating:

- 200 000 high-end service sector jobs in the financial, healthcare and entertainment sectors (based on the use of ICT) – raising GDP by 3% per year
- 500 000 jobs in construction, retail, tourism, and recreation – raising GDP by 2% per year
- 200 000 new jobs in Special Export Zones around the port and airport – concentrating on low-polluting, high-value computer assembly and top-class fashion/apparel industries.
- a consumer city by lowering local taxes and in turn attracting more investment from overseas.

By 2015, Mumbai could be the new 'city that never sleeps'. That is the aim and by then it may resemble Shanghai.

▲ *Shanghai today – an example for Mumbai to follow?*

The world's largest planned city, Navi Mumbai (New Mumbai), built on the mainland opposite the island city of Mumbai, already has 1.2 million residents – most of them middle-class business people moving from Mumbai itself.

Mumbai – global city on hold?

Mumbai is shaping India's future. Its range of skills and services is increasing – from call centres solving computer problems, to film animation – it is all done in Mumbai. In addition, scientific and educational research, publishing, and micro-surgery techniques are being developed for the world in Mumbai.

But the future success of Mumbai is in the balance. If *Vision Mumbai* does not work, the slums, the gridlocked roads and the death trap trains could be its downfall – and it may never be the next Shanghai.

> *Vision Mumbai – good news for everyone?*
>
> **What do you think?**

▲ *Mumbai's stock exchange is helping to push its economic growth*

▲ *Just one of Mumbai's problems – 10 people die on its massively overcrowded railway system every day*

● Over to you

1 Work in groups of three. One of you is a resident of Dharavi, one an international banker, and one a successful Bollywood actor.
   a Read the text on *Vision Mumbai* and look at www.bombayfirst.org/McKinseyReport.pdf
   b In role, discuss the project, and decide on the strengths and weaknesses of the scheme.
   c Which of you would gain and lose from Vision Mumbai? Explain your answer.

● On your own

2 Define these terms: hyper-urbanisation, dual societies, informal work.
3 Explain why shanty towns are common in Asian cities.
4 Use labelled diagrams to compare the causes and processes of urbanisation in Mumbai with those of suburbanisation in Los Angeles.
5 Do you think that Mumbai will follow the same pattern in future as Los Angeles? Explain your answer.
6 Navi Mumbai (New Mumbai) is the world's largest planned city. Research to find out about why it has been built, who lives there and its transport infrastructure.

Exam question: Using examples you have studied, explain the problems that the world's megacities face in trying to be more sustainable. (15 marks)

Environmental consequences of globalisation

In this unit you'll look at how globalisation can contribute to the destruction of the environment.

The price of progress

> There is no chance whatsoever of saving the rainforest unless the debt crisis is resolved. It is entirely impossible to expect a country which is desperately struggling to meet loan interest payments to divert funds to their own long-term needs and the conservation of their natural resources.
>
> Barbara Bramble (International Director, National Wildlife Federation)

Destruction of rainforest in Para State, Brazil, to make room for crops or a cattle ranch ▶

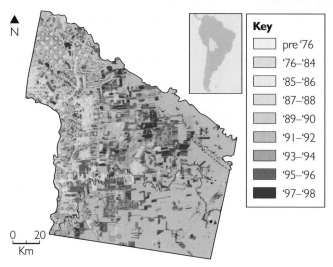

◀N

Key

	pre '76
	'76–'84
	'85–'86
	'87–'88
	'89–'90
	'91–'92
	'93–'94
	'95–'96
	'97–'98

0   20
Km

◀ Deforestation in Bolivia up to 1998. This map has been put together from a series of satellite images

Percentage of rainforest lost, 1990-2005. In the same period, Brazil cleared the largest area (over 42 million hectares, an area the size of California) but, in percentage terms, Comoros (an island nation north of Madagascar) cleared the largest percentage of its forests ▼

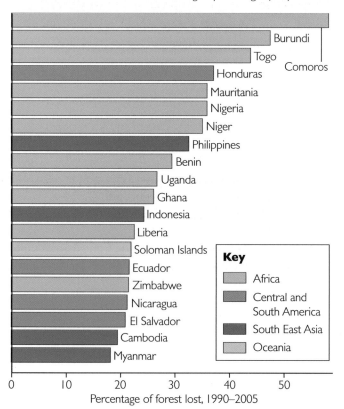

Key

	Africa
	Central and South America
	South East Asia
	Oceania

Percentage of forest lost, 1990–2005

Deforestation, debt and trade

Deforestation, economic growth, trade and debt – they are all linked. Central America once had about 500 000 km² of rainforest cover, but by the late 1980s this had fallen to an estimated 90 000 km². Over 30% of Honduras' rainforest has been lost since 1960, with more than 800 km² being lost every year for ranches, banana plantations, small farms and fuelwood. The expansion of fruit plantations is partly due to the country's need to earn foreign money to repay debt. Exports of hardwood also add to the money earned to repay debt.

20% of Honduras' export earnings is spent on debt repayment every year. In Nicaragua the figure is 39%, and the combined debt of Nicaragua and Honduras is 10.1 billion dollars.

Deforestation and lack of forest cover means that:
- ecosystems are destroyed
- there is more rapid runoff of rainfall
- there is increased soil erosion and the risk of flooding, mud and landslides.

Globalisation, deforestation and climate change

In the 1980s, the plight of the rainforests regularly made headline news. It was recognised that local destruction of rainforests had global effects. Loss of biodiversity became a matter of global concern.

'Debt for nature swaps' were one way of tackling the debt and deforestation issue. They were designed to free up resources in debtor countries for much needed conservation activities. Debt swaps involve purchasing and converting the debt into local currency, and using the proceeds to finance local conservation projects.

Deforestation has major impacts on climate change:
- Burning forests account for 25% of global carbon emissions (the next highest source after burning fossil fuels).

- Deforestation means the loss of a major **carbon sink**.
- One day's burning of rainforest is equivalent to the carbon emissions from 8 million people flying across the Atlantic.
- Indonesia and Brazil became the third and fourth largest emitters of greenhouse gases in 2007, through burning forests.
- The annual global rate of rainforest destruction equals the combined areas of England, Scotland and Wales – releasing 2 billion tonnes of CO_2.

Other environmental costs

Our economic growth is destroying entire ecosystems. At current rates of economic growth, we will need two planets' worth of natural resources by 2050. The Earth cannot regenerate supplies at the rate they are being used. This raises some serious issues:
- As more countries develop, there will be more pressure on resources.
- Emerging economies will demand more of the world's resources.
- Competition for resources will push prices up and undermine development.
- Scarcity of resources creates insecurity.
- How can resources be managed and distributed equitably?

Are the environmental costs of globalisation too high?

What do you think ?

● Over to you

1 In pairs, devise a spider diagram to show the processes by which globalisation affects the environment.
2 Label which processes are deliberate and which are accidental.
3 Label whether the environmental impacts are destructive, neutral or beneficial.

● On your own

4 Define the following terms: deforestation, soil erosion, biodiversity, debt for nature swaps
5 Explain how debt is linked to environmental destruction.
6 a Select one major environmental issue: deforestation, soil erosion, water or air pollution, biodiversity loss.
   b Research the issue in two countries at different levels of development.
   c Assess how serious it is, and why.

In this unit you'll learn about some of the social consequences of globalisation.

Virgin territories

If you visited St Lucia in the Caribbean, you would see the tail fin of a Virgin Atlantic Jumbo Jet planted in the middle of a traffic island as you enter the capital, Castries (see right). Just as early settlers raised their flags in early colonies, the tail fin is a symbol of Virgin Holidays' presence in the Caribbean. In the same way that early colonists repatriated their profits, the profits of transnational corporations go back to the country in which the corporation is based.

Tourism is the world's biggest industry, with 800 million international tourists in 2006. Since the 1970s, tourists have travelled increasingly beyond Europe to America, Asia and Africa – to more exotic and distant locations, as the graph on the right shows.

Cuba

Cuba is a good example of a country switching into the global economy. Cuba became communist in 1959, after a communist revolution led by Fidel Castro. When the Soviet Union and its empire collapsed between 1989 and 1991, Cuba faced a crisis because of the decline of Soviet subsidies. It began a 'special period' of development, accepting foreign capital.

Virgin Atlantic Airways began direct flights to Cuba in 2004. Their 'all-inclusive deals' ensure that visitors from the UK never need to leave their resort hotel. Varadero, a short journey from Havana international airport, receives most of the visitors and has become a **tourist enclave**, completely detached from Cuban life. As tourism increases, locals become dependent on the jobs it provides. Income from tourism now supports economic development.

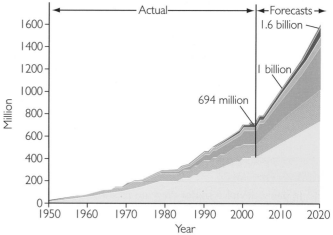

▲ International tourist arrivals, 1950-2020

Key

- South Asia
- Middle East
- Africa
- East Asia/Pacific
- Americas
- Europe

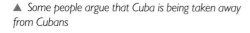

'They come here and look at our country as if it is a park – as if Cubans are animals in a zoo,' said an old man sitting on his porch in the afternoon sun. 'They think that Cuban culture is over there,' he said, motioning toward the hotels, 'believing that Cubans smoke $15 cigars and drink expensive rum. The only Cubans you'll find over there are serving you.'

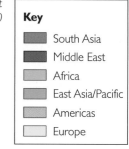

▲ Some people argue that Cuba is being taken away from Cubans

Twenty first century Cuba

Cuba is changing. The influx of tourists, and the influence of television and the Internet are all helping to erode local traditions and values. Globalisation is diluting Cuban culture. It is becoming a commodity and is used in Government tourist advertisements showing local women: 'as sultry, scantily clad, giddy school girls'.

Cuba goes global

Tourism might be big business in Cuba now (see below), but who benefits? Tourism is a global business linking companies in different countries, and they all want to make a profit.

▲ Cuba's communist president, Fidel Castro – not immune to the impacts of globalisation. He is seen here wearing Adidas sportswear in 2007.

- In 1980, there were 130 000 visitors and 5000 available hotel rooms
- In 1990, there were 327 000 visitors and 12 000 available hotel rooms
- By 2005, there were 2.4 million visitors, 266 hotels and 43 000 rooms

Are the social costs of globalisation too high?

What do you think ?

- In 1990, 18 foreign airlines connected to Cuba
- By 2005, 93 foreign airlines connected Cuba to 40 capital cities

- In 2005, China supplied 800 new coaches for tourist excursions and transfers

- 80 000 people work in tourism in Cuba

- 50 hotels with 19 000 rooms are managed by 12 foreign companies
- In 1995–1996, Spanish hotel chains invested US$75 million in Cuban tourism
- In 1996, Canada's Wilton Properties agreed to spend $400 million to build 11 hotels in Cuba

- The new airport is operated by AENA, a Spanish airport-management company

- In 2004, a new tourist currency was devised – the 'Cuban convertible peso', linked to the US dollar

● Over to you

1 Draw a Venn diagram with three circles – social, economic and environmental. Use it to show the consequences of your holidays.
2 Think of ways of calculating the global costs and benefits of these consequences.
3 In pairs, draw up a list of the social and moral consequences of tourism in Cuba. Who benefits from tourism? Who loses out?

● On your own

4 Define the following: repatriation of profits, tourist enclave
5 How and why is tourism a major driving force behind globalisation?
6 Why are international tourists sometimes referred to as twenty-first century colonists?
7 Can tourism ever be sustainable?

2.17 Managing change for a better world – 1

In this unit you'll consider whether globalisation has increased our ecological footprint, and how we can reduce the environmental and social costs of globalisation.

Living beyond our means

People in rich countries, like the UK, have got used to having what they want, from wherever they can get it, at any time of the year – for instance, fresh strawberries in December! Global supply chains fulfil these needs. Our reach has extended across geographical and seasonal boundaries so that developing countries, which depend on food exports to support their economies, end up diverting water and land to grow cash crops and so sacrifice their own needs to supply ours. We now depend on the rest of the world to live beyond our 'environmental means'. The 'veins' on the diagram show resources flowing into the UK.

Ecological footprints

The UK is the world's second largest consumer of natural resources and goods. Only the USA is ahead of us. The table below shows the **ecological footprint** of ten countries at different stages of development. If the whole world consumed resources at the average rate of people in the UK, we would need the equivalent of 3.1 Earths' worth of resources. People in Mauritius currently consume resources at a 'break-even' level.

▲ The 'veins' of the global economy

● The **ecological footprint** is a measure of the amount of land and water that a population needs in order to produce the resources that it consumes, and to absorb its waste, with existing technology.

The USA	5.3
The UK	3.1
France	3.0
Germany	2.5
Russia	2.4
Brazil	1.2
Mauritius	1.0
China	0.8
India	0.4
Malawi	0.3

▲ The number of 'Earths' needed to sustain the whole world at the level of consumption of each country in the table, 2006

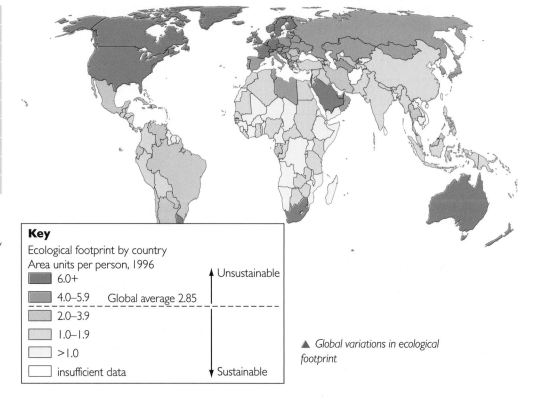

Key
Ecological footprint by country
Area units per person, 1996
- 6.0+
- 4.0–5.9   Global average 2.85
- 2.0–3.9
- 1.0–1.9
- >1.0
- insufficient data

↑ Unsustainable
↓ Sustainable

▲ Global variations in ecological footprint

Globalisation and a global conscience

Globalisation means that economic, cultural, political, and social systems become integrated across geographical boundaries, and countries become less inward looking. Rapid communications mean that millions of people can witness events simultaneously. Over half the world's population (3.9 billion people) watched the 2004 Athens Olympics on televisions across the planet, and many millions in over 130 countries experienced the Live Earth concerts on 7 July 2007 – via television, radio and the Internet. As more people have access to technology, more become aware of the issues of worker exploitation, the impacts of debt, structural adjustment and environmental degradation. A global conscience is emerging as people recognise they have a role in the global village.

▲ *Live Earth concert in Shanghai's new business district*

Some events that have woken our global conscience:
- 1967: The sinking of the Torrey Canyon oil tanker off Cornwall created a 300-km long oil spill and woke up the world to the impacts of development.
- 1970s: Oil crisis – a rapid rise in the price of oil made people realise that resources are finite. The crisis was also a threat to world security.
- 1980s: Debt crisis – led to a realisation of the fragility of capitalism.
- 1980s: Average global temperatures were found to be 0.5 degrees higher than in 1850 and, in 1985, the discovery of holes in the ozone layer showed that our impact on the planet extended beyond the Earth's surface. Events began to lead to action to tackle the human impact on the planet.
- 1985: Live Aid – woke us up to drought and famine in Ethiopia.
- 1987: Montreal Protocol to deal with ozone depletion.
- 1992: Earth Summit in Rio de Janeiro. 178 countries voted to adopt **Agenda 21**.
- 1997-2005: Kyoto Protocol on Carbon Emissions (see page 55).
- 2002: Rio+10 second Earth Summit on key issues of economic and environmental poverty.
- 2005: Gleneagles Agreement to relieve the debt burden on heavily indebted poor countries (HIPCs).
- 2006: Stern Report – polluters producing greenhouse gases impose economic costs on the world (see page 49).
- 2007: Live Earth global concerts to tackle climate change.

● **Agenda 21** is a programme run by the United Nations related to sustainable development. It is a plan of action to be taken globally, nationally and locally by organisations of the UN, governments and major groups in every area in which humans impact on the environment.

● **Sustainable development** is development that meets the needs of the present without compromising the ability of future generations to meet their own needs.

Sustainable development – the final challenge

Rich countries are living beyond their means, and we have seen the environmental and social consequences of this in Units 2.15 and 2.16. The consequences of globalisation must be managed to make the world a better place. Can globalisation and sustainable development exist side-by-side? Can individuals play their part as global citizens and make a difference? The next two pages look at this.

Managing change for a better world – 2

Reducing the environmental and social costs of globalisation

As we become more aware of the consequences of globalisation, the challenge we face as individuals is doing something about it. But we can, at a variety of scales from local to global, as the rest of this unit shows.

Challenge 1: fair trade or free trade?

The World Trade Organisation (WTO) promotes **free trade** by persuading countries to get rid of trade barriers. Free trade normally means that the factory worker or grower of a commodity gets the smallest share of the final price of the goods. **Fair Trade** aims to give a bigger proportion to producers.

Challenge 2: ethical shopping

Retailers are jumping on the ethical bandwagon and organic, Fair Trade and local products are increasing. Marks and Spencer has decided to sell only Fair Trade teas and coffees, naturally dyed clothes, and fabrics made by small businesses overseas. Supermarkets label food with country of origin and use logos to show how it is produced. Local produce is returning to supermarket shelves, and farmers' markets are popping up everywhere.

It is estimated that the total number of miles travelled by all the food eaten in Britain in 2002 was a staggering 30 billion, and the number of miles travelled by all the food in the average person's kitchen in the UK was 41 000 – over twice the circumference of the Earth!

Are there any downsides to ethical shopping?

- Producing food organically can destroy more forests – less use of fertilisers/pesticides means that more land is needed to grow the same amount.
- Buying local produce minimises 'food miles' and carbon emissions – and helps the local economy – but many consumers still use cars to go shopping and, as a result, use more energy than bulk deliveries to supermarkets would.
- Food miles are not all bad. Producing food in Africa is less energy intensive than in the UK, so the emission levels are lower, even if the food is then flown in.
- Buying local produce can undermine Fair Trade and means that poor countries can lose markets.

Banana wars

- Since 1975, each Caribbean country had had a quota of bananas which they could sell to the EU. This protected Caribbean banana farmers from competition from bananas produced by American companies (e.g. Chiquita, Dole and Del Monte) on large-scale plantations in Latin America.
- The USA complained to the WTO and, in 1997, won its case. The WTO ordered the EU to open its markets. The USA got its free trade and banana farmers in the Caribbean rapidly went out of business. By 2000, only one third of the farmers producing bananas in the Windward Islands in 1993 were still growing them.
- Groups such as AgroFair (a Dutch company which first brought Fair Trade bananas into the UK in 2000) say the answer is to produce and sell Fair Trade bananas. Demand for Fair Trade bananas is rising rapidly, guaranteeing farmers a fixed price which reflects production costs.

It's not just our green dyes that won't harm the environment.

Our policy on dyeing clothes is black & white. We've banned all our suppliers from using 56 chemicals in the production of our clothes that put their employees or the environment at risk. We also insist that factories remove residual dyes from their effluents before recycling into the environment.

YOUR M&S
look behind the label

Challenge 3: carbon offsetting and trading

Carbon offsetting is the name given to a credit system called carbon credits. Carbon credits can be voluntary or certified.

Voluntary are payments or projects which offset emissions with equivalent CO_2 savings. For example, Coldplay planted 10 000 mango trees in Karnataka, India, to offset the production emissions caused by their CD 'A Rush of Blood to the Head'. But a year later, few trees had survived Karnataka's dry season and smallholders had lost land in the process.

Certified – these are international exchanges which aim to cut overall emissions. It allows high carbon producers to continue polluting, while purchasing credits from those who do not.

For more on carbon offsetting see page 52.

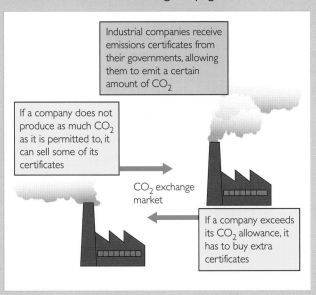

Industrial companies receive emissions certificates from their governments, allowing them to emit a certain amount of CO_2

If a company does not produce as much CO_2 as it is permitted to, it can sell some of its certificates

CO_2 exchange market

If a company exceeds its CO_2 allowance, it has to buy extra certificates

Challenge 4: reduce, re-use, recycle

Londoners produce 3.4 million tonnes of rubbish a year – enough to fill Canary Wharf tower every 10 days! At least 80% of our rubbish could be reused, recycled or composted. 90% of what we buy becomes waste within 6 months of purchase – in London that is a tonne per person per year.

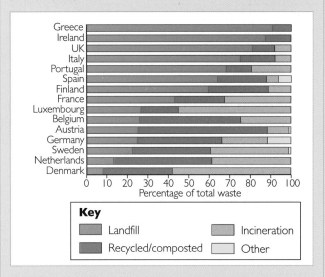

Key

- Landfill
- Recycled/composted
- Incineration
- Other

The challenge of waste:

- Landfill sites are running out, and pose environmental risks. Incineration reduces waste by 75% in weight and 90% in volume. Both incineration and landfill release greenhouse gases.
- Recycling is cleaner, greener and provides new raw materials. But collection, sorting, and processing still uses more energy than simply generating less waste in the first place.
- Composting produces humus that improves soils.

● Over to you

1. **a** In groups of four, devise an investigation into the sustainability of your local area. Design surveys to collect data on recycling, ethical shopping, travel and ecological footprints.
   b Conduct the survey around your local area.
   c Write up a short report of your findings. How sustainable is your local area?

2. Use earthday.net/footprint/index.asp to calculate your own ecological footprint.

3. Why do less developed countries have smaller footprints than more developed countries?

● On your own

4. Define the following: ecological footprint, Agenda 21, sustainable development, free trade, Fair Trade.

5. Choose one of the events listed on page 127, i.e. the Montreal Protocol, Earth Summit, Kyoto Protocol, Gleneagles Agreement, Stern Report or Live Earth concerts. Find out more about the issues they dealt with, their proposals and what has happened since.

6. With ever increasing levels of consumption, are countries likely to succeed in meeting challenges 1-4?

Exam question: Referring to examples, assess some of the strategies that can be used to try to create a more sustainable world. (15 marks)

Unit summary

What key words do I have to know?

There is no set list of words in the specification that you must know. However, examiners will use some or all of the following words in the examinations, and would expect you to know them, and use them in your answers. These words and phrases are explained either in the glossary on pages 297–301, or in key word boxes throughout this chapter.

ageing population	globalisation	quota
Agenda 21	grey pound	remittances
baby booms	greying	replacement level
carbon sink	host nations	replacement migration
centrally planned economies	hyper-urbanisation	rural-urban migration
reverse colonialism	illegal migrant	Seasonal Agricultural Workers Scheme (SAWS)
connections, connected, disconnected	IMF	
	immigration	seasonal workers
core and periphery	informal sector	sheltered accommodation
deindustrialisation	inputs and outputs	social cohesion
digital divide	just-in-time	source nations
dynamic system	megacity	structural adjustment packages
ecological footprint	megalopolis	
economic migrants	immigration	super city
edge city	millennium development goals	sustainable development
enclaves	modernisation theory	tariff
ethnic enclave	natural increase and decrease	three Ds
export processing zones	net migration	Total Fertility Rate (TFR)
Fair Trade	new economy	tourist enclave
fertility rate	outsourcing	trading blocs
First, Third World	periphery	urbanisation
Fortress Europe	population structure	World Bank
free trade	privatisation	world city
	pull and push factors	

Try these questions

1 How and why can international migration create benefits for host countries but problems for source countries? (15 marks)

2 Suggest why most of Africa is *switched off* from development. (10 marks)

3 How and why do different people hold different views about the benefits of global trade? (15 marks)

4 Explain how people can manage the environmental and social costs of globalisation for a better world. (15 marks)

What do I have to know?

This topic is about extreme weather. It focuses on weather events that appear to be 'one-offs', and how and why these occur, together with longer-term processes such as drought.

1 Extreme weather watch: What are extreme weather conditions and how and why do they lead to extreme weather events?

What you need to learn

- Know a wide variety of extreme weather occurrences
- **Fieldwork and research**: Use a weather diary and synoptic maps, to record:
  - air masses, pressure systems, fronts which influence changes in weather
  - contrasting weather events, e.g. depressions or anticyclones
  - contrasting examples of how extreme weather conditions develop, e.g. hurricanes, snow and ice, and drought

This is a research-focused part of the syllabus and emphasizes the importance of keeping your own research and of logging weather events. One or two case studies will be useful.

2 Extreme impacts: What are the impacts of extreme weather on people, the economy and the environment?

What you need to learn

- Extreme weather hazards have different impacts depending on the severity of the event, level of economic development and vulnerability of those affected
- **Fieldwork and research** into the social, economic and environmental impacts of extreme weather, created by:
  - an immediate disastrous event, e.g. tornado
  - a subsequent hazard, e.g. localised flooding
  - a longer term trend, e.g. heat wave or drought

Detailed case study knowledge will arise from your research and studies. You need one study of each of the bullet points above (storm, drought, hurricane or similar).

3 Increasing risks: How are people and places increasingly at risk from and vulnerable to extreme weather?

What you need to learn

- Understand that extreme weather hazards in the UK and elsewhere are more frequent and result from, e.g. climate change, demographics and land management
- **Fieldwork and research** to investigate how a small catchment can suffer increased flood risks resulting from:
  - meteorological causes
  - the physical characteristics of the area
  - growing urbanisation, land use change and attempts at management

Again, detailed examples are needed of the causes leading to the events already mentioned.

4 Managing extreme weather: How can we best respond to and cope with the impacts of extreme weather?

What you need to learn

- **Fieldwork and research** into
  - short- and longer-term strategies for responding to extreme weather events
  - how some management strategies are more successful than others
  - the role of technology in improving preparedness, event forecasting and reducing impacts of disasters
  - ways of managing drought

This should follow on from your studies of extreme weather events – here you will need case studies about how well people have responded to each event, and how well each is managed.

3.1 Extreme heat

In this unit you'll find out about one example of extreme weather and the weather system that caused it.

What is extreme weather?

Usually severe, and often unexpected, extreme weather can bring chaos and misery. Floods, droughts, heat waves and fires are just a few examples of hazards caused by extreme weather. Too much or too little precipitation, temperatures which are higher or lower than normal, extreme winds, and intense low- or high-pressure systems create hazards which affect people all over the world.

2003 – Europe's heat wave

In summer 2003, Europe was in the grip of a heat wave. It began in June and continued until mid-August. While some people took to the beach and sales of ice-cream soared, others were less lucky. Temperatures were 20-30% higher than average over a large part of Europe. Extreme temperatures of 35-40 °C were recorded again and again in July and August in most southern and central European countries from Germany to Turkey.

Impacts of the heat wave

The impacts were wide-ranging. They included:

- A human death toll which exceeded 30 000. The heat wave was the worst natural disaster in Europe for 50 years.
- Farm livestock also died and crops failed – leading to increases in food prices. The heat wave cost European farmers 13.1 billion euros (£7 billion).
- Demand for electricity soared as people turned up air conditioning and fridges, but the heat spelled trouble for French and German nuclear power stations, which are cooled by river water. In some places, river levels were so low that the cooling process was impossible and nuclear plants had to shut down.
- The River Danube in Serbia fell to its lowest level for 100 years. Reservoirs and rivers used for public water supplies and hydro-electric schemes either dried up or ran very low.
- The lack of rainfall meant very dry conditions, with forest fires breaking out across Europe (over 25 000 were recorded). Portugal was the worst hit, with nearly 400 000 hectares burned (an area the size of Luxembourg). The cost to Portugal was over 1 billion euros.
- Tourism increased in parts of the UK, as people stayed at home to enjoy the Mediterranean weather.
- Transport suffered – railway tracks buckled in the heat; speed restrictions were imposed to prevent trains derailing; road surfaces melted.
- Extreme snow and glacier melt in the Alps led to increased rock and ice fall. Total glacier volume loss in the Alps was 5-10% in 2003.

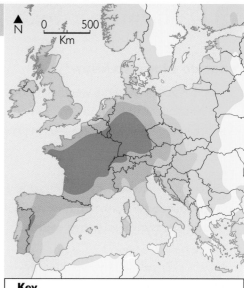

Key

Difference from normal temperatures

| +9 °C | +7 °C | +5 °C | +3 °C | +1 °C | −1 °C |

▲ *Europe hotter than normal – the difference from normal temperatures between 3 and 9 August 2003*

Country	Number of deaths
France	14 082
Germany	7000
Spain	4200
Italy	4000
UK	2045
Netherlands	1400
Portugal	1300
Belgium	150

▲ *Deaths caused by the 2003 heat wave. Elderly people were most affected.*

Fighting a Portuguese forest fire, 2003

Background

Anticyclones

Two large-scale weather systems dominate the weather in the UK – depressions which usually bring wet weather (see page 143), and **anticyclones** which bring periods of dry weather. Anticyclones have high pressure, light winds, clear skies and dry conditions. However, there are differences between winter and summer.

- **Anticyclones** are areas of high pressure, and give clear, calm weather.

Summer anticyclones

In an anticyclone, air is descending. As it descends it warms up, causing any water vapour in the air to evaporate. This prevents clouds from forming. Cloudless skies mean that all the Sun's energy reaches the Earth's surface, raising ground temperatures dramatically. If the anticyclone remains stationary, it produces a heat wave – as in Europe in 2003. That stationary anticyclone blocked the rain-bearing depressions that usually cross the continent from the Atlantic Ocean.

Winter anticyclones

In winter, anticyclones bring different weather. The cloudless skies associated with anticyclones allow heat to escape into the atmosphere, cooling the ground rapidly at night and leading to the formation of frost. Water vapour condenses into droplets, causing fog, which can linger into the day until the heat of the sun evaporates it.

> Heat waves – doesn't everyone prefer a Mediterranean climate?
>
> **What do you think ?**

Synoptic charts

Synoptic charts, or weather maps, are used to plot the weather conditions of certain areas at a single point in time. The synoptic chart here shows the weather at midday on 5 August 2003. It shows an area of high pressure (labelled H) over most of Western Europe. Air is moving around the high in a clockwise direction (the red arrow), bringing a hot, dry, tropical continental **air mass** to the UK.

- An **air mass** forms when air is stationary for several days. The air assumes the temperature and humidity of the area where it originated.

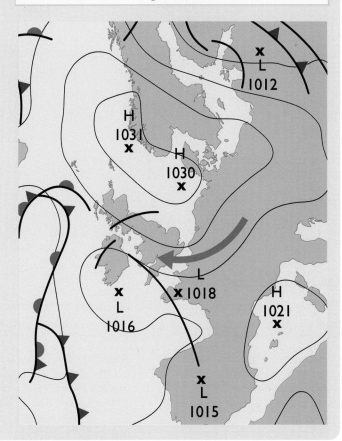

A synoptic chart for midday 5 August 2003 ▶

● Over to you

1 Draw up a table with three columns headed Social impacts, Environmental impacts, Economic impacts. Use the table to classify the impacts of the heat wave (some impacts may go in more than one column).

2 a Who is most at risk from heat waves? Why?

   b Why did more people die in France and Germany than in other European countries?

● On your own

3 Define these terms from the text: heat wave, anticyclone, synoptic chart, air mass.

4 Find two photos, one of the weather associated with a summer anticyclone, the other of the weather associated with a winter anticyclone. Annotate the photos to show the type of weather associated with each.

From one extreme to another

In this unit you'll investigate extreme weather and the links with climate change.

Extremes around the world

Since 2000, extreme weather events have killed tens of thousands of people and caused colossal amounts of damage. In Mumbai, India, 800 people were killed in July 2005 after nearly a metre of rain fell in 24 hours, causing massive flooding.

▼ *Weather-related disasters worldwide, 2000-2005 (droughts, extreme temperatures, floods, landslides, bushfires, windstorms, resulting in ten or more people killed, 100 or more affected, a state of emergency, or a request for international asisstance)*

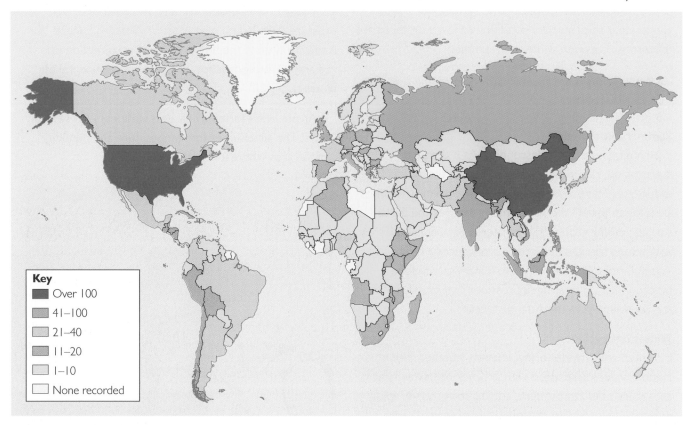

Key
- Over 100
- 41–100
- 21–40
- 11–20
- 1–10
- None recorded

UK weather extremes

The Met Office predicted that 2007 could be the hottest year ever. The first half of the year was highly changeable and noted for its extremes, as the timeline shows.

January *It began with stormy weather in many parts of the UK, with gusts of wind over 70 mph. It was the second warmest January on record, with 17% more sunshine and 21% more rain than normal.*

March *A cold snap in mid-March, with hail, frost, sleet and snow showers, affected much of the UK. Overall, though, it was a relatively mild month.*

May *Above average rainfall in most of the UK. The annual two-day Ten Tors walk on Dartmoor was a wash out and ended early. East Anglia had its wettest May ever.*

January • February • March • April • May • June

February *Heavy snowfall caused widespread disruption, but the winter was still the second warmest on record. It was also very wet.*

April *A record-breaking month – the warmest April in the UK on record, with virtually no rain in some areas. Temperatures reached 26.5 °C on 15 April (about 10 °C above average) and there were about 50% more sunshine hours than average across the UK.*

June *A month of torrential rain and severe flooding in many parts of the UK. A sixth of the annual rainfall fell in some places in 12 hours. The Glastonbury Festival was even wetter and muddier than normal, and Boscastle flooded again (see pages 138–141 for the devastating 2004 Boscastle flood).*

Extreme weather and climate change

Extreme weather at home and abroad – is climate change to blame?

The view from scientists is that it is not possible to link any one particular extreme weather event to climate change. But there is no doubt that the planet is warming up and, at the same time, we are seeing severe weather around the world.

- Records kept by the International Disaster Database indicate that the number of floods and windstorm disasters has increased significantly since the 1960s, plus these events are more intense, last longer and affect more people.
- Most climate models indicate an increase in the frequency and length of extreme events. For example,

in the USA, total precipitation increased between 5–10% from 1997–2007, and rain and snow are falling in fewer, more extreme events.

- Scientists use climate modelling to predict patterns of future heat waves; they predict that in some areas of Europe and North America, heat waves will become more frequent and extreme in the second half of the twenty-first century.

The Intergovernmental Panel on Climate Change (IPCC) is increasingly confident that certain weather events will become more frequent, widespread and intense during the twenty-first century, as climate changes. Some of these are shown in the table below.

▼ *Possible impacts of climate change due to extreme weather and climate events*

Weather, climate and extremes	Impact on agriculture, forestry and ecosystems	Impact on water resources	Impact on human health	Impact on industry, settlement and society
warmer temperatures with more frequent hot days and nights, fewer cold days and nights	increased yield in colder areas; decreased yield in warmer areas	possible decline in some water supplies	fewer deaths caused by cold.	energy used for cooling instead of heating; declining air quality in cities
increase in warm spells/heat waves	decreased yields in warmer regions due to heat stress; increased risk of forest fires	increased water demand; water quality problems, e.g. algal blooms	increased risk of heat-related deaths among the old, sick, very young and poor	reduced quality of life for people without appropriate housing
heavy precipitation events; frequency increases in most areas	damage to crops; soil erosion; water-logged soils cannot be used	adverse effects on water quality; contamination of water supply	increased risk of death, injury or illness	disruption of settlements, business and transport due to flooding; loss of property
area affected by drought increases	land degradation; lower crop yields/failure; increased livestock deaths; increased risk of fire	more widespread water stress	increased risk of: food and water shortage; malnutrition; water- and food-borne diseases	water shortages for settlements, industries; reduced HEP potential; increase in migration
intense tropical cyclone (hurricane) activity increases	damage to crops, trees and coral reefs	power cuts cause disruption of water supply	increased risk of deaths, injuries, water- and food-borne diseases	disruption by flood and high winds; loss of property; withdrawal of insurance cover; possible migration

● Over to you

1 The number of people who died worldwide from weather-related disasters between 2000 and 2005 were: 1171 from drought, 55 279 from extreme temperature, 30 988 from floods, 4537 from land or mudslides and 15 963 from windstorms.
   a Draw a graph to show these figures.
   b Where were people most likely to have died for each type of disaster? Why?

● On your own

2 Continue the timeline for weather events in the UK for July to December 2007. Find out what the weather was generally like each month, and whether any other records were broken. Use www.bbc.co.uk/weather/ukweather/year_review to help.

3.3 Increasing the risk

In this unit you'll find out how people and places are increasingly vulnerable to extreme weather.

Living on the edge – I

One in ten people worldwide (and one in eight city dwellers) lives less than 10 metres above sea level and near the coast (the **at-risk zone**). Low coastal cities attract large populations to areas that are at risk from worsening storms, floods and also rising sea levels. As cities develop, the surrounding ecosystems suffer, e.g. mangroves in Asia and the wetlands around New Orleans (see page 153). These can provide natural barriers against storm surges and floods.

- Asia has about 75% of the people living in coastal areas 'at risk'. The ten countries with the largest numbers of people living in the at-risk zone, are: China, India, Bangladesh, Vietnam, Indonesia, Japan, Egypt, the USA, Thailand, and the Philippines. The map shows that much of Bangladesh is at-risk, and that population density is high within the at-risk zone.
- 21 nations (mostly small islands) have more than half their population in the at-risk zone, e.g. The Maldives.
- In China, people are moving to coastal cities at a rapid rate, and large populations are put at risk.

In England and Wales, 5 million people in 2 million homes live in flood-prone areas. In some areas, the rate of development on floodplains has more than doubled in the past 50 years, putting increasing numbers of people at risk.

In Europe, an increasing number of people are living in flood risk areas, and the number of businesses and industries locating in flood risk areas continues to grow.

Key
Population density (in 2000) inside (red colours) and outside (green colours) a zone of 10 metres or less above sea level.
Persons per km² <25   25–100   100–250   250–500   500–1000   >1000

Largest urban areas ———

▲ *Population density in Bangladesh's at-risk zone*

> *Floodplains are called floodplains for a reason – they flood!*
> **Professor Bob Spicer, Open University**

> *Building houses on floodplains – does that sound like a good idea?*
> **What do you think?**

Living on the edge – 2

Rapid population growth in LEDCS, together with **urbanisation**, means that cities are growing rapidly.

People leave rural areas for the cities for a number of reasons: 'pushes' include landlessness, drought, famine, civil war; 'pulls' include employment, the possibility of education, and a better standard of living. But life in the cities does not always meet expectations, and migrants often end up living in **shanty towns**. Shanty towns usually develop in parts of the city where no-one else wants to live, e.g. on steep hillsides or in swampy areas. Homes are built from any materials that are available – wood, cardboard, metal, plastic sheeting. These settlements are vulnerable to heavy rain, mudflows, and landslides.

▼ *Shanty dwellings in São Paulo, Brazil*

In 2005, at least five people died in São Paulo's shanty towns (or favelas) after southern Brazil experienced some of its heaviest rainfall and flooding on record. The rain caused fatal mudslides. In 2007, flooding devastated South Asia. Residents from Karachi's shanty towns were evacuated as the badly built homes collapsed or were washed away by torrential rain. 200 people died.

Deforestation

The most common reasons for **deforestation** are clearing land for farming, or when the wood is used for fuel. Whatever the reason, the effects are the same. Lack of forest cover means:

- more rapid runoff of rainfall
- increased soil erosion, risk of flooding and risk of mud and landslides.

In Honduras, 30% of the forest has been lost since 1960 – with more than 800 km² lost every year for ranches, banana plantations, small farms and **fuelwood**. The expansion of fruit plantations is partly due to the country's need to earn foreign money to repay **debt**. 20% of the export earnings of Honduras is spent on debt repayment every year.

Clearing the hillsides for farming in Honduras meant that when Hurricane Mitch roared into Central America in October 1998, the slopes were vulnerable to landslides.

Vulnerability

Are we all equally vulnerable to extreme weather events, or are some people more vulnerable than others? Some people live in dangerous places because they have no choice; others do things which make them more vulnerable.

Ultimately, it is today's children (like Ziaul, Raul, and Luke) who will pay the price of extreme weather in the future, whether or not it is linked to climate change.

▲ *Rebuilding a hut destroyed by floods in a Karachi shanty town in July 2007*

Ziaul   *Raul*   *Luke*

***Ziaul**, Bangladesh – catastrophic flooding is likely to feature frequently in his life.*

***Raul**, Nicaragua – hurricanes such as Hurricane Mitch, which devastated Nicaragua's capital Tegucigalpa in 1998, may become an increasingly regular and hazardous fact of life for him.*

***Luke**, Solomon Islands – the sea will feature large in his life, as small islands are threatened with bigger storm surges.*

● Over to you

1 Look at all the ways in which people become increasingly vulnerable to extreme weather events. Draw a Venn diagram with three overlapping circles labelled Social factors, Economic factors and Environmental factors. Put the reasons in the circles. Compare your diagram with a partner and explain any differences.

2 Draw a diagram to show the links between physical and human factors, extreme weather events and vulnerability.

● On your own

3 Define these terms from the text: at-risk zone, urbanisation, shanty towns, deforestation, fuelwood, debt, vulnerability.

4 Explain the link between debt, deforestation and landslides.

Exam question: Using examples, explain how people and places can be at risk from extreme weather. (10 marks)

3.4 One extreme storm

In this unit you'll learn about the impact of the extreme events of 16 August 2004.

Boscastle, August 2004

Some of the worst flooding ever seen in the south-west of the UK devastated Boscastle in north Cornwall on 16 August 2004.

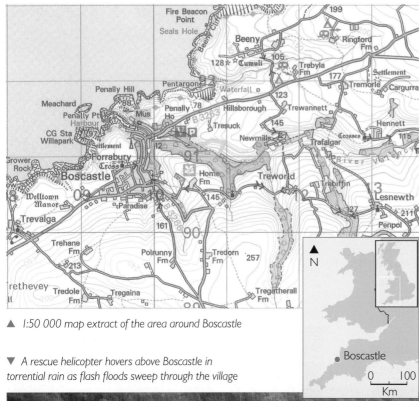

▲ 1:50 000 map extract of the area around Boscastle

▼ A rescue helicopter hovers above Boscastle in torrential rain as flash floods sweep through the village

Andrew Cameron is a crew member of the Port Isaac lifeboat, and was one of the first rescuers to reach the scene of the flash floods in Boscastle. 'We were the first lifeboat on the scene and were greeted by a 10-15 foot wall of water coming down the town, out of the harbour and pushing 30, maybe 50 cars in front of it. There were cars all around us at sea. There was debris everywhere, the air was thick with the stench of fuel.

Then another storm came in as we arrived. Lightning was hitting all around us in a big thunderstorm. The rain was so heavy you couldn't see in front of you.'

Adapted from BBC News Online, 18 August 2004

It was just a normal lousy wet afternoon, it wasn't even raining in the morning.
Resident Steve Lancaster

It had happened before …

Severe weather, and flooding, had hit Boscastle before:

- **28 October 1827** *'One of the most awful days I ever experienced … It rained very heavily in the morning … At Bridge teams of Wagon Horses were saved with difficulty. Pigs also belonging to the Cottagers were taken out of ye Roofs of Houses'* This extract is from *The Journal of Thomas Pope Rosevear.*
- **16 July 1847** Devastating floods swept down from Davidstow Moor.
- **6 September 1950** Torrential rain caused the River Valency to rise rapidly and burst its banks.

- **8 June 1957** There was 203 mm of precipitation in 24 hours (140 mm fell in just 2.5 hours, with hail drifts over 50 cm deep).
- **3 June 1958** The River Valency rose 4.5 metres above normal in 20 minutes after a cloudburst on high ground around the village.
- **6 February 1963** Heavy snow on high ground thawed and flowed into the valleys, flooding Boscastle.

Not far away in Lynmouth, Devon, there was a flood in August 1952 that was so bad that people still talk about it. A surge of water swept through the valley killing 34 people, destroying 93 houses and 28 bridges.

Effects of the flood

Boscastle is located on Cornwall's north coast. It has got a natural harbour, and it evolved as a port and fishing village. The development of the railways in the nineteenth century was both good and bad for Boscastle – the port became less important, but the railway brought the first tourists. And tourism is big business here now – supporting shops, pubs, hotels and other local businesses. Summer is the busiest time of year in Boscastle.

Tourism accounts for 90% of Boscastle's economy. It was estimated that the long-term cost of the flood could reach £50 million.

But it wasn't just businesses that were affected:

- 58 properties were flooded, and 4 were swept away by the water.
- Amazingly no-one died, and only one injury was reported (a broken thumb!).
- Around 1000 residents and visitors were affected.
- 4 footbridges along the River Valency were washed away, and stone bridge parapets collapsed.
- Cars were swept away – 84 were recovered from Boscastle's harbour and streets; 32 were never seen again (probably swept out to sea).
- There was serious damage to Boscastle's infrastructure – water, power and phone services were cut, and roads badly damaged.
- 300 metres' worth of sewer pipes were blocked or washed away. Raw sewage contaminated the flood water and caused health concerns.
- Damage to buildings and services was estimated at £2 million.

The harbour – one of Boscastle's historic attractions ▶

What happened in Boscastle?

The events of 16 August 2004

16 August 2004 was a busy day in Boscastle in more ways than one. There were lots of tourists around, but no-one was aware of what was about to happen. This timeline of events was put together using information collected by the Environment Agency, the emergency services, North Cornwall District Council and others.

> 2004 was not the first time that Boscastle has flooded. Why do people still live in places that flood?
>
> **What do you think?**

▲ Cars are carried through Boscastle by the floodwaters on their way out to sea, crashing into buildings and bridges as they go

12.00 noon
Heavy rainfall begins in parts of north Cornwall.

12.45 pm
The Lesnewth rain gauge records 12 mm in 15 minutes.

1.00 pm
Water flows in the River Valency and its tributaries begin to increase.

1.30 pm
Possible showers are noted on weather radar. Hourly monitoring of rainfall begins.

2.00 pm
The Environment Agency sends two people to check the drains in Boscastle.

3.00 pm
Visitors' cars almost fill the 170-space Boscastle car park.

12.00 noon

1.00 pm

2.00 pm

Background

Flood facts

- The River Valency flows into the sea at Boscastle. Its tributary, the River Jordan, flows into the Valency in the middle of the village.
- The **catchment area** of the Valency and the Jordan is quite small (20 km^2) and steep. It rises more than 300 metres. The steep-sided valleys are known as 'flashy catchments', and act as huge funnels after heavy rain.
- 200 mm of rain fell in 24 hours – most of it in a five hour period on 16 August.
- The peak intensity of rainfall was over 300 mm per hour.

- The storm was very localised. Four of the nearest ten rain gauges (within a few miles of Boscastle) showed less than 3 mm of rain.
- At the peak of the flood (between 5 pm and 6 pm) the flow rate in the Valency was 140 tonnes per second.
- Flood water sped along at over 4 metres per second (10 miles per hour).
- An estimated 2 million tonnes (440 million gallons) of water flowed through Boscastle that day.
- The storm which caused the flooding was classified as extreme.

> - The **catchment area** is the area drained by a river and its tributaries. It is also called the drainage basin.

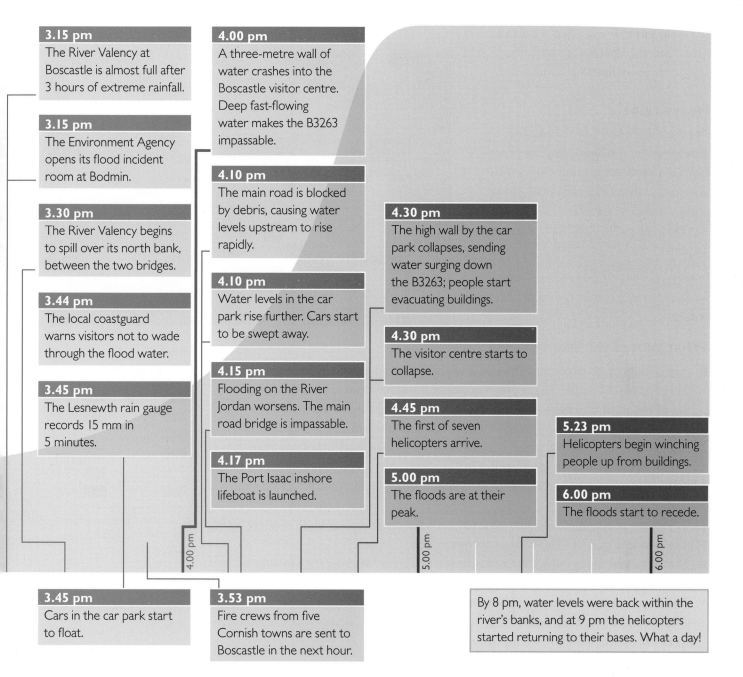

3.15 pm
The River Valency at Boscastle is almost full after 3 hours of extreme rainfall.

3.15 pm
The Environment Agency opens its flood incident room at Bodmin.

3.30 pm
The River Valency begins to spill over its north bank, between the two bridges.

3.44 pm
The local coastguard warns visitors not to wade through the flood water.

3.45 pm
The Lesnewth rain gauge records 15 mm in 5 minutes.

4.00 pm
A three-metre wall of water crashes into the Boscastle visitor centre. Deep fast-flowing water makes the B3263 impassable.

4.10 pm
The main road is blocked by debris, causing water levels upstream to rise rapidly.

4.10 pm
Water levels in the car park rise further. Cars start to be swept away.

4.15 pm
Flooding on the River Jordan worsens. The main road bridge is impassable.

4.17 pm
The Port Isaac inshore lifeboat is launched.

4.30 pm
The high wall by the car park collapses, sending water surging down the B3263; people start evacuating buildings.

4.30 pm
The visitor centre starts to collapse.

4.45 pm
The first of seven helicopters arrive.

5.00 pm
The floods are at their peak.

5.23 pm
Helicopters begin winching people up from buildings.

6.00 pm
The floods start to recede.

4.00 pm

5.00 pm

6.00 pm

3.45 pm
Cars in the car park start to float.

3.53 pm
Fire crews from five Cornish towns are sent to Boscastle in the next hour.

By 8 pm, water levels were back within the river's banks, and at 9 pm the helicopters started returning to their bases. What a day!

● Over to you

1 a Use the OS map extract on page 138 to draw a sketch map of Boscastle. Include Boscastle, the Valency and Jordan rivers, the coastline, the B3266 and B3263.
   b Annotate your map with flood facts (use the information in the box opposite) in one colour, and the impacts of the flood (use the text box on 'Effects of the flood' on page 139) in another colour.

2 Use evidence from the OS map extract on page 138 to explain why flooding affected Boscastle so badly.

● On your own

3 Define this term from the text: catchment area.

4 Write a three-minute news report on events at Boscastle on 16 August. Your report is for the 10 pm news. It must include where Boscastle is, the sequence of events, the impacts of the flood, and facts about the amount of rain and floodwater.

In this unit you'll find out about storms, depressions and other causes of flooding.

Heavy rain

The Met Office had forecast heavy downpours of rain developing over north Cornwall on 16 August. But they couldn't say exactly where it would rain. A string of slow-moving thunderstorms caused localised heavy rain – and that caused the flooding.

It didn't rain heavily everywhere – there were wide variations, as the radar map shows.

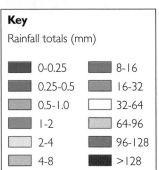

Key
Rainfall totals (mm)

	0–0.25		8–16
	0.25–0.5		16–32
	0.5–1.0		32–64
	1–2		64–96
	2–4		96–128
	4–8		>128

▲ *A radar map showing rainfall totals from 11 am to 6 pm on 16 August 2004*

What caused the storms?

- A large depression dominated the eastern Atlantic Ocean on 16 August, and the storms grew out of this. The depression sucked in warm, moist, tropical air – including the remnants of Hurricane Alex, which had hit North Carolina in the USA.
- The moist air blew in from the sea in the prevailing south-westerly direction. It had been warm and sunny that morning in north Cornwall. The warm, moist air rose – creating clouds and thunderstorms.
- As the south-westerly wind blew over the land, it slowed down and changed direction to south-south-west.
- Where the south-south-westerly wind collided with the prevailing south-westerly wind, it caused strong uplift – this is called **convergence**.
- Storm showers along the north Cornwall coast all started at about the same time – probably due to convergence.
- The storms produced intense rainfall, because of their high moisture content and 'uplift' – the air was being driven upwards, so it cooled giving rain. Cumulonimbus clouds grew to 12 192 metres (40 000 feet) high.
- The storms moved slowly up the coast – dumping millions of gallons of rain as they went.

The storms and the topography of the area were the key to the Boscastle flood. As demonstrated above, rainfall varied widely that day. The rainfall graph here shows the 15-minute rainfall totals for Lesnewth TBR (Tipping Bucket Rain gauge) on 16 August. The rain gauge is just 3 km east of Boscastle. It recorded 184 mm of rainfall that day (7.2 inches). The rain gauge at St Breward (14 km south of Boscastle recorded only 1.5 mm (0.06 inches).

▼ *Lesnewth rainfall graph, 16 August 2004*

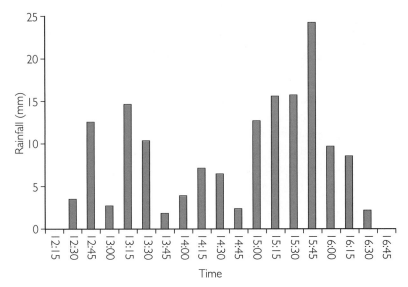

Background

Depressions

Depressions form over the Atlantic Ocean when warm air from the tropics meets cold air from polar regions. Most of the UK's rainfall, and most of our dramatic storms, are the result of depressions.

How do depressions develop?

The diagrams below show different stages in the development of a depression. They are in 3-D.

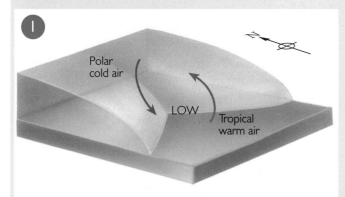

- Cold air flowing down from polar regions meets warm air flowing up from the tropics.
- The warm tropical air is less dense than the cold polar air, so it starts to rise over it, creating low air pressure.

- **Depressions** are areas of low atmospheric pressure.

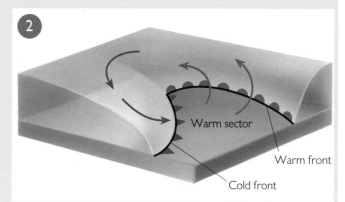

- The warm air is 'sucked' into the low-pressure area, creating a warm sector.
- The cold air is 'sucked' in behind the warm air.
- The whole air mass starts to spiral anti-clockwise.
- Where the warm air rises over the cold air, it is called a **warm front**. As the air rises, it cools and condenses – resulting in clouds and precipitation.
- Where the cold air pushes in behind the warm air, it is called a **cold front**. This forces the warm air to rise – resulting in thick cloud and heavy rainfall.

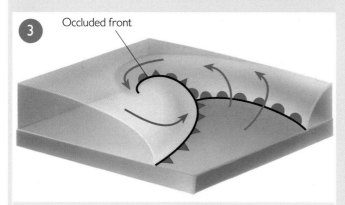

- The warm sector starts to 'climb' above the cold air because it is buoyant.
- Below the rising air the cold front catches up with the warm front. It lifts the warm sector off the ground.
- Where the warm sector is lifted, it is called an **occluded front**.

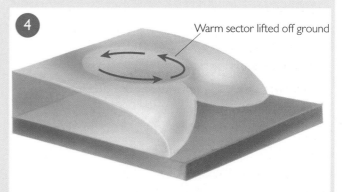

- Cold air has replaced the warm sector on the ground.
- Temperatures even out.
- The fronts disappear, and the depression dies.

What caused the Boscastle flood? – 2

Boscastle – synoptic chart

The synoptic chart on the right shows the weather at midday on 16 August 2004. It shows:

- isobars (lines joining places of equal air pressure)
- cold fronts
- warm fronts
- areas of high pressure (H), and low pressure (L)

On 16 August, the wind was blowing anticlockwise around a low-pressure area. It was approaching Boscastle from a south-westerly direction. It was a warm, moist, tropical maritime **air mass**.

There is a line on the chart labelled TROUGH. This is an area of very localised rain and thunderstorms. On 16 August, it produced very heavy rain.

Boscastle – satellite image

Satellite images allow us to see the cloud cover associated with weather systems, such as anticyclones and depressions. They are used by weather forecasters to tell us what weather is heading towards us. The satellite image on the right shows thick cloud over Boscastle (the white area on the image), and on up the coast, at 1 pm on 16 August 2004. It stayed for most of the afternoon.

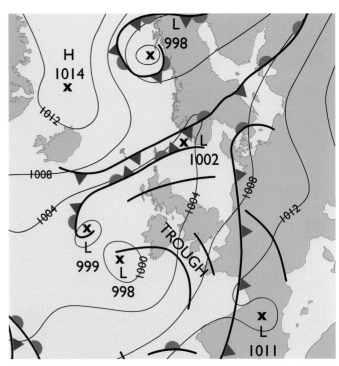

▲ A synoptic chart for midday on 16 August 2004

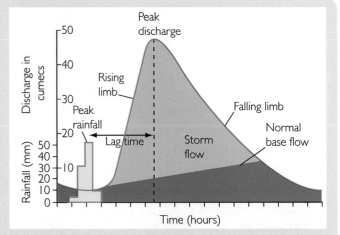

A satellite image for 1 pm on 16 August 2004 ▶

Hydrographs

A **hydrograph** is a graph showing the **discharge** (volume of water) of a river at a given point over a period of time. A flood or storm hydrograph shows how a river responds to a particular storm, and shows both rainfall and discharge. When rain starts, only a tiny proportion will fall into the river, so discharge does not increase straightaway. Most rain falls on the valley sides and takes time to reach the river. As the water makes its way into the river, the discharge increases – shown by the **rising limb** on the graph. The gap between the peak (maximum) rainfall and peak discharge (highest river level) is called the **lag time**.

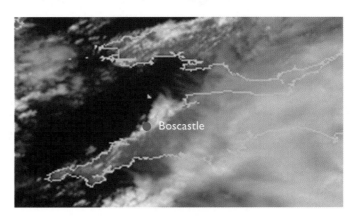

On 16 August 2004, the Valency and Jordan Rivers rose rapidly (a short lag time) as a result of the intense heavy rainfall.

Increasing the flood risk

Did recent changes in land use make the flooding in Boscastle worse? Did they increase runoff? The Environment Agency thought not. The flooding was so severe, and so sudden, that even the facts that:

- the upper part of the village had seen increased development
- hedges had been removed to make fields larger

would not have significantly contributed to the amount of runoff.

In other places, though, human activity can increase the risk of flooding from rivers.

> *Flooding – human activity just makes a bad situation worse.*
>
> **What do you think?**

▲ *Flooding in Tewkesbury, July 2007*

- **Farming** – ploughing up and down hills instead of across them creates channels that can increase the speed at which rainwater reaches a river.

- **Urbanisation** – particularly on flood plains. Impermeable surfaces reduce infiltration and increase surface runoff into drains, and then rivers. The floods in June 2007 in Yorkshire and Humberside were influenced by this.

Increasing the flood risk

- **Destruction of natural environments** – such as deforestation, increases the rate and amount of runoff. Destruction of grasslands and wetlands in South Africa (which absorbed extra water) increased the flooding in Mozambique in 2000.

- **Diverting rivers** – altering and narrowing the channel of the West Lynn River added to the devastation caused by the 1952 Lynmouth flood.

Over to you

1 Use the radar map on page 142 and describe the distribution of rainfall on 16 August 2004. Why was the rainfall so localised?

2 Your teacher will give you a copy of the Lesnewth rainfall graph. Use this as a base to draw a flood hydrograph for 16 August.

   a Add the label discharge to the rainfall axis. You will need to extend the time axis.

   b Use the information on the timeline on pages 140-41 to draw a line for discharge for the Valency and Jordan rivers as they flowed through Boscastle.

   c Add these labels to the hydrograph: peak rainfall, lag time, peak discharge, rising limb, falling limb.

(Hint – you cannot plot discharge accurately, because you do not have the exact figures, but you can plot the shape of the curve against time.)

On your own

3 Define these terms from the text: convergence, depression, warm front, cold front, occluded front, hydrograph, discharge, rising limb, lag time.

3.6 Managing flooding

In this unit you'll learn how we can protect ourselves against the risk of flooding.

Defending Boscastle

As you've already seen, severe weather and flooding has hit Boscastle before – and it happened again in June 2007, but not as seriously. The flood in 2004 was one of the most extreme ever experienced in Britain at the time. It has been worked out that the risk of a similar, or worse, flood happening is 1 in 400 in any one year. So what has been done to help protect Boscastle from major flooding?

▼ *Plans to prevent future serious flooding in Boscastle*

A *Before the August 2004 flood, the Environment Agency had planned to improve the old culvert (a buried pipe) carrying the River Jordan under the Wellington Hotel. After the flood, they decided to lay a larger relief culvert that would carry excess water away after heavy rain. The new culvert was completed in April 2005 and can carry at least twice the floodwater of the old one.*

C *The height of the car park is being raised using stone removed from the river bed and banks during the lowering and widening of the River Valency – with barriers to stop vehicles being washed into the river.*

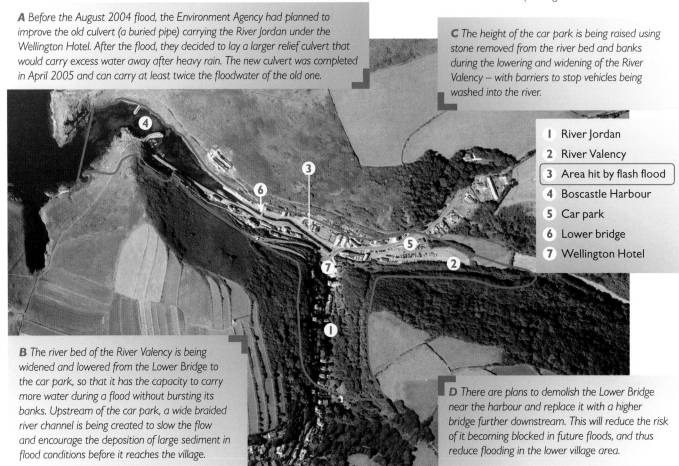

1	River Jordan
2	River Valency
3	Area hit by flash flood
4	Boscastle Harbour
5	Car park
6	Lower bridge
7	Wellington Hotel

B *The river bed of the River Valency is being widened and lowered from the Lower Bridge to the car park, so that it has the capacity to carry more water during a flood without bursting its banks. Upstream of the car park, a wide braided river channel is being created to slow the flow and encourage the deposition of large sediment in flood conditions before it reaches the village.*

D *There are plans to demolish the Lower Bridge near the harbour and replace it with a higher bridge further downstream. This will reduce the risk of it becoming blocked in future floods, and thus reduce flooding in the lower village area.*

It is estimated that raising the car park, lowering and widening the bed of the River Valency, and replacing the Lower Bridge, will cost about £4.6 million. The flood defence plans were approved by the District Council in September 2006. The Environment Agency says that the new planned defence system will reduce the risk of flooding from a flood with a 1 in 75 chance of happening in any one year. However, the new defences will not prevent flooding from an event as severe as that in August 2004 (with a 1 in 400 chance of occurring). The defence work began in Autumn 2006 and is expected to take two years to complete.

The costs and benefits of different schemes to protect Boscastle from future floods had to be weighed up to see what was feasible. Other options considered, but rejected, included:
- digging a tunnel from the car park to a cliff outfall (the high cost meant this wasn't viable)
- floodwater storage (the location made this impractical).

Flood forecasting and warning

It is the Environment Agency's job to work out which places are at risk from flooding. They produce flood maps which show areas at risk. Flood maps are made up of several different types of data, including flood zones – areas which could be affected by flooding from rivers and the sea (see the map below).

The Environment Agency also uses automatic sensors and rainfall radar to monitor rainfall and the water level in many main rivers 24 hours a day. They use this information, together with detailed weather forecasts from the Met Office, to predict flooding.

Flood warnings are issued:
- through the media
- via automated telephone messages to people at risk
- through the Floodline (available by telephone or online). Floodline also advises people about what to do before, during, and after a flood.

But in places like north Cornwall, where rivers can rise rapidly after heavy rain, current technology often means that there is not enough time to issue warnings.

▼ Flood zones around Boscastle. (the areas in blue are at risk of flooding according to the Environment Agency)

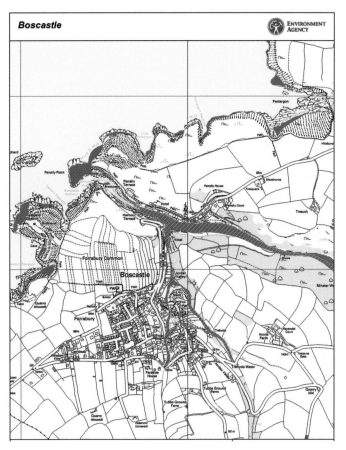

Managing the flood risk

The key players – who does what?

Environment Agency
- Builds, maintains and inspects flood and tidal defences on coasts and some main rivers, including the Valency and Jordan.
- Monitors water levels and flows.
- Issues flood warnings, forecasts and implements major incident plans.
- Has a supervisory role over all flood defences.
- Provides flood risk advice to developers and planners.
- Manages and operates flood defences.

Local authorities
- Build and maintain flood and tidal defences on other rivers and streams.
- Develop and implement major incident plans.
- Deal with some flooded roads and culverts.
- Support police in arranging evacuation and providing rest centres and emergency accommodation.

Police
- Co-ordinate the emergency response in major floods.
- Help save life and property.

Fire services
- Rescue people trapped by floodwater.
- Can pump out buildings

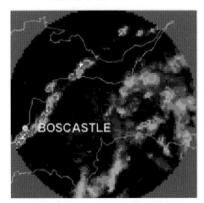

02:00 PM

RAINFALL RATE mm per hour

0 – 0.25	
0.25–0.5	
0.5–1.0	
1 – 2	
2 – 4	
4 – 8	
8 – 16	
16 – 32	
>32	

▲ A rainfall radar image for the Boscastle area at 2 pm on 16 August 2004

Preparing for the future

Sustainable St Ives

The Environment Agency has a new weapon in its fight against flooding – willow!

St Ives (in Cornwall) has suffered a number of floods, including a serious one in November 2002 which damaged 40 properties in the town centre. The flooding happens when the fast-flowing River Stennack overflows, and flood water pours into the town.

As part of the town's flood defence scheme, the Environment Agency demolished a concrete wall and replaced it with a 40-metre-long willow bank in 2007. The bundles of willow were held in place by wooden stakes and covered in soil. Willow cuttings were planted in the bank and grew quickly. The resulting structure is intended to slow down floodwater, prevent erosion and provide a natural habitat for wildlife.

▲ *The willow bank on the River Stennack, St Ives*

Protecting places at risk

Every year the Environment Agency spends about £150 million improving existing flood defences, and creating about 200 new defence schemes. Flood defences can be divided into two types – **hard engineering** and **soft engineering**.

▶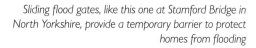
Sliding flood gates, like this one at Stamford Bridge in North Yorkshire, provide a temporary barrier to protect homes from flooding

Hard engineering

- Dams – trap and store water.
- Embankments – effectively make the river channel deeper.
- Flood walls – protect settlements.
- Channelising the river – straightening and deepening it, so water moves more quickly.
- Flood storage reservoirs – water is pumped out of the river, and stored temporarily.
- Flood relief channels – take excess water away from the main river.
- Barriers – e.g. the Thames Barrier. Might only operate when a high tide, or flood, is forecast.
- Weirs, sluices, pumping stations.

Soft engineering

- Washlands, wetlands, salt marshes – areas that provide space for floodwater.
- Land use zoning – land has different building controls depending on how close it is to the river. Land furthest from the river is used for things like hospitals, old people's homes, etc.
- Afforestation – replanting trees in the catchment area.
- Warning systems – warnings issued by the Environment Agency.

As flooding increases – possibly because of climate change – soft defences are increasingly being used. Although they tend to take up more space than hard defences, they are cheaper and need less maintenance.

Flood defences aren't perfect – and they might fail. Flood defence has to take into account:

- how often rivers flood, and how badly
- how much damage a flood does
- how much the flood defences cost
- how sustainable the different methods are.

Boscastle – it happened again

Units 3.4-3.6 focussed on the Boscastle flood in 2004. Boscastle had flooded before, and it flooded again in June 2007. Four inches of rain fell in 8 hours but this time the river did not burst its banks – flooding was caused by overland flow. Debbie Moore, of the Napoleon Inn in Boscastle, said: 'Everyone saw it and thought here we go again'. A spokesman for the Environment Agency said: 'The flood defence schemes put in place since 2004 have worked'.

▲ The flooding in Boscastle, in 2004, cost insurance companies about £15 million

In other parts of the UK, summer 2007 will be remembered for severe flooding in many places, including Yorkshire, Lincolnshire and the Midlands. Rainfall records were broken in 38 locations in the UK in June.

- Hull was deluged in two storms 10 days apart. A fifth of the population was hit by flooding, and a man died after being trapped in the water for 4 hours.
- In Toll Bar, near Doncaster, dozens of people had to leave their homes.
- Engineers were afraid that the volume of water in the Ulley Dam, South Yorkshire, would cause it to collapse and swamp nearby villages.
- In Gloucester, water in the River Severn rose nearly 5 metres above normal – see Unit 3.13 for more on the Gloucestershire floods.

Insuring against the risk

One way in which individual people can protect themselves and their homes is by insurance. Insurance companies will provide insurance for people in areas at risk from flooding – but the amount they will insure people for depends on:

- the chances of flooding, divided into low, moderate or significant (a significant chance of flooding each year would be greater than 1 in 75)
- local flood defences.

The serious flooding of summer 2007 is estimated to have cost insurance companies £1.5 billion. For instance, the flooding in Hull alone caused damage estimated at £200 million. In some places, insurance companies provided caravans for people to live in while they waited for their homes to dry out and be repaired.

In future, people who live in flood risk areas, and who have been flooded in the past, will have to pay more for their insurance.

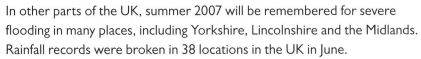

Hard defences are too expensive – we should abandon them.

What do you think?

Over to you

1 How does the Environment Agency manage the flood risk?
2 Which of the flood defences listed opposite are sustainable? Explain your answer.

On your own

3 Define these terms from the text: hard engineering, soft engineering.
4 Research to find out more about the advantages and disadvantages of the hard and soft engineering methods listed opposite. Complete a table to show their advantages and disadvantages.
5 You are going to write a final news report (as on page 141). This one will explain what has been done to protect Boscastle from future flooding. You may want to include a copy of the photo on page 146 to show where flood defences have been put in place (ask your teacher for a copy).

Exam question: Using an extreme weather event that you have studied, assess **a** how well people coped at the time and **b** how effectively protection measures stood up to the test. (10 marks)

Hurricane Katrina, August 2005

In this unit you'll investigate what happened when Hurricane Katrina hit New Orleans, how the hurricane developed, and why it was so disastrous.

In brief ...

Hurricane Katrina hit the coast of Louisiana at 6.10 am on 29 August 2005. It was the third strongest recorded hurricane ever to make landfall on the USA, and the devastation it wreaked was unbelievable.

Infrastructure

- The communications infrastructure failed – many telephones and mobiles did not work; Internet access and local TV stations were disrupted.
- Most major roads into, and out of, New Orleans were damaged.
- The levees and floodwalls protecting New Orleans were breached.
- Much of New Orleans was flooded when the levees broke; by 31 August, 80% of the city was under water.
- 1.7 million people lost electricity across Mississippi and Louisiana.
- There was extensive damage to buildings.

Economic effects

- Nearly everyone who lived in New Orleans was left unemployed – no one was paid, or spent any money, so local government could not collect taxes.
- The cost of repair and reconstruction in Louisiana and Mississippi was running at $10.5 billion in 2006.
- The total economic impact of Hurricane Katrina might well be over $150 billion.
- Hurricane Katrina also affected oil production and refining in the Gulf of Mexico, together with the production of natural gas and the importation of both oil and natural gas. Oil and petrol prices rose. Ten days after the hurricane, oil production was back to 42% of normal.
- 10% of jobs in Mississippi are in forestry – Hurricane Katrina caused significant damage to forests.

Social effects

- 1464 people died in Louisiana; many bodies could not be recovered for days.
- Residents of New Orleans were ordered to evacuate the city on 28 August. Initially, 80% of the city's population left, but 20% remained to face the hurricane because they had no transport or money to enable them to leave. Most of these were residents from poor, black neighbourhoods. The evacuation of the city was finally completed many days later by air and coach.
- Lives were disrupted; hundreds of thousands of people had no access to their homes or jobs, and were separated from their families. One million people were made homeless.
- Displaced New Orleans residents were spread all over the USA, but many went to Houston (Texas).
- There was a lack of clean water, food and toilet facilities in the city, which led to health concerns and the forced evacuation of all remaining residents. There were worries about possible outbreaks of diarrhoea and respiratory illnesses, along with Hepatitis A, cholera, tuberculosis and typhoid fever.
- An estimated 600 000 pets were killed, or left without shelter.
- In the immediate aftermath of the hurricane, looting and civil disturbance became serious problems.

Diary of a hurricane

23 August 2005 Hurricane Katrina begins life as a tropical depression near the Bahamas.

24 August The depression intensifies and is upgraded to Tropical Storm Katrina.

25 August Katrina strikes Florida with 130 km/h (80 mph) winds.

26 August Katrina weakens as it passes over Florida, but then intensifies again as it passes over the warm Gulf of Mexico.

27 August Katrina is graded as a Category 3 hurricane (see page 155).

28 August Hurricane Katrina is upgraded to Category 4, then 5, with wind speeds of 280 km/h (175 mph).

29 August Hurricane Katrina makes a second landfall in Louisiana, having weakened to Category 3. Hurricane-force winds extend 190 km (120 miles) out from the **eye** of the hurricane and graze New Orleans. The hurricane is travelling at 10 km/h (15 mph). A few hours later, Katrina makes a third landfall near the Louisiana/Mississippi border, still as Category 3. Hurricane Katrina causes huge **storm surges**, which smash the whole Mississippi Gulf coast – the surges peak at 8.5 metres. Hurricane Katrina continues northwards.

31 August By now Hurricane Katrina has split in two. It is no longer classified as a hurricane, but it creates further havoc on its journey across the southern USA – bringing heavy rain and gale force winds, creating tornadoes and causing millions of dollars' worth of damage.

▲ *Hurricane Katrina's path. The colours of the dots relate to the colours of the Saffir-Simpson scale (see page 155).*

▼ *Hurricane Katrina heads north*

● A **storm surge** is a rapid rise in sea level caused by the low pressure and strong winds of a hurricane or other intense storm.

Hurricanes – the threats

The three main threats from hurricanes are rain, wind and storm surges.

● The main impact of Hurricane Katrina in Florida was rain – some places had 380 mm, which led to flooding.
Winds caused damage and power cuts.

● Along the Gulf coast, rainfall totals exceeded 200-250 mm along much of Katrina's path.

● Winds gusted over 160 km/h (100 mph) in New Orleans.

● An 8.5-metre storm surge hit New Orleans.

Why did New Orleans suffer so badly?

Location

New Orleans has always been a city at risk. It sits below sea level; has the Mississippi River running through the middle of it; perches underneath a lake twice as big as the city; and to the east and south lies the Gulf of Mexico. New Orleans faces the threat of flooding from the Mississippi, from coastal storms and from heavy rain. A system of **levees** and **floodwalls** around the city provides some protection from flooding (see page 156). A system of pumps and canals is designed to remove water trapped in the city.

Levees and floodwalls

New Orleans' levees were built to withstand a Category 3 hurricane and storm surges of 3-4 metres. As Katrina approached the city, it was classified as Category 5. Katrina's initial storm surge from the south-east overwhelmed floodwalls along the Industrial Canal, flooding part of the city. Although there was a storm surge on Lake Ponchartrain, it was not big enough to flow over the levees and floodwalls – instead, the floodwalls holding back the lake failed and it emptied into New Orleans. Elsewhere, other levees were breached. Poor design, faulty construction and poor maintenance were all possible causes of the failures.

▲ *Lake Ponchartrain pours into New Orleans*

▼ *A cross-section through New Orleans – an accident waiting to happen?*

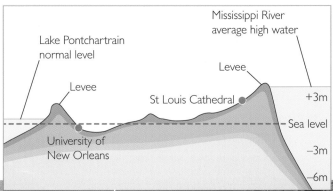

▼ *New Orleans from space*

- **Levees** are broad, earthen structures designed to prevent flooding.
- **Floodwalls** are concrete or steel walls – built on top of levees, or instead of levees.

Disappearing wetlands

The Mississippi delta used to be flooded regularly by the river. Deltas tend to sink, but this was always balanced in the past by material deposited by the river when it flooded. However, widespread building of levees and other engineering works along the Mississippi have put paid to the flooding, and extracting oil and gas from below the delta has speeded up the subsidence. Louisiana's wetlands are disappearing rapidly – at a rate of 25 square miles a year, or the equivalent of a football pitch every 38 minutes! And, in just 2 days, Hurricanes Katrina and Rita turned another 218 square miles of marsh into open water. Why is this important? The wetlands provide a barrier which can absorb the energy of storm surges at the rate of 0.3 m per 2.7 miles when hurricanes hit the coast. In doing so, they provide protection for places inland.

▲ *Hurricane Katrina helped to destroy yet more of the fragile Mississippi wetlands*

Global warming

2005 saw some of the most intense hurricanes ever. But was global warming to blame?

'Probably' said Sir John Lawton, chairman of the Royal Commission on Environmental Pollution. He said hurricanes are getting more intense because of rising sea temperatures, and that increasing intensity is likely to be due to global warming.

'Not sure' said the Met Office's Hadley Centre for climate prediction and research. Hurricane activity fluctuates naturally, and it is only since 1970 that observations have been taken of hurricane intensity. It is difficult to separate human-induced climate change from natural variability.

People know that New Orleans is a city at risk. Why would anyone want to live there?

What do you think?

● Over to you

1 What are the main threats from hurricanes? Why?
2 Draw a spider diagram to show why New Orleans suffered so badly from Hurricane Katrina. You must add explanations to the bubbles on your spider diagram.
3 Complete a table to show the short - , medium - and long-term impacts of Hurricane Katrina. Include economic, social and environmental impacts.

● On your own

4 Define these terms from the text: hurricane, storm surge, levees, floodwalls
5 Research a hurricane in a developing country. List the economic effects, the social effects and effects on infrastructure. Compare them with the effects of Hurricane Katrina. What similarities and differences are there?

Exam question: Referring to an extreme weather event, assess how far its impacts were due to **a** the size and scale of the event itself, or **b** how well people had prepared for it. (10 marks)

All about hurricanes

In this unit you'll learn the theory about hurricanes.

Background

What are hurricanes?

Hurricanes, typhoons, cyclones – are all names for the same thing. They are intense, destructive, low-pressure weather systems – one of nature's most lethal weapons. They have very strong sustained winds of over 120 km/h (75 mph) and torrential rain (250 mm can fall in one day).

How do hurricanes develop?

Hurricanes tend to develop:

- over warm, tropical oceans, where sea temperatures are at least 27°C to a depth of 60 metres (look at the colours on the satellite image and the key)
- in late summer and early autumn, when sea temperatures are at their highest
- in the trade wind belt between latitudes 5° and 20° either side of the equator (trade winds can spiral into a storm due to the Earth's rotation; this spiralling effect is known as the **Coriolis force**)
- where there is low atmospheric shear (wind speeds have to be near constant between ground level and 12 km above ground level, otherwise the hurricane can be pulled apart).

Hurricane Katrina heads for the USA ▶

Aug 27 2005

Sea temperature (°C)

15   20   25   30   35

Geographers are not completely sure how hurricanes form, but it appears to be like this:

- A strong upward movement of air draws water vapour up from the ocean.
- As the air rises, it spirals, cools and condenses – releasing huge amounts of heat energy, which powers the storm.
- Colder air sinks down through the centre of the hurricane to form the eye.
- When the hurricane reaches land – and its source of heat energy and moisture disappears – it rapidly decreases in strength.

▼ *The features of a hurricane*

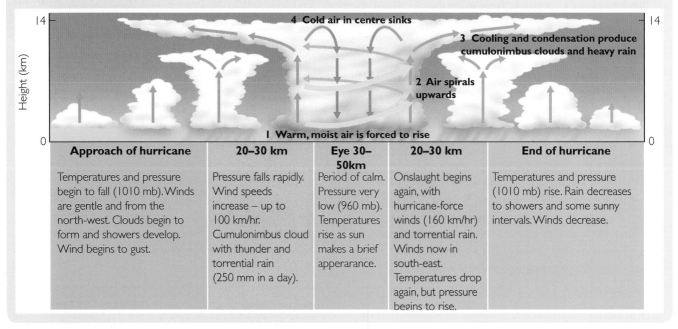

Approach of hurricane	20–30 km	Eye 30–50km	20–30 km	End of hurricane
Temperatures and pressure begin to fall (1010 mb). Winds are gentle and from the north-west. Clouds begin to form and showers develop. Wind begins to gust.	Pressure falls rapidly. Wind speeds increase – up to 100 km/hr. Cumulonimbus cloud with thunder and torrential rain (250 mm in a day).	Period of calm. Pressure very low (960 mb). Temperatures rise as sun makes a brief appearance.	Onslaught begins again, with hurricane-force winds (160 km/hr) and torrential rain. Winds now in south-east. Temperatures drop again, but pressure begins to rise.	Temperatures and pressure (1010 mb) rise. Rain decreases to showers and some sunny intervals. Winds decrease.

Diagram labels: Height (km), 14, 0; **4 Cold air in centre sinks**; **3 Cooling and condensation produce cumulonimbus clouds and heavy rain**; **2 Air spirals upwards**; **1 Warm, moist air is forced to rise**

Where do hurricanes occur?

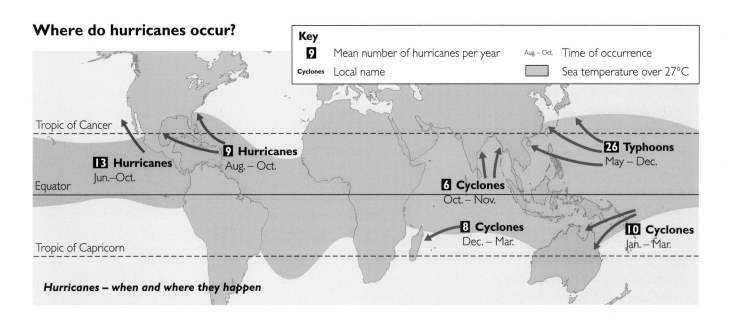

Key

9	Mean number of hurricanes per year
Cyclones	Local name

Aug.–Oct. Time of occurrence

▢ Sea temperature over 27°C

Hurricanes – when and where they happen

Hurricane facts

- Hurricanes are the world's most violent and destructive storms. The hazards associated with hurricanes are:
  - winds (of up to 250 km/h)
  - storm surges (resulting from high winds and low
  - pressure)
  - flooding (caused by torrential rain)
  - landslides (caused by the heavy rainfall)
- The energy produced in a single hurricane could supply the whole of the USA with all of its electricity for 6 months.
- The average lifespan of a hurricane is 7-14 days.
- Hurricanes are named to improve communication.
- A list of names is produced by the National
- Hurricane Centre in the USA – beginning with A and alternating between men and women's names.

How are hurricanes measured?

The Saffir-Simpson scale is a 1-5 rating based on the hurricane's intensity.

Category	Wind speed	Storm surge	Damage
1	119-153 km/h (74-95 mph)	1-2 metres	minimal
2	154-177 km/h (96-110 mph)	2-3 metres	moderate
3	178-209 km/h (111-130 mph)	3-4 metres	extensive
4	210-249 km/h (131-155 (mph)	4-6 metres	extreme
5	>249 km/h (>155 mph)	> 6 metres	catastrophic

Hurricanes are getting more intense. Or are they?

What do you think?

● Over to you

1 Explain where and how hurricanes develop. Use a diagram to help explain how they develop.

2 **a** Describe the global distribution of hurricanes.

 b Why do hurricanes decrease in strength when they hit land?

● On your own

3 Define these terms from the text: Coriolis force, eye

4 **a** Use http://maps.csc.noaa.gov/hurricanes/viewer.html to find the track of 2005 Hurricane Rita.

 b Describe Rita's path and compare it with Hurricane Katrina's (see the photo on page 151).

5 Use the website in **4** to find the track of the hurricane you investigated in activity **4** on page 153. NB: the website shows historical records, so if you have chosen a very recent example, you may have to search somewhere else for this information.

Managing hurricanes

EXTREME WEATHER

In this unit you'll learn how New Orleans' defences are being restored, and how people can manage the hurricane risk.

Defending New Orleans

Hurricanes are a fact of life; they happen every year. New Orleans' defences were in tatters following Hurricane Katrina's onslaught, and they needed to be strengthened as well as repaired. The natural wetlands which had been destroyed for decades also needed restoring.

Repairing the levees

A year after Hurricane Katrina, New Orleans' levees and floodwalls had been repaired to the same standard they were before Katrina struck. Temporary floodgates and pumps were built at the city's four main canal entrances, which had allowed storm surges from Lake Ponchartrain to overwhelm flood defences.

The Mississippi River Gulf Outlet was built in the 1960s to allow ships easier access to the Gulf of Mexico, via the port of New Orleans. Over the years, the original 91-metre-wide channel has been expanded to 914 metres wide. This channel allowed Katrina's storm surge to penetrate right into New Orleans. The US Army Corps of Engineers, who are in charge of the repair work on New Orleans' flood defences, proposed building a system of weirs to regulate the flow of the Mississippi River Gulf Outlet. They also suggested building a system of 8.5-metre-high levees to the east and north of the city.

Restoring the wetlands

The disappearance of Louisiana's natural defensive wetlands has exposed about 2 million people, together with billions of dollars' worth of property and industry, to an increased risk of flooding and damage from hurricanes, storms and even high tides.

Coast 2050 is a $14 billion 30-year plan – which includes flood control, water diversion and coastal restoration programmes – with the aim of recreating the mixture of swamp, marshland and barrier islands that existed in the past, and so help to protect places inland.

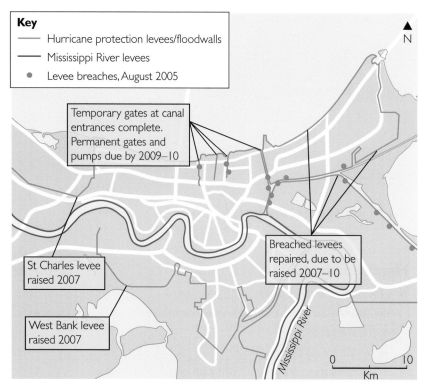

Key

— Hurricane protection levees/floodwalls

— Mississippi River levees

● Levee breaches, August 2005

Temporary gates at canal entrances complete. Permanent gates and pumps due by 2009–10

St Charles levee raised 2007

West Bank levee raised 2007

Breached levees repaired, due to be raised 2007–10

Mississippi River

0   10
Km

N

▲ *Repairing and rebuilding the levees*

▲ *New floodgates being built at the entrance to one of the drainage canals that cut deep into New Orleans*

In the New Orleans area, if all the Coast 2050 strategies are implemented, you stand a pretty good chance of returning to a level of protection similar to 40 years ago.
Denise Reed, a professor of coastal geomorphology at the University of New Orleans

Managing the risk

Preparing for a hurricane

It is possible to predict the general area where a hurricane will strike, but predicting its *exact* strength and location of landfall remain difficult.

- The National Hurricane Centre in Florida provides predictions (including a **hurricane watch**) and **hurricane warnings** to US states and surrounding countries.
- The National Hurricane Centre also aims to make people more aware of how vulnerable they are, and what actions they should take in the event of a hurricane. They run a Hurricane Preparedness Week every year. In 2007, it ran from 20-26 May, and covered:

> • A **hurricane watch** is an announcement that specific coastal areas could experience hurricane conditions within 36 hours.
>
> • A **hurricane warning** is a warning that certain coastal areas can expect sustained winds of at least 119 km/h (74 mph) within 24 hours.

HISTORY	HURRICANE HAZARDS		FORECAST	PREPARE	ACT
Sunday 20 May	Monday 21 May	Tuesday 22 May	Wednesday 23 May	Thursday 24 May	Friday 25 May / Saturday 26 May

- Other organisations run hurricane preparedness programmes, e.g. DHL (an airline courier company) runs a hurricane simulation programme in schools in Florida to help students learn how to plan, prepare for and deal with the aftermath of a hurricane.

Reducing the impact – evacuation

Evacuation is neither easy, nor cheap. A hurricane's path is erratic and it is not always possible to give more than 12 hours' notice of where it will hit, which is not enough time for proper evacuation measures. It is estimated that the cost of evacuating a 500-km stretch of the US coastline is about $50 million.

On 28 August 2005, the Mayor of New Orleans ordered people to leave the city. About one million people evacuated. But New Orleans is a city with a great deal of inequality – 21% of residents did not have cars; many of them were poor and black and unable to leave.

> *New Orleans has always been vulnerable. Should it have been repaired and rebuilt, or abandoned?*
>
> **What do you think?**

• Over to you

1 Why is it important to restore Louisiana's wetlands? Explain your answer.

• On your own

2 Define these terms from the text: hurricane watch, hurricane warning.

3 a New Orleans was being repaired so that people could move back. Describe how New Orleans' defences were being restored.

  b See if you can find out how far the repair and reconstruction work has got, and how many people have moved back.

4 Research to find out how developing countries prepare for hurricanes. Are there differences to the way that developed countries such as the USA prepare? Why?

3.10   The Big Dry

In this unit you'll find out what drought is, and how Australia's first drought of the twenty-first century has affected the country.

Australia's epic drought

By 2007, Australia was in the grip of its worst **drought** on record. The drought began in 2002 and the worst affected area was the Murray-Darling Basin.

Adapted from BBC News Online, May 2007 ▼

Malcolm Adlington has been farming his land for over 30 years, and has never known it so bad. He has a dairy farm close to the Murray River – on what used to be some of Australia's best pasture land.

But then came the 'Big Dry'. Normally the paddocks would be green and the grass 20 cm high. 'Now it looks like scrubland' says Malcolm. He's had to sell two-thirds of his herd and can hardly afford the straw to feed his bony cattle.

For the farmers of Wakool, this has been the driest 12 months in 125 years. The area used to have 19 dairy farms – now there are just six. Farmers have suffered a double blow – a chronic shortage of rain, and now the irrigation water they normally receive from the Murray River has been turned off. Rural communities are being sacrificed to maintain water supplies in the cities.

Malcolm said 'I should have taken my mother's advice and become a plumber'.

The Murray-Darling Basin

The Murray-Darling Basin (MDB):

- is in south-eastern Australia, extending from Queensland to South Australia
- provides 40% of Australia's agricultural produce – and is called Australia's food-bowl
- makes up more than 70% of Australia's irrigated cropland and pastures
- has 55 000 farmers who supply virtually all of Australia's stone and citrus fruits, vegetables, cotton and rice
- is the location of many of Australia's vineyards (60% of Australia's wine grape production relies on water from the MDB)
- is the size of France and Spain combined
- provides 85% of the water used nationally for irrigation, and 75% of Australia's water
- is managed by state governments
- is home to almost 2 million people.

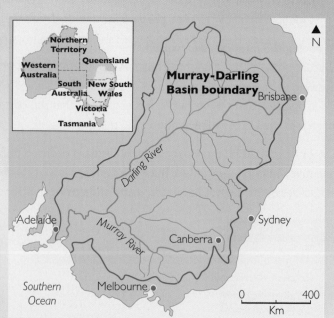

▲ The Murray-Darling Basin

In 2006, just 1317 billion litres of water flowed into the MDB – almost 25% less than the previous minimum in 1902 (during another devastating drought). The annual flow of water in the MDB has been predicted to fall by anything from 10-45% by 2070.

> ● A **river** or **drainage basin** is the area of land drained by a river and its tributaries.

What is drought?

Drought is an acute water shortage. It is associated with long periods of serious or severe rainfall deficiency (lack of rain). Drought exists when there is far less water in a particular area over a period of time compared to what is normal for that same period of the year. What constitutes a drought varies around the world.

▲ In the Sahel, a drought is declared when there is below average rainfall for 2 years. The Sahel has suffered drought for most of the last 30 years – but in Niger it was particularly bad in 2004, leading to food shortages and hunger.

▲ In the UK an official drought is declared when at least 15 consecutive days go by without more than 0.2 mm of rain. The UK suffered drought and water shortages in 1975-76, 1995 and 1996.

Australian drought fact file

- Australia's rainfall pattern is one of the most variable in the world, due to its size, location, geography, climatic range and shifts in global weather patterns. Many areas are subject to the threat of drought.
- Over the long term, Australia has three good rainfall years and three bad rainfall years out of ten.
- There is little chance that all of Australia could be in drought at the same time.
- Some droughts are long-lived, some are short and intense. Some can be very localised, while others affect large areas.
- On average, severe droughts affect some parts of Australia about once every 18 years – but the intervals between severe droughts can vary from 4 to 38 years.
- Drought means a reduction in water storage levels, which leads to water restrictions (limiting the amount of water people can use and the way they can use it).

Australian drought years	Effects
1864-66	all states affected except Tasmania
1880-86	southern and eastern states affected
1895-1903	most of Australia affected, 50 million sheep lost and 40% of cattle
1911-16	19 million sheep and 2 million cattle lost
1918-20	only parts of western Australia were free from drought
1939-45	nearly 3 million sheep lost
1958-68	widespread drought: 1966-68 saw a 40% drop in the wheat harvest, the loss of 20 million sheep and a decrease in farm income of $300-500 million
1972-73	mainly in eastern Australia
1982-83	vast areas affected; total loss in excess of $3000 million
1991-95	production by rural industries fell by 10% (a loss of $5 billion to the Australian economy); $590 million was provided in drought relief
2002-??	ongoing drought – see pages 154–155 for its impacts

Impacts of the Big Dry

Depending on where drought happens, the impacts can range in scale from those that are minor inconveniences to those that have catastrophic consequences. In Australia, it is agriculture that suffers first and most severely when drought hits.

Impacts on agriculture by 2007

- The allocation of irrigation water to some farms was cut. Without water crops cannot be irrigated. Water could only be used for consumption by people and livestock. This ensured that urban areas had sufficient water for drinking and washing.
- Production of irrigated crops, including cotton lint, wine grapes and rice was reduced.
- Harvests were decimated.
- Annual crops were not planted if there was not enough water to support them.
- Farm incomes shrank, leading to financial ruin for thousands of farmers laden with debt after years of drought.
- Farmers were forced to abandon their land.

▲ Hungry and thirsty sheep in a dry river bed

Social impacts by 2007

- Rural communities were suffering. For example, in New South Wales communities became ghost towns as people left because of the drought. Local services such as schools and shops closed because there were no longer enough people left to support them.
- The rural suicide rate soared.
- As less food was being grown in Australia, more needed to be imported and prices rose.
- Prices of energy and water also soared. Water bills were set to rise by up to 20% from 2008, partly to help pay for new infrastructure to prevent future shortages. The drought also reduced the production of HEP in Victoria and Tasmania – more expensive (and polluting) methods were used to produce energy, and consumers had to pay the bill.
- Farmers and small businesses dependent on agriculture were receiving support from the Australian government's drought assistance programme (through income support, interest rate subsidies and free counselling).

Level 5 water restrictions were introduced in Goulburn in 2006 ▶

- There were water restrictions in all major Australian cities. There are different stages of water restrictions – starting at 1 (least restrictive) to 8 (most restrictive). By summer 2007, the highest level reached was Stage 5 for Goulburn in New South Wales, and Brisbane. In Melbourne there were bans on car washing, only extremely limited watering of gardens was allowed, and people were expected to only take four-minute-long showers.

Environmental impacts by 2007

These included:

- loss of vegetation and wildlife
- soil erosion
- declining water quality
- toxic algal outbreaks in depleted rivers, dams and lakes
- increasing numbers of bushfires and dust-storms
- the Snowy Hydro Electricity Scheme was forced to rely on an old polluting gas-fired power station to generate much of its electricity, because it had almost run out of water (its reservoirs had dropped to an average of 12% capacity).

Drought produces a complex web of impacts that affect everyone.

What do you think?

Saltwater haven turns to dust as Big Dry bites hard

The drought has reduced Australia's biggest permanent saltwater lake, Lake Corangamite, in Victoria, to its lowest level for more than 60 years. Conservationists are worried about the impact of decreasing water levels and increasing salinity on native flora and fauna and the thousands of migratory birds that arrive at the wetland to feed on brine shrimp.

Adapted from *The Age*, April 2007

Economic impacts by 2007

- The drought wiped 1% off the Australian economy in 2007 – affecting government spending.
- Nearly 60% of businesses in Victoria had been affected by drought, and rural areas were hardest hit. Over one third (34%) of rural businesses reported a major impact, compared to 10% in urban areas. In the Alpine shires up to 80% of income is from tourism – and few people were visiting.

- The resource base of farming was being eroded. For example, once a fruit tree dies it takes 4-5 years for replanted trees to produce fruit. This affects supplies of stone fruits, grapes, avocadoes and almonds for years.
- The 2008 vintage wine was crippled due to lack of water in 2007. The 2007 vintage suffered a 40% drop in the amount of wine grapes picked and crushed.

Over to you

1 Draw a spider diagram to show the importance of the Murray-Darling Basin to Australia.
2 Read the text boxes on this spread on the impacts of the drought. Are they minor, moderate or severe? Explain your answer.
3 The supply of irrigation water to some farms was cut during the drought, so that urban areas had sufficient supplies. How do you think the affected farmers would feel? Do you agree with this strategy?

On your own

4 Define these terms from the text: drought, river (or drainage) basin.
5 Research a drought in a developing country. How long did it last? What were the impacts? Divide them into social, environmental and economic. How do they compare with the impacts of the Australian drought?

Exam question: Referring to a drought that you have studied, assess how far its biggest impacts were economic, social or environmental. (10 marks)

EXTREME WEATHER

In this unit you'll find out about the possible causes of the drought in Australia.

Climate patterns

Global climate patterns are mainly due to the circulation of air in the atmosphere, but ocean currents can have a major influence on weather patterns and climate. This unit looks at what caused the Big Dry.

Apart from the seasons, **El Niño** is the single largest cause of variability in the world's climate from year to year. In order to understand El Niño, you need to know about air and water circulation in the Pacific Ocean.

> • **The Walker cell**
> is the circulation of air, where air in the upper atmosphere moves from west to east, and surface air moves from east to west, as the **trade winds**. The Walker cell is named after Sir Gilbert Walker who first identified it in the 1920s.

Climatic variation in the Pacific

Normal years

In 'normal' years, winds in the Pacific circulate around the **Walker cell**. Winds travel westwards along the surface of the Pacific Ocean taking warm surface water with them. The cold Humboldt current, which flows northwards along the west coast of South America is drawn into the circulation and flows westwards. As the current flows west it is heated by the tropical sun. Warm, moist air rises over Indonesia, creating a low-pressure area, tropical cumulo-nimbus clouds and heavy rain. The air then circulates east in the upper atmosphere and sinks into the cooler high-pressure area over the west coast of South America, giving the dry conditions that created the Atacama Desert in Peru.

▼ *The Walker circulation cell*

a) Atmospheric circulation

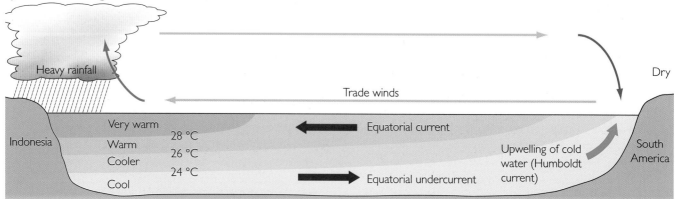

b) Section through the Pacific Ocean

La Niña

At certain times, the 'normal' situation intensifies and is known as La Niña. During a La Nina event, the low pressure over the western Pacific becomes even lower, and high pressure over the eastern Pacific even higher. As a result, rainfall increases over South-East Asia, and South America suffers drought. Trade winds become stronger due to the increased difference in pressure between the two areas. La Niña can occur just before, or just after, El Niño.

El Niño

During **El Niño** years, pressure systems and weather patterns reverse. Warmer waters develop in the eastern Pacific, with temperatures rising by up to 8 °C. Low pressure forms over the area, drawing in westerly winds from across the Pacific. Warm, moist air rises, creating heavy rainfall over the eastern Pacific. The air circulates west in the upper atmosphere. Around Northern Australia and Indonesia the descending air gives drier conditions than usual and can cause drought.

● **El Niño** is Spanish for 'boy child'. It appears every 3-8 years, around Christmas (hence the name) and can last for 14-22 months.

▼ El Niño

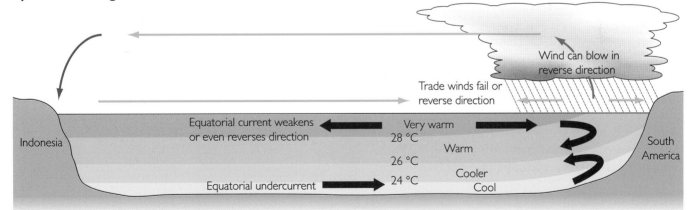

a) Atmospheric circulation

b) Section through the Pacific Ocean

The Southern Oscillation Index

The change between 'normal' years and El Niño is called the Southern Oscillation. The strength, direction and speed at which it takes place is known as the Southern Oscillation Index (SOI). High pressure dominates the eastern South Pacific and low pressure dominates the west. If pressure rises in the east, it falls in the west. Meteorologists record air pressure at Easter Island (west of South America) and subtract it from that at Darwin in northern Australia to calculate the SOI. A sharp drop in the index indicates that El Niño is on the way.

Many droughts affecting eastern and northern Australia are a direct result of the Southern Oscillation and El Niño.

Are El Niño events increasing?

El Niño events have been occurring for 15 000 years. But scientists are investigating whether climate change is leading to an increase in their intensity and duration.

What caused the drought? – 2

Climate change

2002 was the fourth driest year on record in Australia, while maximum temperatures across the continent were the warmest on record.

A report produced by the UN Intergovernmental Panel on Climate Change (published in 2007) made it clear that global warming would bring severe drought with disastrous economic consequences. There is a growing drought problem in the Horn of Africa – probably caused by global warming – but is Australia's long drought the first in a developed country to be caused by climate change?

Australia's climate is changing as the maps show – but it is not the same everywhere.

Melbourne records driest 12 months

Melburnians have lived through their driest year on record. In the previous 365 days, Melbourne received 316 mm of rain – less than half its yearly average. The Melbourne area had a very severe drought year on top of the extended dry period over the last decade. In 2006, Australia as a whole had above-average rainfall, but the Murray Darling Basin was very dry.

Adapted from *The Age*, May 2007

New South Wales January 2006 – warmest month on record

In January, New South Wales experienced its warmest month ever recorded – with a mean temperature of 3.5 °C above the average January temperature of 24.7 °C. It is the tenth consecutive month of above-average temperature for the state. The temperatures were generally highest inland, with a large proportion of the Murray-Darling Basin recording the highest mean temperatures on record.

Australian Bureau of Meteorology, February 2006

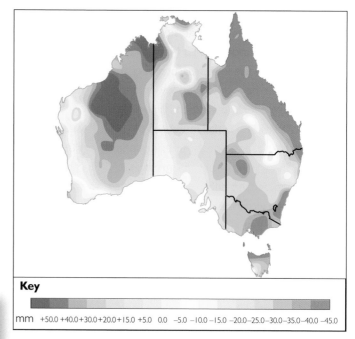

Key

mm +50.0 +40.0 +30.0 +20.0 +15.0 +5.0 0.0 –5.0 –10.0 –15.0 –20.0 –25.0 –30.0 –35.0 –40.0 –45.0

▲ *Australia – annual total rainfall trend, 1970-2006. Notice the difference between eastern and north-western Australia.*

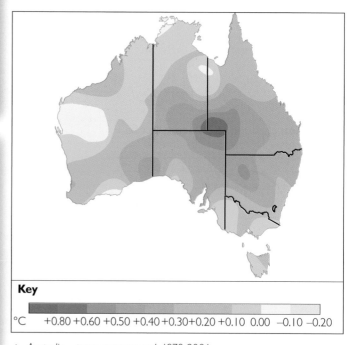

Key

°C   +0.80 +0.60 +0.50 +0.40 +0.30 +0.20 +0.10 0.00 –0.10 –0.20

▲ *Australia – temperature trend, 1970-2006*

● Over to you

1 Draw two simple diagrams to show atmospheric circulation in 'normal years' and El Niño years. Annotate your diagrams to explain what is happening.
2 Describe the temperature trend and rainfall trend from 1970-2006 shown on the two maps on this page.
3 What caused the Big Dry? Explain your answer.

● On your own

4 Define these terms from the text: El Niño, Walker cell, trade winds, La Niña, Southern Oscillation Index.
5 Research other Australian droughts, e.g. those mentioned on page 159 to find out which of those were related to El Niño.
6 What were the causes of the drought you researched on page 61?

Managing Australia's water – 1

In this unit you'll find out how Australia is managing its scarce water resources.

Australia's Prime Minister, John Howard, announced a $10 billion national water management plan. The plan includes the biggest modernisation of irrigation in Australia's history, boosting water efficiency on farms and addressing the over-allocation of water in the drought-ravaged Murray-Darling Basin. The government plans to take control of the river system. Mr Howard said the 'old way' of managing the basin had 'reached its use-by date. Our goal is to increase agricultural production by using less water.'

Adapted from the *Sydney Morning Herald*, January 2007

A National Plan for Water Security

It is predicted that the amount of water flowing into the Murray-Darling Basin will decrease further, and there is evidence that rainfall is declining. Bushfires, the growth of farm dams and boreholes, plus reforestation, are all reducing available water resources. The National Plan for Water Security (The Plan), announced in January 2007, is designed to ensure that the use of rural water becomes sustainable. It builds on work begun under two other programmes – the Living Murray Initiative and the Australian Government Water Fund. The government will invest $10 billion over ten years to improve water management, and it will focus on the Murray-Darling Basin.

▼ *The Plan*

Objective/Area	Why is it needed?	Action
Modernising irrigation methods	In an average year, irrigated agriculture uses 14 000 gigalitres of water (about 70% of all water use in Australia). Up to 65% of this water is lost either before it reaches the farm, on the farm, or because of over-watering.	• Piping or lining channels and adopting more efficient watering methods. • Adopting more accurate water meters to improve measurement and reduce over-watering. • Improving river operations and storage management, e.g. reducing evaporation at lakes in the lower MDB.
Addressing over-allocation in the MDB	Over-allocation is where more entitlements to irrigation water have been issued than can be sustained. Water availability is declining.	• Parts of irrigation schemes which are not viable, e.g. in salt-affected areas, will be identified and retired. • Irrigators (farmers) will be given help to leave the industry. • Water saved will be managed to restore wetlands and rivers in the MDB.
Reforming the management of the MDB	The MDB covers 14% of Australia (1 million km²); is home to over 2 million people; a further 1 million people rely on its water; the land and water of the MDB are of huge economic importance in terms of agriculture and tourism.	• The government plans to take control of the MDB; it was previously jointly managed by New South Wales, Victoria, Queensland, South Australia and Australian Capital Territory. • The new MDB Commission will control and be responsible for water management in the basin.
Upgrading water information	As water becomes scarcer, and demand increases, it is vital to monitor the resource and its use – at a national scale, a basin scale and for individual farms.	The Bureau of Meteorology will have a key role in a number of areas, including: • helping water data collecting agencies to modernise and extend their measurements (e.g. stream flow, groundwater monitoring) • maintaining a national database and reporting system for all water information • providing a forecasting service.
Northern Australia	There is an opportunity for the northern land and water resources to be developed – but it must be done in a sustainable way.	• A taskforce will identify the key ecosystems, communities and indigenous people and their water needs to ensure that they are sustained. It will look at whether the drainage basins can support further use.
Great Artesian Basin	The Great Artesian Basin is one of Australia's most important water resources, but uncontrolled water flow from boreholes and drains threatens farmers' livelihoods and those ecosystems which depend on the groundwater.	• The repair of uncontrolled bores and replacement of drains ('capping and piping') has improved farm and environmental management. The job has not been finished – but will be completed as part of The Plan.

Managing Australia's water – 2

The Australian Government Water Fund

While the National Plan for Water Security concentrates on the MDB, Australia has other programmes which are nationwide. The Australian Government Water Fund is a $2 billion investment in water solutions for current and future generations. It focuses on water infrastructure and improved management of Australia's scarce water resource. It is made up of three programmes: Water Smart Australia; Raising National Water Standards; Community Water Grants.

Lake Brewster Water Efficiency Project

This project was announced in February 2007, and aims to improve Lake Brewster for storage on the Lachlan River in New South Wales. The project is a joint initiative between regional natural resource managers, water delivery managers and water users on the Lachlan River to provide a more sustainable, efficient and secure river operation. The project will:

- save up to 10 gigalitres of water
- improve water security by 4%
- create 1100 hectares of managed wetlands to improve water quality and habitat in Lake Brewster and water discharged to the Lachlan River
- manage river flows in a more natural way
- improve the river environment downstream of Lake Brewster.

Water Smart Australia

This programme aims to accelerate the development and uptake of 'smart' technologies and practices in water use. It is targeted at large-scale projects. The Lake Brewster project is just one of many Water Smart projects.

▼ *The Lake Brewster plan*

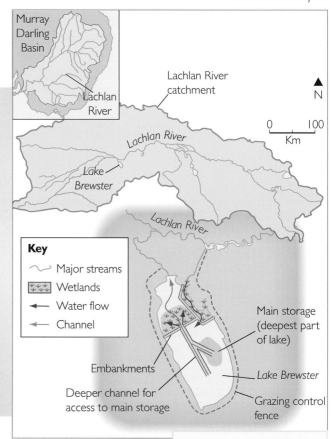

The project includes:

- the construction of an embankment to divide the lake into two parts
- the deepening of the outlet channel to access more water from the deepest part of the lake
- the creation of two wetlands to improve water quality.

If farming practices do not change, Australia can wave goodbye to home-grown food.

What do you think?

Changing farming

Drought has focussed attention on the need to develop farming systems better suited to Australia's environment than some of the current practices. There are a number of ways in which Australian farming is changing:

- **Drought-tolerant plants** are being developed. The Australian Commonwealth Scientific and Industrial Research Organisation (CSIRO) is looking at different ways to breed drought-tolerant and water-efficient crop varieties. CSIRO has bred a wheat variety 'Drysdale' which yields 5% more than other varieties under dry conditions.
- CSIRO is also researching **crop and pasture management** technology for conserving water in different regions.
- **Precision farming**. CSIRO uses satellite information to estimate biomass and deliver this to farmers' computers. Farmers can then make informed decisions on the sustainable management of their land.
- **Global positioning systems (gps)** are being used for accurate sowing of seed. A gps system guides a tractor and seeder to an accuracy of 2 cm when cutting rows in the earth and sowing seeds.

- **Sustainable farming** is being adopted. For the last two centuries, many farmers have felled trees to clear land for livestock or crops – leaving it vulnerable to drought and flood. Some farmers are now changing their methods, e.g. reducing livestock numbers; planting native trees and shrubs (which provide protection for livestock, reduce wind speeds and the risk of soil erosion, and lower soil moisture evaporation); reducing the amount of ploughing, leaving stubble in the fields and planting in the stubble (which helps to shield young plants, reduces soil erosion, prevents moisture loss and the stubble breaks down to improve soil structure).

▲ *Tree planting to restore eroded land in Victoria, Australia*

Other ways to secure Australia's water supply

Desalination plants

In 2007, the government in Victoria announced plans to build one of the world's largest desalination plants near Melbourne. The plant is expected to be completed by 2011 and will help to 'drought-proof' water supplies to Australia's second largest city.

Urban dams and disused reservoirs

There are suggestions in Sydney that small dams could be built near parks, playing fields and open spaces in order to store water to use on the parks, etc.

Sydney could capture and re-use millions of litres of storm water by using disused reservoirs and storage spaces around the city, e.g. cable and railway tunnels to store water. This water could be recycled to irrigate parklands which currently use precious drinking water.

Pray

When the report on the state of the MDB was released, John Howard (Australia's Prime Minister) said 'It is a grim situation, and there is no point pretending otherwise. We must all hope and pray there is rain.'

▲ *Desalination plants may produce more water, but can also increase greenhouse gas emissions and damage coastal environments*

● Over to you

1 Draw two large outline maps of the Murray-Darling Basin (use the map on page 158). On one, list the impacts of the drought (use different colours for impacts on agriculture, social impacts, etc.). On the other, list the ways in which Australia is managing its water supplies. There is a lot of information to fit on the maps, so you may need to be selective (choose what you think are the most important things).

● On your own

2 The Great Artesian Basin is one of Australia's most important water sources? Why? Find out, and write 150 words about it.

Exam question: Using examples of drought that you have studied, assess how well people managed the resources to cope with it. (10 marks)

3.13 Gloucestershire under water

In this unit you'll learn about the impacts and causes of the floods which hit Gloucestershire in July 2007, and about how Cheltenham manages flooding.

Flooding – 2007

Part of Longlevens in Gloucester is typical of one of the UK's new high-density housing developments. A local builder told the council about the area's vulnerability to flooding when the development was proposed in 2000. It went ahead anyway and, in July 2007, it was under a metre of water.

What were the impacts of the flooding?

In late July 2007, some areas of England and Wales were deluged by three times the average rainfall for July – which fell in just 24 hours. (The average July rainfall is 5-6 cm and, for example, Pershore in Worcestershire had 14.5 cm of rain in 25 hours.) The heavy rain caused massive flooding.

In Gloucestershire:
- three people died.
- thousands of homes were flooded.
- people had to be evacuated from their homes and rescued from trapped cars. Some were airlifted to safety. Gloucestershire Fire and Rescue Service received nearly 2000 calls in 12 hours (25% of the amount they normally get in a year).
- the damage to roads was estimated at £25 million – the County Council's total annual budget for road repairs.
- thousands of drivers were stranded overnight in their cars on the M5.
- the River Severn in Gloucester peaked nearly 5 metres above its normal level.

Floods also affected other parts of the UK:
- Hundreds of rail and bus services were cancelled. Cars were stranded. Flights were cancelled or delayed.
- 50% of some vegetable crops were ruined, and prices in the shops were likely to rise as a result.
- Fetes, fairs and shows were cancelled. The plug was pulled on the annual Game Fair in Leeds, causing a loss of £50 million to the local economy.
- The cost of flood damage throughout the UK was estimated at £3 billion.

But perhaps the biggest surprise was the effect that the flooding had on power and water supplies in Gloucestershire.

- The Mythe water treatment plant at Tewkesbury produces 120 million litres of drinking water a day, and supplies 350 000 people. It was deluged by floodwater and shut down. Severn Trent (the local water company) was overwhelmed by demand for water. It supplied over 50 million litres of bottled water to people, and put 1300 bowsers (large mobile water tanks) on the streets of Gloucestershire, so that people had safe water to use.
- After the flood, Severn Trent immediately installed additional flood defences at Mythe.
- Severn Trent estimated that the flooding cost them between £25 and £35 million.
- Castle Meads electricity substation in Gloucester was also flooded, leaving 48 000 homes in Gloucester, Cheltenham and parts of Herefordshire without power.
- Walham electricity substation in Gloucester supplies electricity to 500 000 homes and was saved by temporary flood barriers.

The Walham substation was saved from flooding by temporary barriers, but firefighters still had to pump out 18 inches of rain water ▶

◄ *Flooding is a regular event in Tewkesbury. Floodwater usually flows onto a large meadow called 'The Ham', which means the town stays dry – but not this time.*

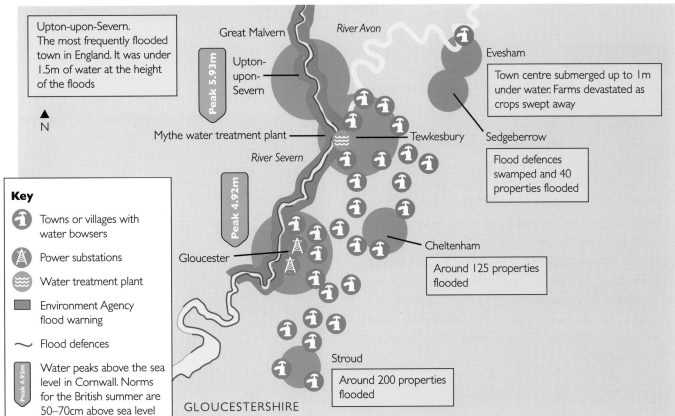

Upton-upon-Severn. The most frequently flooded town in England. It was under 1.5m of water at the height of the floods

N

Great Malvern

River Avon

Upton-upon-Severn

Peak 5.93m

Evesham

Town centre submerged up to 1m under water. Farms devastated as crops swept away

Mythe water treatment plant

Tewkesbury

River Severn

Sedgeberrow

Flood defences swamped and 40 properties flooded

Peak 4.92m

Gloucester

Cheltenham

Around 125 properties flooded

Key

- 🛈 Towns or villages with water bowsers
- ⚠ Power substations
- 〰 Water treatment plant
- ▇ Environment Agency flood warning
- ~ Flood defences
- Water peaks above the sea level in Cornwall. Norms for the British summer are 50–70cm above sea level

Stroud

Around 200 properties flooded

GLOUCESTERSHIRE

▼ *Mythe water treatment plant had to be shut down.*

▼ *Water, water, everywhere … but with a flooded water treatment plant, people had to rely on bottled water and mobile bowsers.*

What caused all the rain?

There was above-average rainfall in May, and torrential rain and severe flooding in June and July. The map on the right shows how parts of the UK suffered from heavy rain during the summer of 2007. But why was this? Meteorologists blamed our wet summer on the jet stream.

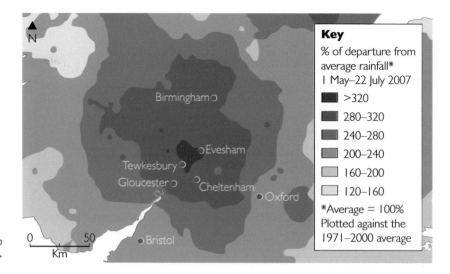

Heavy rain fell across Central England in the summer of 2007 ▶

Key

% of departure from average rainfall*
1 May–22 July 2007

- ■ >320
- ■ 280–320
- ■ 240–280
- ■ 200–240
- ■ 160–200
- ☐ 120–160

*Average = 100%
Plotted against the 1971–2000 average

Background

36 000 feet above the Earth is the polar **jet stream**. It separates cold polar air to the north and warm subtropical air to the south. Jet streams can be thousands of miles long, hundreds of miles wide and a few miles deep. During most summers, the polar jet stream passes to the north of Scotland, which allows the Azores High – an area of high pressure in the mid-Atlantic – to bring us warm and settled weather. In the summer of 2007, however, the jet stream was further south than normal, and passed directly over the UK. This brought us depressions, storms and heavy rain instead of our normal weather. It is not unusual for the jet stream to move around, but some scientists think its movement could be linked to La Niña, or to the warming of the Atlantic.

● **Jet streams** are fast flowing 'rivers' of air.

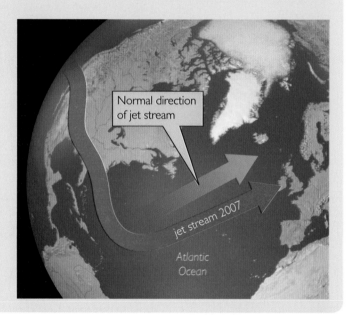

The jet stream ▶

Managing flooding

Cheltenham was just one of the places flooded in July 2007. (It had also been flooded in June.) Although the River Chelt, which flows through Cheltenham, is far smaller than the River Severn, Cheltenham still needs protecting from the threat of flooding.

The River Chelt has a small and steep catchment. Rain started falling early in the morning of 20 July 2007. The catchment became saturated before rainfall peaked (between 12.00 and 3.00 pm). The intensity of the peak rainfall resulted in peak riverflows occurring before the rain stopped. Flooding was caused by the increased river flow, combined with surface water runoff.

No running water in the twenty-first century? Why weren't our utilities (water and power) better protected?

What do you think?

The flood defences on the River Chelt had been upgraded between 2000 and 2006. They were designed to protect the town centre from floods up to, and including, a predicted 1 in 100 year flood (i.e. a flood with a 1% chance of happening in any one year). The July flood was a 1 in 120 year event (i.e. it had a 0.8% chance of happening) so some flooding was expected. The map below shows some of the flood defences on the River Chelt.

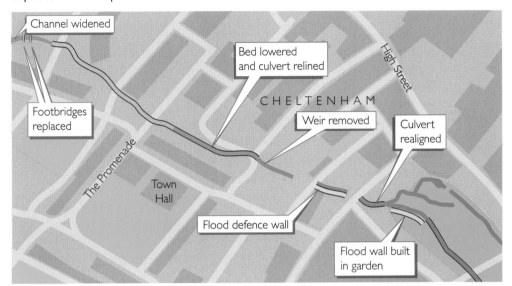

◄ River Chelt flood defences in the centre of Cheltenham

A swale is a shallow depression, like a ditch. It might not look much but it can provide temporary water storage, and is just one example of SUDS ▼

SUDS

The flooding in Cheltenham was made worse by surface water. The drainage system could not cope with the amount of rain falling on to hard surfaces such as tarmac, roofs, etc. A long-term solution to this problem could be to implement SUDS – sustainable urban drainage systems. SUDS use a variety of techniques which allow water to infiltrate into pavements, trenches and so on, or provide temporary water storage. But although new developments can build SUDS into their design, it can be hard to build them into existing urban areas, such as the centre of Cheltenham.

Swale

● Over to you

1 Make a table with three columns headed Social impacts, Economic impacts, Environmental impacts. Use your table to classify the impacts of the flooding in 2007. Which column is longest? Explain why this is.

2 Explain how the jet stream caused the UK's wet summer in 2007.

● On your own

3 Define these terms from the text: jet stream, sustainable urban drainage systems.

4 In September 2007, Gloucestershire County Council launched an inquiry into the floods. Find out the results of the inquiry (use www.gloucestershire.gov.uk) and write your own 500-word report on what happened, how various agencies responded, and how Gloucestershire can prepare for future flooding.

5 Research the flooding in South Asia in the summer of 2007. Focus on Bangladesh and find out about the causes of the flooding, impacts, flood protection and warning. Compare what happened in Bangladesh with either the flooding in Gloucestershire or the Boscastle flood in 2004.

Unit summary

What key words do I have to know?

There is no set list of words in the specification that you must know. However, examiners will use some or all of the following words in the examinations, and would expect you to know them, and use them in your answers. These words and phrases are explained either in the glossary on pages 297–301, or in key word boxes throughout this chapter.

air mass
anticyclone
catchment area
cold front
convergence of air
  masses
coriolis force
deforestation
depression
discharge of a river
drought
El Niño
eye of a hurricane
flood walls
hard engineering

hurricane watch
hydrograph
jet stream
landslides
levees
occluded front
rising limb
shanty towns
soft engineering
storm surge
trade winds
urbanisation
Walker cell
warm front

Try these questions

1 Examine the success of strategies used to manage **one** type of extreme weather event. (10 marks)

2 Describe and explain **a programme of fieldwork and research** that you would use to investigate the links between precipitation and flooding along a stretch of river. (15 marks)

3 Using a named extreme weather event, assess to what extent its impacts were economic, social or environmental. (10 marks)

4 Crowded coasts

What do I have to know?

This topic is about the ways in which coasts attract people, and the consequences of that growth. It focuses mainly on two UK stretches of coast.

1 Competition for coasts: Why are coastal zones so favoured for development?

What you need to learn
- How physical factors create varied coastal environments, e.g. geology, landforms, ecosystems
- Why some coasts are experiencing rapid population growth:
  - Physical factors, e.g. flat land, equable climate
  - Human factors, e.g. fishing, tourism, industry and ports
- **Fieldwork and research**: how these factors have shaped the development of contrasting coasts.

This is a general question and does not need specific knowledge. In-depth case studies come later. Examples are needed but in little depth.

2 Coping with pressure: How do coastal developments create competition and conflict? How can these pressures be resolved?

What you need to learn
- How development leads to zoning and pressure on coastal environments
- **Fieldwork and research** into coastal pressures between development and conservation, e.g.
  - overuse of resources, pollution
  - threats to high-value coastal habitats
- How coastal development brings economic benefits and environmental costs and how people's views about this may vary

Here, detailed case study knowledge is needed of the Dorset coast. You need examples from Bournemouth, Studland, the World Heritage Coast, the Solent (Dibden Bay and Fawley), and the surf reef at Boscombe.

3 Increasing risks: How is coastal development at risk from and vulnerable to physical processes?

What you need to learn
- Risks to coastal development from:
  - rapid coastal erosion along vulnerable coasts
  - rising sea levels and flooding in areas of dense population
- **Fieldwork and research** into rates of coastal erosion **or** coastal flooding and impacts on people, environment, and local development

Detailed examples are needed from Holderness (coastal erosion) and the Thames Gateway and estuary (for coastal flooding). The syllabus requires knowledge of a few areas – not wide coverage.

4 Coastal management: How is coastal management adapting to new ideas and situations?

What you need to learn
- Different management strategies from hard engineering to 'do nothing' and the evolution of Shoreline Management Plans (SMP)
- **Fieldwork and research** into:
  - coastal defence schemes
  - strategies used to manage a high-value coastal environment
- Management strategies for the future including sustainable and integrated approaches, e.g. SMP

This section is about knowing how to evaluate different ways to manage coasts in future – you need examples to help illustrate what is or is not sustainable.

4.1 The coast: not all sand and sea

In this unit you'll look at different coasts and think about how and why they vary.

A

How and why do coasts vary?

Coast – the UK is surrounded by it. Coasts vary enormously, and are always changing. Some are low-lying with long sandy beaches, while others have steep rocky cliffs. Some have human uses, such as ports, fishing, industry, or tourism. Coasts are like magnets attracting people, and, for many, they are increasingly a place to live. In wealthy countries and developing economies alike, coastal populations are increasing.

The shape of the coast – its **topography** – is related to the rock type – or **geology** – from which it is made. The **relief** – or height and slope of the land – is also affected by its geology.

More-resistant rocks, such as granite and chalk, usually form steep high cliffs, while less-resistant rocks, like clays, erode easily. Low, flat land often has wide beaches, e.g. the east coast of England. Often, coastal relief helps to determine how the coast is used.

- On sandy, low-lying coasts, tourism can be important; coastal resorts develop where there is space.
- Where there are deep-water harbours, coasts may be used for shipping and industry.

The photos shown in this unit are all from the UK coastline. They show how variable coasts can be, and how they can be used in different ways.

B

C

A   Studland Bay, Dorset

B   Blackpool Pleasure Beach

C   Holyhead Port, Wales

D   Lulworth Cove, Dorset

E   Newquay, Cornwall

Are some coasts special and worth saving from human activity?

What do you think ?

● Over to you

1 **a** Work in groups and study the photos in this unit. Discuss the physical features shown and then draw a spider diagram to show how coasts vary physically.

  b Now discuss what value each place has for people – perhaps for the jobs it offers, or for tourism. Draw a separate spider diagram to show the variety of human uses of coasts.

  c Feed back your ideas, and complete the spider diagrams using others' ideas.

● On your own

2 Define the following words: topography, geology, relief.

3 Research one of the places shown in the photos, or another coastal place you know. Find out about:

  a its physical features (use the words topography, geology, relief). How have these affected the coast?

  b its human activities (e.g. industry, tourism, port). What were the reasons for these activities developing there?

D

E

4.2 The shift to the coast

In this unit you'll learn how people are increasingly moving to live at the coast.

What is coastalisation?

Just as urbanisation means the movement of people and human activities to urban areas, **coastalisation** means the movement of people to coastal areas. This has some major implications, with more homes and jobs required by the new residents. In attractive areas, this new development can lead to conflict – because people become concerned about too much development spoiling the location. It may also mean that more people are at risk from erosion and rising sea level if global warming is really happening.

Coastalisation is happening in many parts of the world, including Australia, Spain and Florida.

Australia

Australia is the most urbanised country in the world; 90% of the population live in urban settlements, and 60% live in the five largest cities – all of which are located on the coast. The main ports and industries here bring employment and, therefore, urban growth. Rainfall is also much greater near the coast, so most farming occurs within 300-400 km of the coastline.

Coastalisation in Australia has been referred to as **sea change**. There is a movement of people to smaller coastal towns, as well as to the five big cities. There are four main factors causing this sea change:

- Some people are leaving farming in inland rural areas and moving to coastal towns. Recent droughts (see Units 3.10–3.12) have made farming in Australia very difficult, and young people are leaving the land to work in service jobs on the coast, e.g. in tourism.
- New immigrants usually move straight to one of the coastal settlements.
- Coastal towns offer an outdoors lifestyle, as well as urban attractions, e.g. culture, restaurants, theatre.
- House prices are cheaper in smaller coastal towns. Young families often choose to move there.

While the coastal economies are booming, some people are concerned about whether Australia can provide food and water for itself without harming the environment. The country is huge – but the amount of available freshwater and good farmland is not.

Coastline crowding as sea change grows

Australia's coastline is becoming more crowded. Figures from the Australian Bureau of Statistics show that the tide of people moving to coastal communities surged by 2% last year, compared with a national growth rate of 1.2%.

'Nearly half of Australia's population increase in 2004 occurred in smaller coastal communities, even though only 28% of Australians live in these areas,' Joe Natoli, Mayor of Maroochy (Queensland), said. 'Migration to the coast continues to gather momentum, placing demand on council services in sea change areas.'

Mr Natoli said that housing affordability appeared to be driving the migration. '79% of sea changers are under the age of 50. Many have been priced out of capital city housing markets and are looking for affordable housing by the coast which has the benefit of a more enjoyable lifestyle.'

Adapted from an Australian Bureau of Statistics article, March 2005

Coastalisation in Australia ▶

Spain's Costa Geriatrica

65% of the entire Mediterranean coast is urbanised, and, by 2025, there are expected to be 135 million people living on just the northern coast of the Mediterranean. The population on Spain's Mediterranean coast has increased by 1.2% a year since 2000, caused by **inward migration**. Most new migrants are either:

● families moving from large inland cities, or
● international migrants retiring to Spain from countries such as the UK. In 2005, 22% of people along the Spanish coast were over 65. They move for cheaper housing and living costs, together with the warm Mediterranean climate.

But can the environment cope with this influx of migrants? There are many concerns: Is there enough water? What will happen to the sewage? What will happen to habitats and biodiversity?

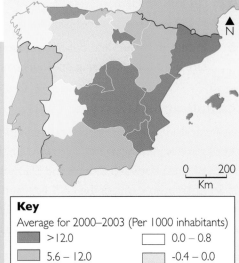

Key
Average for 2000–2003 (Per 1000 inhabitants)
- >12.0
- 5.6 – 12.0
- 2.4 – 5.6
- 0.8 – 2.4
- 0.0 – 0.8
- -0.4 – 0.0
- -1.2 – -0.4

▲ *Average population change in Spain per 1000 people, 2000-2003*

Florida

75% of people in Florida live along the coast, and the coastal population is increasing rapidly. The value of property along the Florida coast is estimated to be $1.9 trillion. Like Spain, the greatest number of inward migrants consists of families and retired people moving from northern US cities to Florida's warm climate. But population growth comes at a cost:

● The Florida Everglades wetland area has already shrunk by 80%, and providing freshwater for Florida's residents is a problem in some years.
● 9% of Florida lies less than 2 metres above sea level, in an area which regularly experiences hurricanes. Coastal flooding and hurricane damage is a problem, and Florida's population growth means that the number of people now at-risk is increasing.

Florida lifestyle – Adam's blog

I just moved down to Florida from New York. We LOVE it! I have lived in many places in this country, and could not think of a better place to raise a family. 15 minutes from some of the most beautiful beaches/ sunsets ... 20 minutes from Sarasota ... 40 minutes from Tampa ... only an hour and 45 minutes from Orlando ... now it is 75 degrees and not a cloud in the sky! There are endless things to do ... my five-year-old son plays sport outside all year. I live in Bradenton (Lakewood Ranch) and it is fairly pricey, but you can still find cheaper housing, with many very good schools in Manatee County.

● Over to you

1 Draw a spider diagram with three bubbles labelled: social, economic and environmental. In three different coloured pens, write down the factors that have led to coastalisation in Australia, Spain and Florida.
2 Have people largely been moving for social, economic, or environmental reasons?
3 Now discuss and add in the social, economic and environmental issues that could arise as a result of further population growth in these places.

● On your own

4 Define the following: coastalisation, sea change, inward migration.
5 Research one of the issues arising from coastalisation in **one** of the countries included in this unit, e.g. water shortages, environmental loss, increased risk.

Can the current rate of coastalisation be maintained?

What do you think?

In this unit you'll find out how coastalisation is occurring along the Hampshire and Dorset coast, focussing on Bournemouth.

Coastalisation UK-style

Like Australia, Spain and Florida, coastalisation is happening in the UK, and many coastal counties are experiencing rapid growth. Units 4.3 to 4.6 focus on issues affecting the stretch of coast between Dorset and Hampshire. It has several towns and cities, and – in Southampton – one of the UK's largest ports. However, it also contains the UK's first stretch of World Heritage Coast, as well as a popular beach at Studland Bay (where the sand dune ecosystem is under huge pressure). These units look at how the different pressures and conflicts can be managed.

Key
- New Forest National Park
- East Devon and Dorset World Heritage Coast

◄ *The Dorset and Hampshire coast*

Bournemouth – the place to be?

Bournemouth is Dorset's largest city, and growing rapidly. In 1851 it was a village of 695 people. By 2001, its population was 163 600. It has the highest population density in the UK's south-west, at 3543 people per km^2. Its population grew by 6.4% between 1995 and 2005, and Bournemouth Council expects a further increase of 11% by 2011. But why is it growing so quickly?

The reason is not **natural increase** – Bournemouth's birth rate is actually lower than its death rate, and its *natural* population decreased by 300 people in 2005. The reason for Bournemouth's population growth is **inward migration** – mostly of over 50 year olds from other parts of the UK. Like Australia, Spain and Florida, Bournemouth attracts people because of its:

- **climate** – Dorset's coast is the UK's second sunniest (after Eastbourne), and its mild winters have few days of frost or snow.
- **high environmental quality** – the coast is an attractive environment to live in, situated only a few miles away from the UK's first area of World Heritage Coast.
- **accessibility** – local urban amenities are easily available, and the area attracts commuters because it is less than two hours by train to London.

> ● **Natural increase** is the difference between birth rate and death rate per 1000 people. If birth rates exceed death rates, the population increases, if the reverse is true then it decreases.
> ● **Inward migration** is the movement of people into an area.

Climate data for Bournemouth ▼

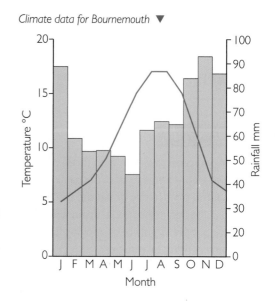

The coast at Bournemouth ▶

The growth of UK seaside resorts

Most UK seaside resorts date from the Victorian period, as a result of the growth in the railway system – which made more of the country easily accessible. In 1859, the railway was built between London and Bournemouth to attract summer holidaymakers. Average wages were low at this time and the first 'tourists' were generally wealthy. Like most seaside resorts, the earliest hotels were large, built directly on the seafront, and were given names like the 'Grand Hotel'. In 1875, Bournemouth's Winter Gardens were built for the town's symphony orchestra.

Bournemouth developed in ways that are typical of UK seaside resorts. In the 1880s, its pier was built to attract more people (the mass market). As average wages increased, more people visited – including day-trippers from the big cities (Bournemouth and Brighton attracted visitors from London, while Blackpool attracted those from Manchester and Liverpool). Day-trippers, and those holidaymakers staying for a week or two, helped to shape these towns, and cheaper guesthouses and accommodation spread inland away from the seafront.

It was all change in the 1960s and 1970s, because people began to go on package holidays to the Mediterranean – and UK resorts deteriorated as a result. The populations of resorts declined – Bournemouth's fell from 154 296 in 1961 to 144 803 in 1981. As the resorts became run down, they had trouble attracting people back. You can read more about this in Unit 6.7 on Walton-on-the-Naze.

The 1980s saw another change, as regeneration took place. The railway was upgraded, meaning that Bournemouth was now less than 2 hours journey from London. A determined effort by the Council to attract new employment in the 1980s and 1990s gave the city an economic boost that few other resorts have achieved. Its airport linked people with the Channel Islands, and later Europe, attracting major employers to the town.

An idealised model of how resorts such as Bournemouth grow and change over time ▼

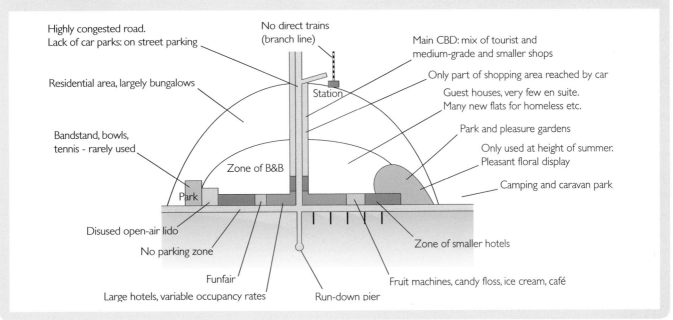

Bournemouth's economic boom

In recent years, the service sector in Bournemouth has seen increasing levels of employment, particularly in banking, finance and tourism. The biggest growth has occurred in the number of financial institutions. In 1999, 14 889 people worked in banks and other financial institutions, growing to 18 300 by 2003. One international bank alone, JP Morgan Chase, has a large office on the edge of the city, with smaller offices in the centre, and employs over 4000 people there. Barclays Bank has its IT Centre in nearby Poole, the Portman Group and Abbey Life also have large offices there.

These industries are **footloose** – they can locate anywhere, because they are not tied to raw materials. This means that companies can locate in a place because it offers a good quality of life. However, economic reasons are also important, and companies also need help from local councils. Companies have been attracted to locate in Bournemouth because:

- land is available for development, for which the council has given planning permission (together with financial and advisory support)
- the university provides a supply of skilled graduates
- there is easy access to London and other European cities (via the local airport)
- wages are lower than in South-East England, especially London
- house prices are also lower than in most of the South-East, so people can afford a bigger property there.

Bournemouth's retirement boom

Bournemouth's population is ageing. It attracts retired people. Most come from the South-East, **downsizing** their property in order to release money to invest. The city has many facilities for elderly people, including day centres, clubs, and social activities. However, unlike some retirement areas (e.g. Cornwall – see Unit 6.8) Bournemouth retains its younger population. The University has expanded in recent years, and attracts younger people who create a demand for vibrant nightlife around the University.

▲ *The Portman/Nationwide offices in Bournemouth, part of the recent financial boom there*

Bournemouth's population structure in 2001 (right-hand side), compared with the average for England and Wales (left-hand side) ▶

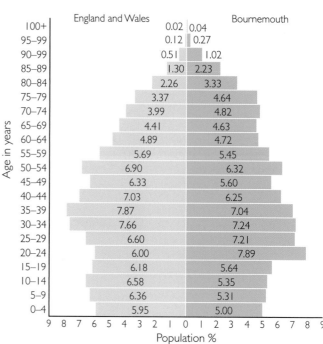

Age in years	England and Wales	Bournemouth
100+	0.02	0.04
95–99	0.12	0.27
90–99	0.51	1.02
85–89	1.30	2.23
80–84	2.26	3.33
75–79	3.37	4.64
70–74	3.99	4.82
65–69	4.41	4.63
60–64	4.89	4.72
55–59	5.69	5.45
50–54	6.90	6.32
45–49	6.33	5.60
40–44	7.03	6.25
35–39	7.87	7.04
30–34	7.66	7.24
25–29	6.60	7.21
20–24	6.00	7.89
15–19	6.18	5.64
10–14	6.58	5.35
5–9	6.36	5.31
0–4	5.95	5.00

Population %

Planning for Bournemouth's future growth

Bournemouth's population growth has led to urban expansion. Victorian buildings have been surrounded by early twentieth century housing to the north and east, with later housing estates to the west, and infilling between villages. The city is now squeezed between the coast on one side and rural areas inland, outside the city boundaries. This is called **coastal squeeze**, where development has to fit into less space. The city's **green belt** puts further pressure on this.

> • The **green belt** is a zone around an urban area where further building cannot take place and on which there are strict planning controls.

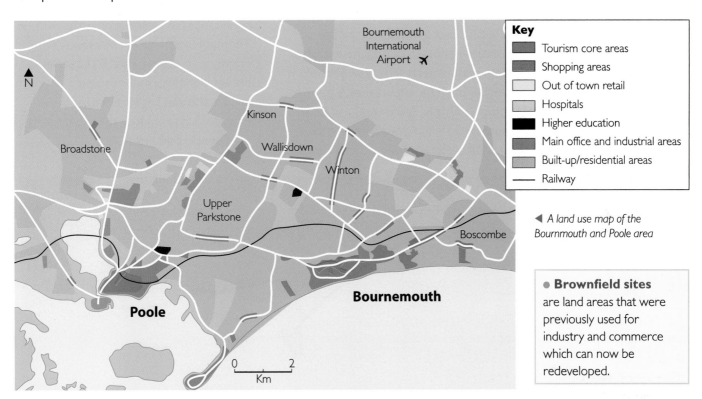

◄ A land use map of the Bournmouth and Poole area

Key
- Tourism core areas
- Shopping areas
- Out of town retail
- Hospitals
- Higher education
- Main office and industrial areas
- Built-up/residential areas
- ── Railway

> • **Brownfield sites** are land areas that were previously used for industry and commerce which can now be redeveloped.

Bournemouth Council now encourages people to renovate existing buildings, rather than build new ones, and to use **brownfield sites**. It aims to have 99% of new residential development on brownfield sites, including 12 000 new homes. The seafront has also been redeveloped, with many old hotels being converted into flats or student accommodation for the University.

> *Can coastal places such as Bournemouth be 'over-developed'?*
>
> **What do you think?**

● Over to you

1 In pairs, read about *where* Bournemouth developed and *how*: **a** in the nineteenth century, **b** in the early twentieth century, **c** in the later twentieth century, and **d** recently. Draw a labelled sketch map to show this, and shade in the different periods in different colours.

2 Now compare your diagram with the idealised model of seaside resorts. List the similarities and differences.

3 Draw a greenbelt around your diagram, then label the city boundary around it. Discuss where any redevelopment should take place.

4 Using Google Maps, identify new sites where development could occur. Justify your choices.

● On your own

5 Define the following: natural increase, inward migration, footloose, downsizing, coastal squeeze, green belt, brownfield sites.

6 Research one other seaside resort, e.g. Brighton or Blackpool. What similarities are there **a** in its development, **b** in the issues that it faces?

Exam question: Describe how physical and economic factors may have made one area that you have studied a crowded coast. (10 marks)

4.4 Conflict on the coast – 1

In this unit you'll learn about the special nature of the Dorset coast, and how using it for tourism can create conflict.

The Jurassic Coast

Dorset's coast has some of the UK's most stunning coastal scenery. This, together with its wildlife, nature reserves and great beaches makes it very popular. In 2001, UNESCO made it the UK's first coastal World Heritage Site. Places are only awarded World Heritage status if they are of global importance. The Dorset and East Devon Coast World Heritage Site (known as The Jurassic Coast) was awarded this status because it depicts a 'geological walk through time', which spans several geological periods. It puts Dorset alongside places in the UK of similar importance, e.g. Ironbridge, Shropshire (for its historical industrial significance) and Kew Gardens (for its global botanical importance) – and globally gives it the same status as the Daintree Rainforest in northern Queensland, Australia.

History and culture
The Dorset coast is also historically significant for its fishing villages and small ports, such as Lyme Regis. Culturally, the area's greatest significance is its association with Thomas Hardy – whose Wessex novels, such as *Tess of the D'Urbervilles*, are set in the Dorset landscape and heathlands. English Heritage is interested in promoting this aspect of the Dorset countryside.

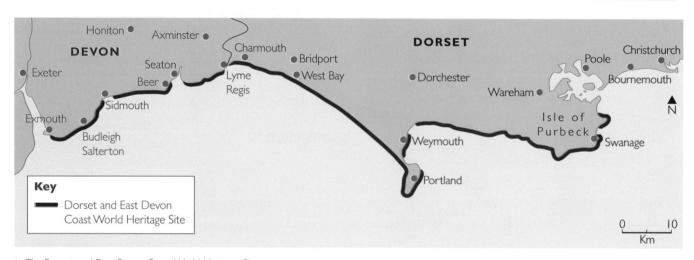

▲ The Dorset and East Devon Coast World Heritage Site

Ecology
Ecologically, the Dorset coast is very significant. Studland beach and its sand dunes were formed by the deposition of sand. The dunes vary in height, and much of the area is a Site of Special Scientific Interest (SSSI), protected by law from human interference. The dunes are home to rare plants (e.g. marsh gentian), insects (e.g. ladybird spider), birds (e.g. Dorset heath nightjar) and reptiles (e.g. sand lizard, smooth snake). As you go inland, the sand dunes change to heath vegetation – a combination of low shrubs such as heather and gorse, and trees such as birch. Heath vegetation and fauna are increasingly rare in the UK. Heaths are often used for cattle grazing, or planning permission might be given for housing, but ecologists try to protect dunes from further development.

Studland sand dunes ▼

182

(A) Durdle Door, where a small remaining band of Purbeck limestone has been eroded to form a coastal arch.

A geology map of the Isle of Purbeck ▼

Geology, landforms, and scenery

The **geology** (or types of rock) on Dorset's coast is very varied. It has a major impact on landforms, especially on the Isle of Purbeck – where the combination of different rocks has led to varied scenery. Rock types vary between resistant Purbeck limestone, which forms steep cliffs, and less-resistant clays and sands. These rocks alternate along the coast, so that where a resistant rock is eroded (e.g. at the entrance to Lulworth Cove), and the sea breaks through to the less-resistant rocks behind, erosion follows more quickly. At Lulworth, this has led to the formation of a small bay or cove.

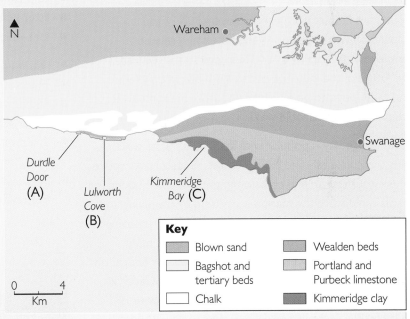

N

Wareham

Swanage

Durdle Door (A)

Lulworth Cove (B)

Kimmeridge Bay (C)

0   4
Km

Key

Blown sand	Wealden beds
Bagshot and tertiary beds	Portland and Purbeck limestone
Chalk	Kimmeridge clay

(B) Lulworth Cove – the resistant Purbeck limestone at the entrance has been eroded by the sea, and subsequent rapid erosion of the less-resistant clays behind it has opened up the coast into a cove or bay.

(C) The cliffs of Kimmeridge Bay are less resistant, and fossils can easily be found there.

Conflict on the coast – 2

The need for protection

One of the benefits of World Heritage status is that it helps to protect the coast. Dorset County Council has to reassure UNESCO that certain types of development will not be given planning permission. But this can lead to conflict. World Heritage status increases tourism – and tourists need places to eat, places to stay, and roads to travel on. Increasing tourist numbers has impacts on:

- Studland's dunes, where vegetation can be trampled
- cliff footpaths, which become eroded
- villages and roads, where traffic congestion can damage the very peace and quiet visitors come for.

Conflicts are bound to happen. These can be recorded on a conflict matrix.

Key:

+++	Strong agreement	– – –	Strong disagreement
++	General agreement	– –	General disagreement
+	Slight agreement	–	Slight disagreement

▼ *A conflict matrix*

	Dorset County Council	English Heritage	Conser-vationists	Ecolo-gists
Dorset County Council				
English Heritage				
Conser-vationists				
Ecolo-gists				

Tourism in Dorset

Tourism is Dorset's biggest employer (30% of the population work in catering and hotels and related employment). Gaining World Heritage status secured the future of the industry. But tourism brings problems, and these need to be managed if tourism in Dorset is to continue in a sustainable way.

Kimmeridge Bay is popular with fossil hunters; its shale cliffs crumble easily and expose fossils. It has a rocky shore where tourists can find crabs or limpets. People hunting for fossils can disturb the marine ecosystem by trampling and stone turning, as well as degrading the cliff and removing fossils from the area. To protect the cliffs, the Limpet Protection Zone at Kimmeridge provides visitor education about how to reduce impacts on the shore. However, it is voluntary, and relies more on the good behaviour of tourists than on having any enforceable rules.

Studland Bay – too popular?

Studland Bay, owned by the National Trust, is one of the UK's best sand dune areas – situated on a peninsula beside Poole Harbour. It is close to both Poole and Bournemouth, and the popularity of the beach means that 1.5 million people visit each year, to sunbathe, picnic, sail, jet ski or to walk in the sand dunes. The area can be reached by car or a ferry. In the summer, queues of cars can extend the full length of the Studland peninsula. On summer Bank Holidays, 35 000 people can visit in one day, bringing:

- cars which cause congestion
- litter which can prove dangerous to birds and people
- feet which trample the dune plants.

At sea, water- and jet-skiing bring particular problems:

- The noise disturbs wildlife, both on land and at sea.
- People with boats and skis cause traffic jams along the minor road leading to Studland.
- The swash from fast jet skis and boats damages seagrass beds.
- Both swimmers and dolphins are threatened by injury from outboard motors.

▼ *Studland on a summer's day*

Surfers' paradise?

Even though the Dorset coast is protected by its World Heritage status, other tourist developments nearby can cause conflict. Boscombe's new surf reef (2.5 km from Bournemouth Pier), which is due for completion in 2008, has many supporters – but it also has opponents. The reef consists of large sand-filled geotextile bags, built over a hectare of seafloor. It works by forcing waves to break out to sea, and with greater height. Costing £1.4 million, developers estimate that the surf waves that it will produce will be 4 metres high. The reef is part of a development of flats, restaurants, cafes and shops.

Why build a reef? It will double Boscombe's current average of 77 good surfing days per year, and could attract 10 000 surfers a year. Surfers spend, on average, 8% more than other tourists. The council expects it to earn £10 million annually, and create 60 full- and 30 part-time jobs.

> It will put Boscombe on the map as a surfing resort. There are no artificial surf reefs in the northern hemisphere. There are two in Australia, one in New Zealand and one in America.
> **Andrew Emery, service development officer for the seafront**

> To spend such a large percentage of resources on a minority sport cannot be justified. Where will the maintenance funds come from every five to six years?
> **Parry Brooks, Chair of Boscombe Cliff Residents' Association, quoted in** The Dorset Echo

> Building 169 flats and a surf reef is not going to bring tourists back to Boscombe. A surf reef is not going to fill up hotel rooms. Surfers spend a lot of money on their gear – not on accommodation or restaurants. Such a major housing scheme will look absolutely appalling. Car parking will be a joke – finding a space is already a problem.
> **Debbie Utteridge, owner of the Rosscourt Hotel, quoted in Bridport news online**

▲ Playing in the waves at Boscombe before the artificial reef is built

World Heritage Sites - a good idea or not?

What do you think?

▲ Views on the new surf reef at Boscombe

● Over to you

1 a Make two copies of a conflict matrix. On one, label the various groups that would have an interest in Studland Beach, and on the other the groups for the Dorset World Heritage Coast as a whole.

 b Using a scale from +++ (strong agreement) to – – – (strong disagreement), complete both matrices to show which groups would agree and which disagree with how each area might be managed.

 c Explain your decisions.

2 a Now complete a third matrix to show the people with interests in Boscombe's new artificial surf reef, e.g. surfers, developers.

 b Explain who would and who would not agree with the scheme, and why.

● On your own

3 Research and design a PowerPoint to show the key features of Dorset's World Heritage Coast.

4 Write a report of about 750 words entitled 'The difficulties faced in managing Dorset's Heritage Coast'. ●

4.5 Industry on the coast: Southampton Water

In this unit you'll investigate how one stretch of coast is used for industry, and the environmental impacts that this has.

The Solent and Southampton Water

Southampton Water and the Solent together form one of the UK's best natural harbours.

- Southampton Water is part of the **estuary** of the rivers Test and Itchen.
  It is a **ria**, with Southampton lying at its northern end.
- Southampton Water flows into the Solent, a stretch of water separating the Isle of Wight from the mainland.

The area has huge advantages for industry:

- It is sheltered from storms in the English Channel.
- It has deepwater channels for large ships.
- Around the estuary is a broad area of flat land for development.

Pressures on Southampton Water

Four issues affect the area – creating competition for land and conflict between different land uses.

The growth of employment in Southampton, including Fawley, has led to the **expansion of suburbs and villages** close to the estuary, e.g. Hythe and Fawley. The growth is squeezed between the New Forest National Park and Southampton Water. As new homes are built here, developers face problems trying to build either along the shore or into the New Forest.

Along its western edge is a large area of salt marsh, just where **industrial development** has taken place for oil refining at Fawley. There is also a large container port at Southampton Docks.

- An **estuary** is the mouth of a large river where it meets the sea.
- A **ria** is a flooded or drowned river valley formed at the end of the last Ice Age as glaciers melted.

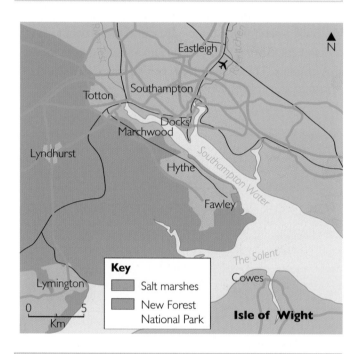

New housing and development creates problems of **sewage disposal**.

Southampton Water is the most important area of the UK for **sailing and leisure craft**; thousands of leisure boats moor there, and boatyards occupy the waterfront.

Southampton's container port requires a huge amount of space for development ▶

Fawley oil refinery

Fawley oil refinery is the largest refinery in the UK – 3000 people work there. Owned by Esso, it handles 2000 ships and 22 million tonnes of crude oil every year, and has 1.5 km of docks. Its location on the Solent means that it is convenient for tankers from Europe. Tankers from the rest of the world reach it via the English Channel. It opened in 1951 and, since then, has expanded massively.

Fawley has three main environmental impacts:

- There are some great environmental pluses, such as the 50 000 trees and shrubs planted when the refinery first opened – but expansion of the refinery has meant that the surrounding area of **salt marsh** has been greatly reduced in size. This salt marsh is designated a Site of Special Scientific Interest (SSSI).
- **Effluent** (liquid waste) from the refinery is the main environmental problem. The effluent can be as warm as 30 °C, much warmer than the estuary water. This affects the estuary ecosystem – some species mature more quickly in warmer water and become more abundant, so altering the food web. In Southampton Water, scientists think that the increase in the population of hard clams is caused by water temperature change. They only breed in water over 23 °C, and breed more often if water is warmer. They then consume more algae, leaving less for other species.
- **Metal pollution**. Although refinery effluent is monitored to check it for things such as metals (cadmium, lead, mercury) and phosphates, these are still emitted, though at lower levels than those legally permitted.

▼ *Fawley oil refinery*

Background

Salt marsh

Salt marshes form in sheltered coastal estuaries, which have a gently sloping seabed and a large supply of sediment. Most sediment is fine clay deposited by rivers. If the wave energy does not erode it, sediment builds up to the surface of the water and is exposed at low tide. This allows colonising plants to establish, such as grasswort and cord grass. These trap more sediment and add nutrients as old leaves decay, changing the habitat so that new plant species can invade. This process is known as **plant succession**.

Salt marshes are hugely valuable in a number of ways:

- **Ecologically**, because they contain huge numbers of marine species, and are feeding grounds for migrating birds.
- **Economically**, because they offer sheltered water for shipping and sailing boats, and beds for oysters and other shellfish.

Most British salt marshes have now disappeared – used for oil refineries, chemical plants, and port developments.

Industry on the coast – impacts

Oil spills

Although the number of oil leakages from the Fawley refinery has fallen, oil spills still happen. On 1 October 1989, a ship offloading crude oil spilt 20 tonnes into the water, which spread downstream from Fawley to Calshot Spit.

Clearing up the oil involved skimming it off the surface of the water and spraying the remainder with detergent to break it up. Beaches had to be cleaned, and the salt marsh was particularly badly affected. Experts believe that trying to clean up salt marsh causes even more damage, so it was left alone. Eventually the oil disperses or is covered by sand. 800 birds were affected by the oil spill.

▲ *Salt marsh near Fawley polluted by the 1989 oil spill*

Sewage and industrial discharge

Although discharging treated sewage into the sea is legal, sewage pipes discharge 300 million litres of treated sewage every day into the Solent, which is a huge amount for a relatively small area. As population in the area increases, so does the sewage discharge.

Raw sewage is treated to reduce the numbers of bacteria and viruses entering the seawater. This is important for swimmers, and for food sources such as shellfish. Clams and oysters are harvested from Southampton Water, and they could be dangerous to eat if contaminated. Oysters feed by filtering seawater for micro-organisms, and so could ingest any bacteria and viruses present in the sea.

Run-off from farmland is high in nutrients, such as nitrates and phosphates. The Solent has a higher nitrate level than other UK estuaries, because its daily input of fresh water is low, and it takes nearly a week for the tides to flush it out with 'new' seawater. If nitrate and phosphate concentrations increase, the estuary ecosystem becomes **eutrophic** – or artificially enriched.

> *Should industry be allowed to locate wherever it wants to, whatever the environmental consequences?*
>
> **What do you think?**

▼ *Sewage outfall and industrial outlets into the Solent*

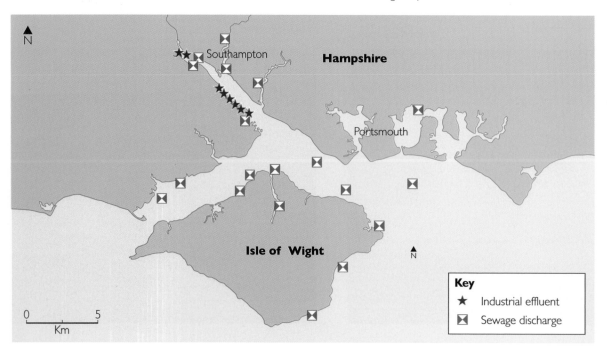

Hampshire

Southampton

Portsmouth

Isle of Wight

N

0   5
Km

Key
★ Industrial effluent
⊠ Sewage discharge

Eutrophication

Eutrophication is the artificial enrichment of water by fertilisers or sewage. It encourages the growth of algae and waterweeds. This causes:

● algal bloom, which produces toxins that kill off plants
● sunlight to be blocked out.

When algae die, their decomposition uses up oxygen. A lower level of oxygen in the water reduces the number of fish and insect species. If the water becomes **anoxic** (deficient in oxygen), then bacteria like *clostridium botulinum* grow rapidly, and are deadly to fish, plants and animals (including humans).

Waste

Even treated sewage still contains solid waste, such as sanitary towels. These can end up in the sea or littering beaches. Marine litter also includes waste from ships and the rubbish that tourists leave behind on beaches. A survey around the Solent found coal, brick, glass, old shoes, rubber gloves, sanitary towels, plastic bottles, sweet wrappers and bottle tops on the beaches.

▲ *Washed up waste in the Solent*

Metal pollution

Metal pollution is a problem in Southampton Water. Paints used to stop organisms such as barnacles attaching themselves to the underside of a boat (called anti-fouling paints) contain a chemical called TBT, which contains a lot of tin. Although these are banned from being used on small vessels, they can be used on the large container ships which use Southampton's container port. Tin levels are above those for other UK estuaries, and much of it is trapped in sediments.

▲ *The dogwhelk is now extinct in Southampton Water due to pollution*

● Over to you

1 In pairs, identify the Strengths, Weaknesses, Opportunities, and Threats of Southampton Water and The Solent as a location for **a** environmentalists, **b** economic activity.
2 Can a compromise be reached between environmental concerns and economic activities?
3 Copy and complete the table below to show the advantages and disadvantages in adopting the following policies.

● On your own

4 Define the following words: estuary, ria, salt marsh, effluent, plant succession, eutrophication, anoxic.
5 Explain why Southampton Water faces so many threats to its environment.

Exam question: Referring to a stretch of coast that you have studied, explain how and why some coasts are under threat from economic development. (10 marks)

Statement	Advantages	Disadvantages
Industry should be free to develop as it wishes.		
Industry should be regulated to control pollution and environmental impacts.		
For environmental reasons, Southampton Water should never be developed.		

In this unit you'll find out about the conflict arising from proposals to build a new container port at Dibden Bay on Southampton Water.

Save Dibden Bay

People living at Dibden, near Southampton in Hampshire, were stirred up in 2001. Associated British Ports (ABP) announced proposals to build a **container port** at Dibden Bay, beside Southampton Water. The plan involved constructing a new deepwater port for container ships – comprising 2.1 km of docks – spreading over an area of 350 hectares (equivalent to 1600 football pitches). This would have transformed an undeveloped area of estuary opposite the existing Southampton Docks into a large port facility.

Some people felt that it offered a great economic opportunity and the chance to expand Southampton as a port, while others feared that it would transform the landscape into a huge area of concrete, with cranes visible from miles away, year-round noise and light pollution – not to mention increased lorry and rail movements in the surrounding area.

▲ The Friends of the Earth campaign to save Dibden Bay

Existing container port

Proposed new container port

Southampton's main docks

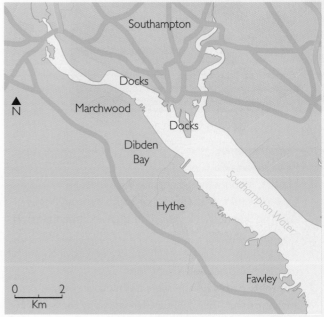

◄ The proposed Dibden Bay container port

● A **container port** is a place where goods are imported and exported in large metal containers, which are then transferred between ships and lorries or trains.

Why did ABP want the container port?

ABP wanted to make Southampton more competitive as a container port. Containers can be easily transferred between lorries or trains and ships – no goods are loaded or unloaded from the containers at the actual port, so container ports are cheaper to operate than traditional ports, where people handle the goods. ABP was worried that other ports in the UK (like Felixstowe) were bigger and more competitive than Southampton, and that, without the new container port, Southampton would decline.

- The new port would cost £700 million, take nine years to build, and be capable of handling six container ships at a time.
- Dredging the estuary bed to make it deeper would allow larger ships further inland, and could also accommodate the large cruise ships which often use Southampton.
- 3000 jobs would be created.
- A new access road to the A326 and a new rail link would be built.

How did people react?

There were many protests against the proposal, including from Friends of the Earth, the Ramblers Association, English Nature and some local residents. They believed that the port should be sited somewhere else because:

- Dibden Bay is a Site of Special Scientific Interest, as a result of its salt marsh ecosystem. It supports over 50 000 water birds each winter and migrating birds each spring.
- traffic (especially large container lorries) would increase on the A326, causing congestion, noise and air pollution
- local homes would lose their views of Southampton Water
- the port would impact negatively on the New Forest, a National Park, e.g. through increased traffic
- increased fuel spills by ships, and contamination by anti-fouling paints on their keels, would risk the ecosystem and water quality.

> If the government turns this down, it would be a bad decision for UK plc. We would lose business to Europe and consumers would pay more for goods in the shops.
> **Mike Stacey, ports project director at Dibden**

> I'm in favour of the new port at Dibden. This is an investment in Hampshire's economy. Despite the view of environmentalists, the Waterside is not tranquil countryside. It houses Europe's largest oil refinery, a power station, chemical industries, army bases and industrial estates. It's the right place for a new port, especially on land that has no useful purpose.
> **Keith Legg, local resident**

> Dibden Bay consists of 240 hectares of open grazing and mudflats. The area forms part of an international wildlife haven noted for its diversity and number of birds. It also forms a gap between Hythe and Marchwood, providing a wildlife corridor to the New Forest from the Waterside, and it is one of the few remaining undeveloped areas on Southampton Water
> **Friends of the Earth**

▲ *Reactions to the proposals at Dibden Bay*

> *Which is more important – jobs and the economy, or protecting the environment?*
>
> **What do you think?**

● Over to you

1 Copy the table at the bottom and assess the main impacts of the Dibden Bay proposal.
2 In pairs, rank the five bulleted objections to the Dibden proposal to decide which you think is the most important and which the least. Explain your reasons.

● On your own

3 In 2004, the Transport Secretary announced that the scheme had not been given planning permission because it was felt that potential environmental damage was greater than economic benefits. Do you think this was the right decision? Justify your answer in 500 words.

	Short term	Medium and long term
Economic		
Social		
Environmental		

CROWDED COASTS

In this unit you'll learn about how and why some coasts erode rapidly, and the impacts this has.

Coastline at risk

For centuries, erosion has been a problem along the coast of the East Riding of Yorkshire, known as Holderness, which stretches between Bridlington and Flamborough Head in the north and Spurn Head in the south. It has the fastest eroding coastline in Europe. On average, nearly two metres of its coastline are lost every single year. Since Roman times, it has eroded by 4 km and 29 villages have been lost. Other villages are disappearing now. During the winter of 2006–2007, five metres vanished from parts of the area, including around the small coastal resort of Skipsea. In the village of Barmston, 1 km south of Bridlington, a single storm in 1967 caused the coast to retreat by 10 metres, leaving 12 bungalows so near to the cliff edge that they had to be demolished. No wonder that some people have to move out and lose their homes and livelihoods.

Peter Johnson's story

In 2007, erosion victim Peter Johnson, 63, said he would have to move sooner rather than later from his home in Seaside Road, Aldbrough, near Hornsea. He said: 'I fully expected to live out my days here, but the land on which my property stands has eroded at a greater rate than I ever expected. I pay my rates to the council, so I would expect some form of compensation package if I have to leave, because it is not something of our making. The least I would expect is compensation to cover the cost of demolishing my home and moving to another property.' Mr Johnson, who has lived in his cliff-top home for 11 years, said he did not think people who had to move would get the full value of their properties.

Adapted from the *Hull Daily Mail,* January 2007

▼ Cliff erosion has nearly reached this house in Holderness

So, the coast is eroding ... and?

Coastal erosion has economic impacts. Apart from deciding whether the area could, or should, be protected from erosion, and the cost of that protection, continued erosion and loss of land has a considerable effect on the economy.

Local boulder clay produces good, fertile soils for arable farming

◄ There is a gas pipeline terminal at Easington, where North Sea gas supplies are piped ashore. It provides hundreds of jobs, and would be very expensive to re-locate further inland if coastal erosion threatens it.

The **chalk** at Flamborough Head is fairly resistant, and produces features such as cliffs, caves, arches and stacks. ▼

Why is erosion such a problem?

The three main reasons for the rapidly eroding coastline at Holderness are:

- geology (rock type)
- the fetch
- longshore drift and beach material.

Geology

The Holderness coast has two rock types: chalk and boulder clay.

Bridlington

Flamborough Head

Great Driffield

Hornsea

Mappleton

Beverley

North Sea

Hull

Withernsea

River Humber

Spurn Head

Key

☐ Chalk

▨ Alluvium

☐ Boulder clay (glacial tills)

0   10
Km

N

◄ The geology of the Holderness coast

◄ The rest of the coast is **boulder clay** (also known as glacial till or drift), a mixture of fine clays, sands and boulders deposited by glaciers after the last Ice Age. It has little resistance to erosion, especially when wet. Boulder clay produces shallow sloping cliffs, between 5 and 20 metres high, and it is these cliffs that are rapidly eroding, on average by 2 metres a year.

The Holderness coast ... going ... going ... – 2

Fetch

One of the main factors affecting how quickly the sea erodes is wave energy. This, in turn, depends on how far waves have travelled – known as the fetch. Holderness is exposed to winds and waves from the north-east, which have a small **fetch** of about 500-800 km across the North Sea. This is not far, compared to some of the world's large oceans, but the size of the waves attacking the Holderness coast is increased by four other factors:

- Currents – or **swell** – which circulate around the UK from the Atlantic and into the North Sea. The Atlantic fetch is 5000 km or more, and its currents add energy to waves in the North Sea. Therefore, there are often powerful **destructive waves** along this coastline, eroding the cliffs.
- Low pressure weather systems passing over the North Sea are often intense, and locally produce very strong winds and waves.
- Small, almost **enclosed seas**, like the Mediterranean or North Sea, often generate huge waves during storms. Waves move within the sea but cannot disperse their energy – rather like water slopping up against the side of a washbasin.
- The **sea floor is deep** along the Holderness coast. Therefore, the waves reach the cliffs without first being weakened by friction with shallow beaches.

As a result of these four factors, waves along the North Sea coast can be huge.

▲ *North Sea storm waves at Scarborough, north of Holderness*

Longshore drift and beach material

The beaches at Holderness are its main problem. Boulder clay erodes to produce mainly clay particles, which are easily transported out to sea, rather than accumulating close to the cliffs as beach sand. Although there **are** beaches, there is never enough sand to stop the waves reaching the cliff base at high tide. The beaches are narrow and offer little friction to absorb the wave energy and protect the cliffs. The sand that is produced is taken southwards by **longshore drift**, leaving the Holderness cliffs poorly protected against wave attack. Eventually, a small amount of beach material reaches the spit at Spurn Head, where it accumulates.

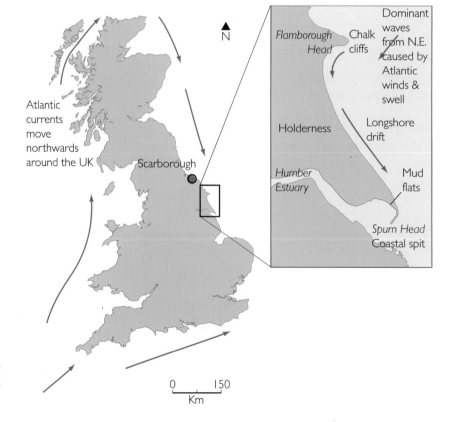

The fetch of the Holderness coast, and the additional swell circulating from the Atlantic ▶

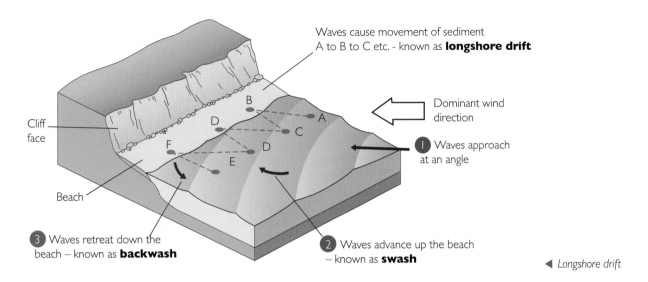

Waves cause movement of sediment
A to B to C etc. - known as **longshore drift**

Cliff
face

Dominant wind
direction

B

D

A

C

D

F

E

Beach

① Waves approach
at an angle

③ Waves retreat down the
beach – known as **backwash**

② Waves advance up the beach
– known as **swash**

◀ *Longshore drift*

Cliff-erosion processes

Two types of processes are involved in the rapid erosion of the clay cliffs on the Holderness coast:

- **Cliff-foot** erosion, caused by wave action at the base.
- **Sub-aerial** processes, which affect the cliff face.

Cliff-foot erosion

Waves attack the base of the cliff, eroding the cliff foot:

A Waves advance, picking up sand and pebbles from the seabed. As they break, the material is hurled at the cliff, chipping away at the base. This is known as **abrasion or corrasion** – both terms are correct.

B Advancing waves trap air inside cracks in the cliff, increasing air pressure. As the wave retreats, the air explodes, weakening joints and cracks and causing pieces of rock to break away. This is known as **hydraulic action**.

C Where cliffs are formed from alkalis such as chalk or limestone, or an alkali cement bonds the particles together, solution by weak acids in sea water dissolves them, causing **corrosion**.

Each process causes the cliff to become undercut and unstable, leading to collapse.

Cliff undercutting ▶ ▲

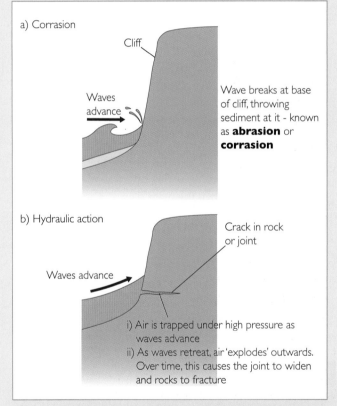

a) Corrasion

Cliff

Waves
advance

Wave breaks at base of cliff, throwing sediment at it - known as **abrasion** or **corrasion**

b) Hydraulic action

Crack in rock or joint

Waves advance

i) Air is trapped under high pressure as waves advance

ii) As waves retreat, air 'explodes' outwards. Over time, this causes the joint to widen and rocks to fracture

Background

Cliff-face processes

When exposed, the cliff face is affected by two sets of sub-aerial processes:

- **Weathering**, which breaks down the solid rock fragments into smaller fragments, or dissolves them chemically.
- **Mass movement**, which is the movement of fine weathered material downslope by gravity.

Weathering processes

Three types of weathering processes affect the Holderness cliffs:

- **Physical weathering**, including any kind of mechanical action. The best example along this coast is **freeze-thaw**. Water enters cracks when it rains and freezes during cold weather. Freezing water expands in volume by 10%, putting pressure on the rock and enlarging the crack. If this is repeated many times, the crack enlarges and eventually fragments of boulder clay fall from the cliff face.

- **Biological weathering**. As seedlings germinate, fine root hairs find their way into microscopic cracks in the rock. As the roots grow, they prise the cracks apart and enlarge them. Fully grown plants can be seen growing out of solid rock, and their roots break the rock up. Once the plant dies, the rocks fall apart.

- **Chemical weathering**. Rainwater combines with CO_2 in the atmosphere to form weak carbonic acid, or with SO_2 to form sulfurous acid. These acids dissolve the alkali chalk at Flamborough Head, weakening it, but chemical weathering is less important than physical or biological weathering along the Holderness coast.

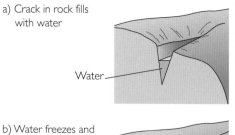

a) Crack in rock fills with water

Water

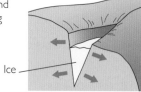

b) Water freezes and expands, pushing rocks apart

Ice

c) Repetition of process prizes off rock fragment which falls

▲ Freeze-thaw

● Over to you

1. In pairs, identify **a** the people who might be concerned at the erosion of the Holderness coast, and **b** why they might be concerned.
2. Classify the reasons in your list as **a** economic, **b** social, and **c** environmental. Which list is longer, and why?
3. Why are some of these people likely to come into conflict in deciding whether or not something should be done about erosion?
4. In pairs, draw a large map of Holderness, showing its main physical and human features. Add details to it about:
   a the geology of Holderness, and where most erosion is taking place
   b how erosion and the size of waves is affected by fetch
   c the direction of longshore drift and the features that it produces along the coast.
5. Research photos of **a** how erosion is affecting specific places, **b** tourism along the coast. Stick these on to your map. Annotate them to show where the photos were taken and what they show.

Mass movement processes

Slumping is the main form of mass movement affecting the clay cliffs along the Holderness coast. Alternate wetting and drying of the clay causes it to expand when wet and shrink when dry. This causes cracks to form, especially during long dry periods. The next time it rains, water rapidly enters the clay and percolates into the cliff. The clay then becomes lubricated or slippery, and gains extra weight. As a result, the cliff cannot support the heavier clay and gravity leads the clay to slide downslope. This movement is known as slumping, or landslip. The slumped material collects at the base of the cliff, and is then washed away by the sea.

▲ *Slumped cliffs at Holderness*

Slumping in action ▶

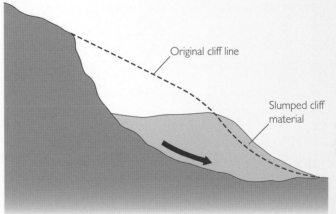

Original cliff line

Slumped cliff material

Should governments pay compensation to people whose homes are threatened by erosion?

What do you think?

● On your own

6 Produce a PowerPoint presentation to show the processes and effects of erosion at the cliff foot. You should mention the key processes on page 195.

7 Now do the same for sub-aerial processes (weathering and mass movement). Include photos of slumping and draw diagrams to explain how this occurs.

8 Define the following terms: boulder clay, fetch, swell, longshore drift (including swash and backwash), cliff-foot processes, abrasion, corrasion, hydraulic action, corrosion, weathering, mass movement, sub-aerial processes, physical, biological and chemical weathering, freeze-thaw, slumping.

9 Most processes in physical geography occur in logical stage-by-stage steps. Produce a series of flow charts showing step-by-step stages of **a** cliff-foot erosion, **b** weathering on the cliff face, **c** cliff slumping.

10 Research one other area of the UK where erosion is a problem; good areas for research include Christchurch Bay in Hampshire and Dorset, and the Norfolk coast. Explain how **a** geology, **b** the fetch, and **c** longshore drift contribute to their erosion problems.

Exam question: Referring to a stretch of coast, explain the factors that have led to this coast suffering from erosion. (10 marks)

In this unit you'll find out about different methods of tackling erosion at Holderness, and the effects they have.

Can erosion on the Holderness coast be stopped?

Coastal erosion can certainly be prevented, but at a cost. Who should pay? The cost of coastal protection is often controversial. Most people want their own stretch of coast protected, but do not always see why their taxes should pay to protect someone else's! Until the 1990s, it was usual for local councils to tackle erosion by designing **hard engineering** structures. Most are expensive, so now more **soft engineering** techniques are used. But which, if any, work best?

Hard engineering techniques in Holderness

Wooden groynes at Hornsea

Coastal resorts usually have one agenda – protect the beach! At Hornsea, wooden groynes have been built to trap sediment carried by longshore drift. These keep the beach in place at Hornsea, and in turn this protects both the cliffs and the town. Advancing waves – or swash – have to travel further over the beach, and the friction caused absorbs much of their energy. Less energy results in less erosion.

However, what happens in one place almost always affects other places. South of Hornsea, the village of Mappleton was being starved of the sediment trapped by Hornsea's groynes. While longshore drift continued to remove the sediment from Mappleton's beach, none of it was being replaced. Rapid wave attack eroded the cliffs so much that by the 1990s nearly 4 metres of cliff per year were being eroded – and the village was in danger of disappearing completely.

- **Hard engineering** involves structures built along the coast, usually at the base of a cliff, or on a beach, e.g. sea walls, groynes and revetments.
- **Soft engineering** is designed to work with natural processes in the coastal system to try to manage – and not necessarily prevent – erosion.

▼ Wooden groynes at Hornsea

Rock groynes at Mappleton

After a campaign by residents and local councillors to protect Mappleton, Humberside Council built two rock groynes in 1991, costing £2 million. These consist of boulders of Norwegian granite, laid out like groynes. Like those at Hornsea, their aim is to prevent the removal of beach sand by longshore drift, and so reduce cliff erosion. The cliff face was also **re-graded** – or shaped – to make it less steep and therefore more stable.

It is not difficult to predict the impact of the rock groynes at Mappleton. At Cowden, three kilometres south of Mappleton's rock groynes, sediment starvation has caused increased erosion of the cliffs. This ripple effect – where erosion increases just past the last groyne – is called **terminal groyne syndrome**, and happens wherever groynes trap beach material and starve other areas. The village of Cowden has seen its erosion rate increase from an average of 2.5 metres a year to 3.8 metres a year between 1991 and 2007.

▲ *One of Mappleton's rock groynes and the re–graded cliff*

Revetments at Easington

A rock revetment has been built to protect Easington's gas terminal. Like stone groynes and **rip-rap**, a revetment consists of large granite boulders. The boulders are placed like a wall in a line of defence along the shoreline. It works by absorbing – not reflecting – wave energy using large air spaces between the boulders and a broad surface area. Revetments are very expensive but long lasting. Some people think they are unattractive.

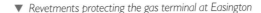

▼ *Revetments protecting the gas terminal at Easington*

Managing erosion on the Holderness coast – 2

Sea walls at Withernsea

Sea walls are usually built at coastal resorts. They can be either straight or recurved.

- **Straight walls** reflect wave energy out to sea, rather than absorbing it, and form a barrier to prevent undercutting of cliffs. Withernsea had a straight wall built in 1875. It worked well in protecting the holiday resort, and also provided a promenade for tourists. However, over time, wave energy eroded – or **scoured** – the base of the wall, undercutting it, and causing it to collapse.
- **Recurved walls** dissipate – or spread out – some of the wave energy, thus reducing the erosion at the base of the wall and extending its life. However, they are more expensive to build because of their shape.

Withernsea replaced its old straight wall with a recurved wall costing £6.3 million (£5000 per metre). The expense was justified following a **cost-benefit analysis**, which showed that the benefits outweighed the construction costs. The wall protects Withernsea – a strategy known as 'holding the line' (see page 202) – and has prevented further decline in the value of local property, and saved tourism employment.

However, it has created new problems. Waves are noisier when they break against the wall and there is a smaller promenade. Views from seafront hotels have also been restricted. At the base of the sea wall is rip-rap, also known as **rock armour**. Like rock groynes, rip-rap consists of large granite boulders at the base of the sea wall. These are designed to dissipate energy at the base, so preventing scouring. They are effective in reducing erosion, but are expensive and some tourists find them unattractive.

(see page 202)

Background

Cost-benefit analysis (CBA)

A cost-benefit analysis will be carried out before a coastal management project is given the go-ahead. Costs are forecast (e.g. a sea wall – its design, building costs, maintenance, etc.) and then compared with the expected benefits (e.g. value of land saved, housing protected, savings in relocating people, etc.).
Costs and benefits are of two types:

- Tangible – where costs or benefits are known and can be given a monetary value (e.g. building costs)
- Intangible – where costs may be difficult to assess but are important (e.g. the visual impact of a revetment).

A project where costs exceed benefits is unlikely to be given permission to go ahead.

The new recurved sea wall at Withernsea, with its protecting rip-rap ▼

Soft engineering techniques at Holderness

Beach nourishment at Hornsea

Beach nourishment has added sediment to the beach at Hornsea, by dredging and pumping from offshore straight onto the beach. In some parts of the world, sand is imported from more distant locations. The purpose of beach nourishment is:

- to create a wider beach to protect the cliff line
- to add sediment into the coastal system so that areas downdrift benefit.

Its disadvantage is that much of the sand can be removed in a single storm and, therefore, the sediment has to be renewed every year or two.

Coastal zoning

This process is also known as red-lining, and is a device used by planners to divide stretches of coast into land use zones. Red-lining identifies those zones at risk of erosion where the costs of protection exceed the possible benefits. Decisions are then made not to protect the area and planning permission is refused for anything inside these zones. It is a way of reducing costs and targeting coastal management spending into areas where there is economic benefit, e.g. from tourism.

Managed retreat along the Holderness coast

Managed retreat works hand-in-hand with coastal zoning. Decisions are made by councils that the coast should be allowed to erode in some areas. Those zones at greatest risk of erosion are refused planning permission for new permanent buildings or other activities, e.g. caravan sites, and a policy is adopted whereby existing residents have to 'roll back' away from the cliff. Existing local residents at risk can apply for planning permission to be allowed to move their homes or caravans inland. East Yorkshire Council now encourages people to relocate to one of the existing towns, especially Bridlington, Hornsea and Withernsea. Residents agree to demolish their existing home and build a new one in the new location that is no larger than the original. In return, they receive help with the costs. Farm buildings and farmhouses can still be built in rural areas so long as the owners still use the buildings for farming and do not sell them for other purposes.

▲ Beach nourishment in East Anglia

How managed retreat works; caravans like this can be moved inland as erosion undermines the caravan site ▼

Managing erosion on the Holderness coast – 3

Integrated coastal management

There is a move now to adopt **integrated coastal management** strategies, which means that sections of coast are managed as a whole, rather than by individual towns or villages. Coastal engineers now realise that acting in just one place affects other places along the coastline. This is because sediment moves along the coast in **sediment cells**. In each cell, sediment moves around between the beach, cliffs and the sea. The cell operates between physical barriers, which prevent the sediment moving further, e.g. a headland or river estuary. The coast of the UK is divided into a number of sediment cells, and these form the basis of management plans for the coast.

The Holderness coast is part of a cell which extends from Flamborough Head to the Wash. Anything affecting sediment **input** (e.g. preventing cliff erosion) or transfer (e.g. groynes) affects the rest of the coast in that cell.

Cells can be broken into smaller sub-cells which work almost independently. When thinking about coastal management, engineers now try to consider a whole cell or sub-cell and devise plans that apply to the whole stretch of coast. These schemes are called **shoreline management plans** (SMPs). In a SMP, all local interest groups are consulted and provide engineers with background information about that stretch of coast. Four options are considered:

- Do nothing, e.g. let existing defences collapse.
- Hold the line, i.e. keep the coastline where it is by using hard engineering.
- Advance the line, i.e. build coastal defences further out to sea, e.g. artificial breakwaters.
- Retreat the line, i.e. allow the coast to erode back to a defined line.

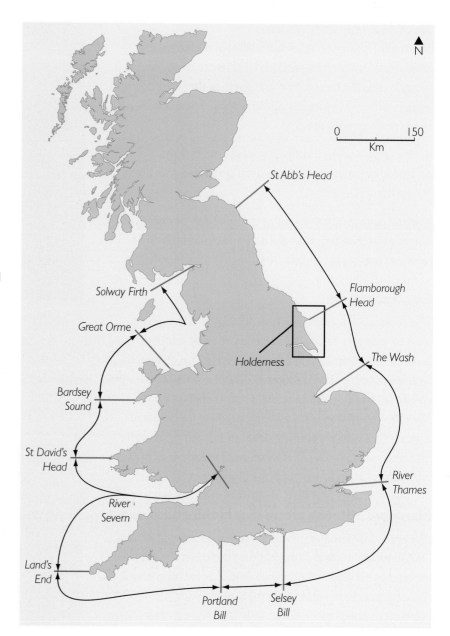

▲ Sediment cells in the UK

Hard engineering is just throwing good money after bad.

What do you think?

A cost-benefit analysis and an environmental impact assessment (EIA) are carried out to decide which is the best option. An EIA decides whether environmental quality will improve or worsen as a result of each option. Planners then select one of these options, and all schemes from then on have to fit into the overall plan. The SMP can be valid for up to 50 years.

The SMP for the Holderness coast comprises a mixture of:

- hold the line in places of economic value, e.g. the gas pipeline terminal at Easington, or towns such as Bridlington and Hornsea
- do nothing, in areas where it is not worth protecting the coast from erosion.

Eventually, unprotected areas erode inland and protected areas form small headlands. This will alter the shape of the coastline into a series of bays. Erosion might then reduce inside the bays because they are sheltered, so eventually erosion should stabilise. These areas would then be referred to as 'stable bays'.

● Over to you

1 Complete the table at the bottom to show the advantages and disadvantages of each of the methods of hard and soft engineering.
2 In pairs, followed by class feedback, discuss the advantages and disadvantages of managed retreat as a policy. How might different people be affected in different ways?
3 In class, discuss whether you think the combined policies of 'hold the line' and 'do nothing' are right for Holderness.
4 Who do you think should be responsible for coastal management in Holderness – local district councils for each town/area, the East Yorkshire County Council, the UK Government, or the EU? Draw up a table of the pros and cons of each, and explain which you think would be best.

● On your own

5 Define the following: hard and soft engineering, terminal groyne syndrome, cost-benefit analysis (including tangible and intangible), environmental impact assessment, integrated coastal management, sediment cells.
6 a Research what the following methods of coastal management involve, and explain their advantages and disadvantages: cliff drains, cliff regrading, vegetation planting, offshore reefs.
   b Would you recommend that any of these methods should be tried at Holderness? Explain why.
7 Draw a table to show the impacts of managed retreat socially, economically, environmentally and politically.
8 Is integrated coastal management better or worse than managed retreat? Explain your answer.

Exam question: Referring to a coastline you have studied, assess the effectiveness of management strategies used to protect this coast from erosion. (10 marks)

Method	Advantages	Disadvantages
Wooden groynes		
Rock groynes		
Sea wall – straight		
Sea wall – recurved		
Revetments		
Beach nourishment		
Coastal zoning		

In this unit you'll learn about how and why the Thames estuary is at risk from rising sea levels and flooding.

Disaster scenario

Imagine a one-metre storm surge racing along the Thames and into London. Imagine the impacts that it would have, especially on London's:

- banking and finance industry – which creates up to 40% of the UK's GNP
- insurance industry – with huge claims for damaged property
- transport – if both surface and underground rail networks flooded.

A storm surge like this would have major impacts on the economy, not just of London but of the UK as a whole. So, erosion is not the only threat to coastal areas – flooding is a major issue too. No wonder London's Thames Flood Barrier is important, because it helps to protect the capital from storm surges like the one described above.

Key
◻ Area at risk of flooding
--- 10 metre contour line

▲ *Flood risk map of London*

Coastal flooding in the UK

Scientists are predicting that sea level is likely to rise by one metre by 2100 – and coastal flooding is almost certain to increase as a result. Unless we begin planning for this now, the impact on London could be like that shown below. It is not just London and the UK at risk, many other parts of the world are low-lying (e.g. most of the Netherlands and Bangladesh), and they face sea level rises of the same amount. During high tides and storms, tidal rivers would flood even further inland.

Coastal flooding is a serious threat to the UK. Many British cities are situated on or near the coast, on low-lying land, together with 40% of the UK's manufacturing industry.

London's Thames Flood Barrier at work ▶

In East Yorkshire:

- Holderness is expected to experience a 30 cm rise in sea level by 2050. This would bring waves closer inshore, with less friction from the seabed to absorb their energy, increasing the already serious cliff erosion.
- increasing wave energy would increase the volume of material removed by longshore drift – affecting places such as Mappleton.
- flooding is expected to increase around Spurn Head

and the Humber Estuary, inundating salt marshes and threatening the calm, sheltered Humber River – used for shipping.

In Dorset:

- the Studland peninsula is already losing 2-3 metres of land per year due to rising sea level, with increased beach and dune erosion.
- Poole Harbour is vulnerable to sea level rises – the main railway line, ferry port and industrial areas could flood.

Background

Why does sea level rise?

Sea level varies over time. It is measured relative to land, so the relative sea level can change either if the land falls or the sea rises. There are two types of sea level change:

- Where sea level itself changes, known as **eustatic change**.
- Where land rises, known as **isostatic change**.

Eustatic change is global. During the last Ice Age, ice sheets stored water normally held in oceans, so sea level fell. At the end of the Ice Age, the ice sheets melted and retreated, water flowed once again into rivers and the sea, and sea level rose. River valleys flooded to form **rias**, and steep-sided glacial valleys filled with water to form deep troughs, known as **fjords**.

Isostatic change occurs locally. It is caused by a change in the level of the land, e.g. the 2004 Indonesian earthquake raised parts of Southern Asia by up to 6 metres. However, isostatic change also occurs as ice melts. In the UK, two different types of isostatic change have occurred since the last Ice Age.

- During an Ice Age, the weight of ice on the Earth's crust causes it to sink. Once the ice retreats, the reduced weight on the Earth's crust causes it to readjust and rise. In these circumstances, sea level appears to fall, as land rises. This has affected Scotland in the last 10 000 years.
- In South-East England, rivers pour water and sediment into the Thames estuary. The weight of the sediment causes the crust to sink, so sea level there appears to rise.

Therefore, South-East England faces inundation from isostatic change, as well as from sea level change as a result of global warming.

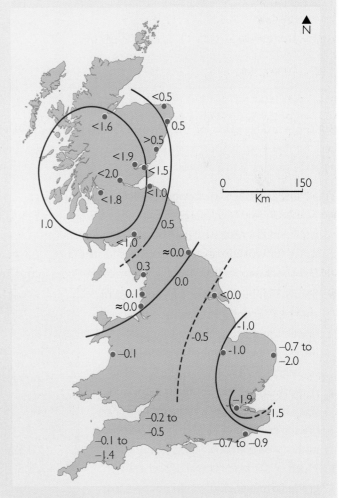

▲ Isostatic change in the UK. On the diagram, the lines show how much parts of the UK are either rising (plus figures) or falling (minus numbers) in metres. Scotland is rising most while South-East England is sinking.

Thames estuary at risk – 2

The Thames Gateway

One area which is expected to be affected by sea level rise, is the land around the Thames estuary, known as the Thames Gateway. Rising sea levels and storm surges threaten the area. While the Thames Flood Barrier protects London, the Thames Gateway is downstream from the Barrier and has no protection.

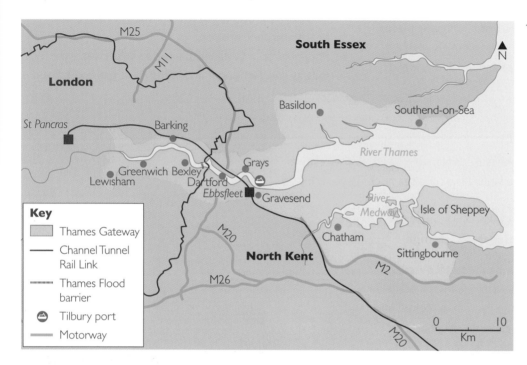

◀ The Thames Gateway area

But there are big plans for the Thames Gateway areas. It will provide 160 000 new homes and, further east at Southend-on-Sea, there will be a new university campus, retail and nightlife venues, and a new airport.

There are a number of reasons for these developments:

● The economy of South-East England is continuing to expand rapidly, both in London itself and the area within an 80-km radius. As a result, there is a demand for more space for workplaces, such as offices, distribution depots, manufacturing sites, transport depots, retail and government services in this area.

● South-East England has a severe housing shortage. Economic expansion has driven the demand for housing well beyond supply – especially affordable housing.

● Almost uniquely in South-East England, the Thames Gateway area had over 10% unemployment in 2007, so job creation in this area is vital.

● Much of the land is regarded as low-grade farmland, and would not be a loss to the farming industry, although the salt marshes around the estuary are of considerable environmental value.

The Thames Gateway area – one of the locations for expansion ▼

What about the flood risk?

The Thames Gateway presents planners with a problem. The region needs houses desperately, and the estuary is probably the only area where large development could take place, yet there are huge flood risks! As part of the planning process for the Thames Gateway, the Environment Committee for the London Assembly researched the state of flood defences along the estuary. It found that:

- defences near the coast were patchy and in a poor state
- even without population expansion, 1.25 million people are already at risk of flooding.

The London Climate Change Partnership estimates that sea levels will rise in the estuary by between 26 and 86 cm by 2080. The Association of British Insurers is pessimistic too. It believes that, by 2100, sea level rise could increase flood frequencies by between 8 and 12 times.

The Thames Gateway itself would make the situation worse:

- A proposed airport is planned for marshland in the estuary. The estuary salt marsh and water meadows close by provide some flood defence in their natural state, because they act like a sponge, holding and storing water.
- Urban expansion would create more impermeable surfaces, and lead to increased surface run-off and a greater risk of flooding.

How to cope? The UK Environment Agency is currently planning new flood defences to be built along the Thames estuary between 2007 and 2016. To deal with the increased risk:

- floodwalls are planned for the riverbank to protect built-up areas.
- there are plans to set aside 'reserves' of land, where flooding will be permitted and in which floodwater would be held naturally like a sponge.

Decisions are also needed now about longer-term flood protection and how the estuary can best be protected – whether the planned Thames Gateway developments go ahead or not.

> *Should housing be built in areas where there is a severe flood risk?*
>
> **What do you think?**

● Over to you

1 In pairs, produce a spider diagram of the impacts of a major London flood on **a** London, **b** the UK as a whole.
2 Select either Holderness or the Dorset/Hampshire coast, or an area of coast known to you. Using atlases and OS maps, assess the potential risk from coastal flooding on **a** settlements, **b** industries, **c** the environment.
3 In pairs, draw up a list of social and environmental benefits of developing the Thames Gateway. Then list any potential problems or costs. Which is greater – costs or benefits?

● On your own

4 Define the following: eustatic and isostatic change.
5 Research development plans (e.g. housing, jobs) for the Thames Gateway, using, for example, the London Development Agency website (www.lda.gov.uk). Include:
   a planned flood defences
   b whether these defences will be sufficient.
   Use this to prepare a presentation either in favour of or against the developments.

Exam question: Explain how coastal flooding presents environmental, social and economic risks for the UK. (10 marks)

In this unit you'll learn about pressure on salt marshes and how they can be managed sustainably.

Why are salt marshes under pressure?

Coastal estuaries are in demand. Sheltered from the open sea, they make ideal ports, particularly for industries such as oil refining which need deepwater access. As ports and industries develop, cities around them expand, putting pressure on land and estuary salt marshes. Many British salt marshes have been reduced in size, e.g. along the Thames estuary, while others have been lost to projects such as the tidal barrage in Cardiff Bay. Now, the developments in the Thames Gateway pose further threats.

Pressure on the Thames estuary

Along the Thames estuary, two-thirds of the salt marsh has been lost since 1945. Apart from building and industry, the main threats are grazing, landfill, fertilisers, drainage and tourism.

- *Summer grazing* by livestock leads to selective removal of grass species such as red fescue.
- The proximity of London and its demands for waste disposal has led to the loss of marshland for **landfill**.
- The use of **chemical fertilisers** on surrounding farmland alters salt marsh plant communities as surface run-off changes soil acidity.
- Marsh soils are rich in nutrients and tempt farmers to **drain and plough them for crops**. The change to arable farming has caused a major loss of natural marsh pasture. As the land dries out, it shrinks and needs protection by sea walls to prevent flooding.
- *Recreational uses* of the estuary (e.g. boating, jet skiing, yachting) demand space, reduce water quality and disturb wildlife.

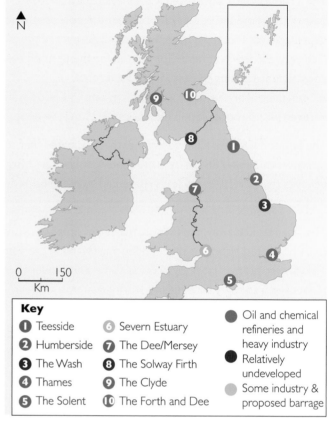

Key
1. Teesside
2. Humberside
3. The Wash
4. Thames
5. The Solent
6. Severn Estuary
7. The Dee/Mersey
8. The Solway Firth
9. The Clyde
10. The Forth and Dee

● Oil and chemical refineries and heavy industry
● Relatively undeveloped
● Some industry & proposed barrage

▲ *The UK's major estuaries and their uses*

▼ *The Thames estuary and its environmental pressure points*

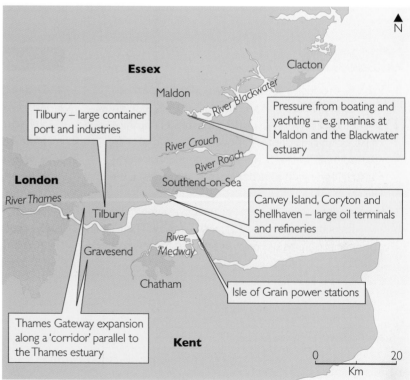

Essex
Clacton
Maldon
River Blackwater

Tilbury – large container port and industries

River Crouch
River Roach
Southend-on-Sea

Pressure from boating and yachting – e.g. marinas at Maldon and the Blackwater estuary

London
River Thames
Tilbury

Gravesend
River Medway
Chatham

Canvey Island, Coryton and Shellhaven – large oil terminals and refineries

Isle of Grain power stations

Thames Gateway expansion along a 'corridor' parallel to the Thames estuary

Kent

0   20
Km

Protecting salt marsh – Abbot's Hall Farm

Although there are plans to develop much of the Thames Gateway, there are also efforts being made to protect its salt marshes and, in one case, extend them. At Abbot's Hall Farm in Essex, part of the Blackwater estuary, the sea has been allowed to breach the seawall to convert 84 hectares of farmland back into salt marsh. The new marsh acts as a natural defence against flooding for the land behind it. If, as expected, sea levels rise in the future, the salt marsh will migrate inland naturally. The plan brings benefits for marsh birds (like Brent geese and wigeon), and salt marsh plants (such as samphire and sea lavender), and is supported by several groups, e.g. English Nature and The Environment Agency.

Different interest groups along the Essex coast
- Local councils, e.g. Essex County Council, local District Councils and parish councils
- Environmental charities, e.g. National Trust, Essex Wildlife Trust, English Nature, RSPB, Friends of the Earth
- National Government, e.g. Countryside Agency, Environment Agency

Should we give up valuable farmland to re-establish salt marshes?

What do you think?

▼ *The plan for Abbots Hall Farm*

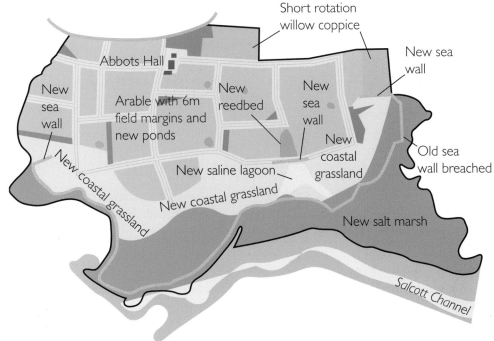

Short rotation willow coppice

New sea wall

Abbots Hall

New sea wall

New reedbed

New sea wall

Arable with 6m field margins and new ponds

New coastal grassland

Old sea wall breached

New saline lagoon

New coastal grassland

New coastal grassland

New salt marsh

Salcott Channel

● Over to you

1 In pairs, complete the table at the bottom to explain the issues that the Thames Estuary faces by 2050.
2 a Explain why each of the interest groups along the Essex coast would be interested in how the Thames estuary is managed.
   b How might some groups conflict? Give examples.

● On your own

3 In 400 words, assess the benefits and problems caused by allowing the re-establishment of salt marsh at Abbott's Hall Farm.

Issue	Why it presents problems for salt marsh and estuary environments
Climate change / global warming	
The urban development of the Thames Gateway	
The need for larger ports for oil tankers, e.g. on Canvey Island	
Water and sewage treatment for the Thames Gateway	
The further growth and expansion of London	

Unit summary

What key words do I have to know?

There is no set list of words in the specification that you must know. However, examiners will use some or all of the following words in the examinations, and would expect you to know them, and use them in your answers. These words and phrases are explained either in the glossary on pages 297–301, or in key word boxes throughout this chapter.

abrasion
anoxic
boulder clay
brownfield sites
coastal squeeze
coastalisation
container port
corrasion
corrosion
cost-benefit analysis
  (including tangible and
  intangible)
downsizing
estuary
eustatic and isostatic
  change

eutrophication
fetch
footloose
freeze-thaw
geology
green belt
hard and soft engineering
hydraulic action
integrated coastal
  management
inward migration
longshore drift (including
  swash and backwash)
natural increase

plant succession
relief
ria
salt marsh
seachange
sediment cells
Shoreline Management
  Plans
slumping
swell
terminal groyne
  syndrome
topography
weathering

Try these questions

1   Using **named** examples, examine the **environmental** costs of coastal development. (10 marks)

2   Describe and explain **a programme of fieldwork and research** that you would use to investigate the impacts of **either** coastal erosion **or** coastal flooding, along a stretch of coastline. (15 marks)

3   Hard or soft engineering? Using a stretch of coast you have studied, justify which you think is best. (10 marks)

5 **Unequal spaces**

What do I have to know?

This topic is about unequal spaces. It looks at how inequalities exist within countries, and within cities and rural areas.

1 Recognising inequality: What are unequal spaces and what causes them?

What you need to learn

- Define inequality and what it means in different areas
- How inequalities arise, e.g. of environmental quality, wealth and poverty, and quality of life
- **Fieldwork and research**: research the pattern of spatial inequality in one rural and one urban area using primary surveys and secondary data

Although this is a general question, and does not need specific knowledge, it is blended in with the case studies in this book to help you understand it better.

2 Inequality for whom? What impact do unequal spaces have on people?

What you need to learn

- Inequality and social and economic exclusion; denying opportunities and access to services in urban and rural areas
- How inequality creates marginalised groups in rural and urban areas
- Understand effects of inequality by gender, age, ethnicity, health employment, education, income and disability
- **Fieldwork and research** into:
  - criteria to identify the pattern of 'haves' and 'have-nots' in rural and urban areas
  - know how to evaluate schemes to tackle inequality

Again, detailed case study knowledge is not needed – but the concepts have been woven into the main case studies to help you understand them.

3 Managing rural inequalities: How can we manage rural inequality and improve the lives of the rural poor? How successful have particular schemes been?

What you need to learn

- Know the social, economic and environmental problems creating rural inequality that need to be overcome
- **Fieldwork and research**: e.g. the success of ways to reduce rural inequalities, e.g.
  - appropriate technology
  - community involvement and empowerment
  - improving access to transport and services
  - local employment
  - sustainable solutions

You need to know detailed examples about schemes which help to reduce rural inequalities. The schemes chosen are not important, but the criteria that make them successful or not must be clear.

4 Managing urban inequalities: What strategies can be used to combat inequality in urban areas? How successful have particular schemes been?

What you need to learn

- Social, economic and environmental issues associated with urban inequalities
- Key players are involved in delivering solutions
- **Fieldwork and research** into the success of specific ways to reduce urban inequalities, e.g.
  - self-help schemes
  - traffic and public transport
  - town planning initiatives
  - business initiatives
  - crime and policing

Like rural examples, you need to know detail about schemes which help to reduce urban inequalities. The schemes chosen are not important, but the criteria that make them successful or not must be clear.

In this unit you'll learn about different types of inequality across the UK.

The North-South divide

The UK is a divided nation. The economy has changed from one based on primary and secondary activities, to one where over 70% of our jobs are in the tertiary sector. And London dominates.

> The North-South divide has widened so much in the past 10 years that the UK is now split in two, with the entire southern half – from Lincolnshire to Cornwall – dominated by the ever-expanding London metropolis. While London attracts the young and well-educated, who enjoy wealth and employment, northern communities are slowly sinking demographically, socially and economically.
>
> 'Increasingly, those who can, move south,' says a report from Sheffield University. 'The South is London and London is the South'. But, while the young and the educated flood into the metropolis, it is also disgorging people who are older, sick or less skilled, who find there is no room for them.
>
> Adapted from *The Daily Telegraph*, 2004

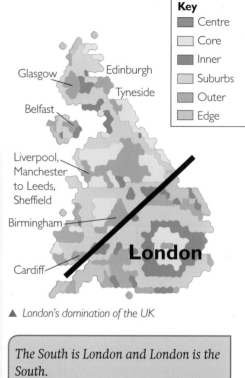

Key
- Centre
- Core
- Inner
- Suburbs
- Outer
- Edge

Glasgow, Edinburgh, Tyneside, Belfast, Liverpool, Manchester to Leeds, Sheffield, Birmingham, Cardiff, London

▲ *London's domination of the UK*

> *The South is London and London is the South.*
>
> **What do you think?**

Spatial inequality

We all have an image of what makes a place urban or rural, prosperous or poor, but can we always measure what makes places different and say that one place is better or worse than another? Spatial inequalities exist where the distribution of wealth, resources and opportunities are not evenly spread. Rural and urban environments offer different opportunities and obstacles to different people.

- **Environmental magnetism** is where geographical characteristics of local surroundings attract or repel.
- **Positive externalities**, such as open spaces, clean air and tranquility, attract people.
- **Negative externalities**, such as high-density buildings, pollution and noise, repel people.

For several decades, hundreds of thousands of people in the UK have been swapping their urban homes for a place in the country. Movements of people affect property and land values. As a place becomes more appealing or accessible, it is likely that a spatial inequality will emerge. While some people win, others lose.

Indicators of wealth and prosperity, e.g. average household income, can show spatial patterns and regional inequalities. Often the best-paid jobs are in London and the South East. In 2005, Surrey was the UK's highest-earning county (average household income was £33 400), while Cornwall was the lowest (average household income was £17 700). Quality-of-life indicators, such as general well-being and happiness, produce different spatial patterns – with rural and remote counties perceived as being better places to live. Quality of life may be more important than economic wealth, but it is difficult to measure.

Environmental inequality

The map on the right was produced by the Campaign to Protect Rural England. New developments, noise and physical surroundings enhance or spoil people's experience of tranquility (calm, quietness). Many people move into 'tranquil areas' because of the positive impact that they think the move will have on their health and quality of life. Not surprisingly, as more people move, house prices rise and communities change. The map shows that remote areas like Cumbria, Yorkshire, Northumberland, Dartmoor and Exmoor remain truly peaceful, but people living there have some of the lowest average incomes in the UK. Increasing numbers of urban migrants and owners of second homes create local inequalities in such areas.

Green belts around urban areas and National Parks preserve a degree of tranquility, but again they encourage higher property values. People on lower incomes who depend on rural employment may be squeezed out of the local housing market as those able to buy their piece of the countryside move in. A process of social and economic filtering occurs, which reinforces spatial divisions.

Socio-economic inequality

The decline of the UK's traditional primary and secondary industries, along with the reduction in the numbers employed in farming, means that some regions now depend on the government for more than 20% of their jobs. These include jobs in education, healthcare, and the Civil Service.

> The 2001 Census showed 'widening inequalities in all areas of life in Britain. Areas with the highest levels of poor health also tend to have the lowest numbers of doctors, dentists and other health professionals'. In remote rural areas with second and holiday homes, 'more local people are still renting their homes at ages when they would traditionally be expected to have entered the housing market.'
>
> From the Joseph Rowntree Foundation

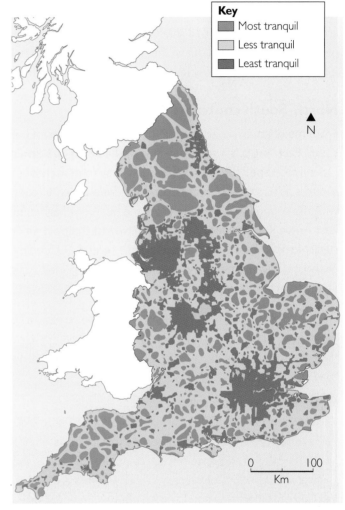

Key
- Most tranquil
- Less tranquil
- Least tranquil

N

0   100
Km

▲ *Tranquility levels in England, 2007*

● Over to you

1 **a** What are the obstacles and opportunities faced by people living in urban and rural areas? Set them out in a table.
 b How might these differences lead to inequalities?
2 Why might green belts and National Parks distort the values of property and land, and then create spatial inequalities?
3 Draw a spider diagram and complete it with examples of positive and negative externalities in your local area, town or city.

5.2 East Anglia: emerging inequalities

East Anglia

In this unit you'll find out about the changing characteristics of East Anglia and how places are becoming more unequal.

Rich estates and cheap streets

East Anglia boasts some of the UK's richest landowners and business people (e.g. Bernard Matthews), and also has six of the UK's cheapest streets. While the former are scattered across the region, the six streets can be found together in the community of Jaywick. East Anglia's six counties are in crisis, with a shortage of affordable rural housing (young people are being forced out of the market); a decline in services; limited job opportunities; and low levels of child-care and health provision.

North-South contrasts

From the northern coastline of Norfolk to the fringes of London in the south, East Anglia is a good example of the **rural-urban continuum**. Open landscapes and small villages give way to densely populated suburbs as you move towards London – where city functions merge with the countryside. For fifty years the number of agricultural jobs in East Anglia has declined and villages have lost their old functions. Central rural Norfolk, East Suffolk and The Fens receive EU funding to help with economic development, and several areas have gained Assisted Area Status to promote small businesses.

▼ *A street in Jaywick*

● A **rural-urban continuum** is where there is no clear division between town and country.

▼ *Distinctive patterns across East Anglia*

Northern counties
● Northern Cambridgeshire, Norfolk and Suffolk are sparsely populated and relatively remote.
● 8.3% of the working population in North Norfolk, and 3.9% in Suffolk, are employed in agricultural activities.
● Between 1991 and 2005, farm incomes declined by 75%.
● Farmers often have other part-time jobs to boost their income.
● Older people dominate the age structure (5% above the national average), and 49% of primary sector workers are over 45.
● There are declining employment opportunities for young people.
● Traditional rural industries are disappearing and, with them, opportunities for people with poor literacy and/or numeracy skills.
● Young people are leaving farming because of low pay and long working hours.
● Norfolk and Suffolk have the greatest concentration of low earners. In 2005, 39% of full-time employees earned less than £250 per week.

Southern counties
● Essex, Hertfordshire and Bedfordshire are more densely populated than the three northern counties.
● They have strong links with London.
● They have a high number of commuters.
● They have greater affluence than the northern counties (15% earn less than £250 per week).
  ● There is pressure on land for development.
  ● There is a loss of rural communities.
  ● There is severe house price inflation.

Common features
● The perceived good quality of life and the growing importance of tourism.

Norfolk

Cambridgeshire

Suffolk

Bedfordshire

Hertfordshire Essex

Greater London

0 ___ 50
Km

N

The data in the following tables is from the 2001 Census.

	Essex	Herts.	Beds.	Cambs.	Suffolk	Norfolk	UK
Population growth (% p.a.)	0.5	0.5	0.6	1.1	0.4	0.5	0.3
Age 0-15 (%)	20	20	22	21	20	19	20
Age 65+ (%)	16	16	13	14	18	19	16
Primary employment (%)	5	5	5	7	8	8	6
Secondary employment (%)	24	24	31	25	26	26	25
Tertiary employment (%)	71	71	64	68	66	66	69

◀ East Anglia – population and employment

	Essex	Herts.	Beds.	Cambs.	Suffolk	Norfolk	UK
Unemployed male/female (%)	7/3	6/3	7/3	6/2	5/2	6/3	8/3
Average weekly incomes (£)	539.9	552.60	548.0	560.5	455.2	446.2	522.6
% earning less than £250 per week (male)	8.2	7.6	8.1	6.9	12.4	10.4	11.3
% earning less than £250 per week (female)	24.1	21.3	23.8	17.4	27.2	32.3	24.4

◀ East Anglia – income

Percentage of rural parishes lacking …	Essex	Herts.	Beds.	Cambs.	Suffolk	Norfolk	UK
a GP	76	78	87	81	88	86	83
daycare for the elderly	89	83	89	85	95	92	91
a pharmacy	70	74	83	77	83	82	79
a general store	64	62	58	67	79	72	70
a post office	35	40	36	33	48	40	43
a daily bus service	62	77	65	72	80	84	75
a public house	12	9	14	21	33	38	29
a village hall	13	14	12	22	23	26	28
a primary school	39	26	30	50	64	54	50

◀ East Anglia – service provision. In addition to what the table shows, over 90% of rural parishes (95% in Suffolk) do not have a bank, dentist or public nursery.

Changing communities

The decline of traditional industries and the invasion of an urban population means that few areas in rural East Anglia now depend on the primary sector. Since 1981, the population has increased from 4.9 to 5.5 million, with 12% of the increase consisting of migrants from Greater London. Where barns, maltings, breweries and grain warehouses once clung to river estuaries and town harbours, exclusive apartments and second homes now stand. The loss of workplaces and jobs means that local spending is reduced, and shops and services close. Converting old buildings into new homes may have a positive effect, but a lot depends on who lives there.

East Anglia has experienced the loss or closure of many community amenities in the last twenty years, such as village primary schools, post offices and pubs. Increasing numbers of elderly and young people have become disadvantaged by government policies of centralising rural health services and by the loss of public transport (caused by deregulation). These policies have added to the feeling of exclusion for some sectors of society.

The suburbanisation of East Anglia

In the mid 1980s, the mainline railway between London and Norwich was electrified. Towns north of Colchester were now within easy commuting distance of the capital. The increase in commuting and car-ownership, and the appeal of living in the country, meant that **counter-urbanisation** increased throughout East Anglia. Settlements along the rail and road corridors in East Anglia have experienced significant population growth as a **ripple effect** has developed. The countryside has become increasingly urbanised.

> • **Counter-urbanisation** is the movement of people out of cities to smaller towns and villages.

Emerging inequalities

Commuters contributed £11 billion to the East Anglian economy in 2006, up 2.3% on 2005. The increase was greatest in Suffolk. Added to this, house prices in Suffolk have risen markedly since 2003 (values rose between 19% and 26% a year in Bury St Edmunds and Sudbury). A web-like pattern of inequalities emerges, where the 'urban haves' (the commuters) live near the transport routes, leaving the 'rural have-nots' isolated.

Newcomers to the countryside tend to be middle-aged people, whose children have either left home or are over primary school age. They tend to use cars to travel to work and shop, and may only spend weekends in the village – and so contribute little to the community. As a result, a dormitory settlement develops.

▲ A web of inequalities – towns like Colchester and Ipswich have become detached suburbs of London. In turn, villages around these towns have become their suburbs, creating an urban and rural web-like pattern.

The ripple effect

As people move out of cities, they have a ripple effect on house prices – each wave of people forces house prices up, so the next wave has to move further out. House prices increase so much that local young people cannot afford them and are excluded from their own community. Once wealthier newcomers have bought their home in the countryside, they do not want their view destroyed by the building of affordable new homes for local young people. **Affordable homes** are often small and built to a high density. By trying to block plans for new housing, the fabric of the community is threatened.

▲ Local services are what people need.

▲ But when newcomers move in, secondhand bookshops and art galleries open, and local services for the **residual population** decline.

Rural young lose out as rich move in

The trend of rich people buying up the available housing in the countryside and forcing poorer young families to move to cities is increasing dramatically. The sale of rural council houses, together with a lack of housing association homes, means that there is no accommodation for people setting up home for the first time. Four out of ten new households in rural areas are unable to buy a home.

Mark Shucksmith, of Aberdeen University, told a geographers' conference in Plymouth: 'Rural areas are now ruled by market forces, so the rich will live in the countryside and the poor in the urban areas. The social glue that holds rural communities together is falling apart. Rural communities had an age balance of all generations – now these social networks are increasingly being fractured. The old will have no-one to look after them, and the young will be without the support of their parents.'

Adapted from *The Guardian*, 2001

▲ *Properties in rural areas close to good transport links attract people moving out of cities and create property hotspots, like this one in Dedham on the Essex/Suffolk border.*

● Over to you

1 On an outline sketch map of East Anglia, showing the six counties, draw the network of major roads and railways. Annotate the map to show:
   a which areas have been under most pressure from commuters
   b those areas considered to be remote.

2 Study the tables on page 215. In pairs, draw up evidence to prove and explain why the following statements are correct:
   • East Anglia possesses its own north-south divide.
   • The more remote rural areas experience lower levels of income.

3 Use the tables on page 215 to decide, with reasons:
   a the most deprived county in terms of services
   b the least deprived county in terms of services
   c the links between income and number of services

4 Who should be responsible for improving life in rural communities and why?
   a Local communities?
   b Local councils?
   c Regional government?
   d National government?
   e The EU?

● On your own

5 Define the following terms from the text: rural-urban continuum; counter-urbanisation; ripple effect; residual population.

6 Briefly outline the recent changes in East Anglia. Use the terms in **5**.

7 Why do you think so many people choose to leave the towns and cities to live in the countryside?

8 Who do you think suffers most when services and amenities decline in rural East Anglia?

9 Find out about a village that has experienced counter-urbanisation close to a major line of communication near your home or school. Write a 500-word illustrated case study.

Exam question: Referring to examples, explain how the influence of urban areas can lead to rural deprivation. (10 marks)

> *The rural young are losing out as the rich move in.*
>
> **What do you think?**

In this unit you'll learn about the ways in which two areas are attempting to manage spatial inequalities in East Anglia, and about the problems faced in a third area.

Rural housing crisis

'*Shocking State of Rural East*' ran the front-page headline of the *East Anglian Daily Times* when it reported on the findings of the *State of the Countryside Report, 2004*. Rural communities in East Anglia were in crisis, with low wages, limited skill levels, a declining birth rate, a shrinking agricultural economy, and a shortage of housing. (Limited access to affordable housing forces young adults to leave an area, has serious impacts on local employment prospects, and undermines service provision.)

The report concluded that the rural housing crisis was due to: in-migration of people from London, and retirees; second-home ownership; a shortage of social and rented housing; and increased life expectancy.

Communities at risk

Changes in the local economy, such as a decline in local jobs, can have a negative effect on a place. Conversely, an increase in investment can have a positive effect, as the multiplier diagram shows. This unit looks at spatial inequalities in three areas: an Area of Outstanding Natural Beauty (AONB – North Norfolk); a declining traditional holiday resort (Jaywick in Tendring); and an area with increasing numbers of second homes (Southwold and Walberswick in Waveney).

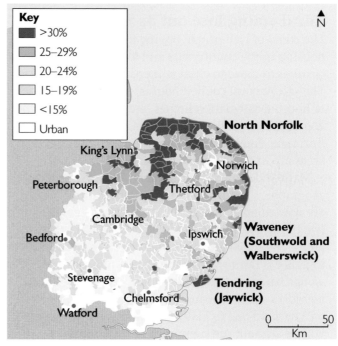

Key
- ■ >30%
- ▨ 25–29%
- ▨ 20–24%
- ▨ 15–19%
- □ <15%
- □ Urban

North Norfolk · King's Lynn · Norwich · Peterborough · Thetford · Cambridge · Bedford · Ipswich · Waveney (Southwold and Walberswick) · Stevenage · Chelmsford · Tendring (Jaywick) · Watford

0 — 50 Km

▲ *The proportion of households with an income below £10 500 per year*

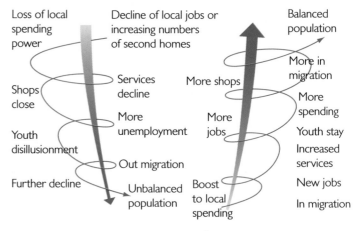

Negative multiplier
Loss of investment

Loss of local spending power · Decline of local jobs or increasing numbers of second homes · Services decline · Shops close · More unemployment · Youth disillusionment · Out migration · Further decline · Unbalanced population

Balanced population · More in migration · More shops · More spending · More jobs · Youth stay · Increased services · New jobs · Boost to local spending · In migration

Investment
Positive multiplier

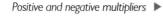

Positive and negative multipliers ▶

	Tendring Jaywick	Waveney Southwold	Waveney Walberswick	North Norfolk Brancaster	East Anglia	England
Total population, 2001	4665	4025	1995	1484	5.3m	49m
% aged 0-4	3.86	3.3	3.66	2.43	5.97	5.96
% aged 5-15	9.11	10.83	12.13	7.28	14.13	14.2
% aged 16-19	2.62	3.95	4.06	3.1	4.66	4.9
% aged 20-44	19.19	19.45	21.85	20.28	34.25	35.31
% aged 45-64	31.25	24.27	30.98	32.01	24.54	23.75
% aged over 65	33.98	38.19	27.32	34.91	16.45	15.89
Ethnic group – % white	98.35	98.31	99.35	99.73	95.12	90.92

◀ *Population characteristics*

		Southwold	Walberswick	Jaywick	Brancaster
Weekly household income	April 2001–March 2002	£460	£580	£380	£500
Disability living allowance claimants	August 2006	115	60	565	35
Income support claimants	August 2006	55	20	355	20
Jobseekers allowance claimants	August 2006	25	15	185	5
Pension credit claimants	August 2006	195	85	635	85
% aged 16-74 with no qualifications		31.24	27.98	59.61	32.81

◄ Indicators of deprivation

North Norfolk – problems in an AONB

The North Norfolk coast AONB is home to 40 000 people in the most rural and remote part of East Anglia. Elderly people make up more than 30% of the population; 16-29 year olds have been leaving (down 22% since 1991) as 45-69 year olds have been increasing (up 24%). Up to 30% of properties in some villages are second homes.

	Binham	Worstead	Brancaster	Great Massingham	North Norfolk
Housing stock					
% terraces and flats	21.9	11.7	19.0	14.9	24.8
% owner-occupied	60.3	71.2	62.5	66.4	71.4
% rented	30.2	24.1	32.5	29.4	25.0
% second homes	28.7	3.4	34.6	6.1	7.2
Average house price, 2003-4	£254 500	£162 986	£351 458	£136 943	£170 747
Population					
% of households with at least one pensioner	55.7	35.2	61.8	41.1	48
% of households with an annual income of under £20 000	36.3	39.0	37.4	36.2	38.2

◄ The scale of the housing problem in North Norfolk

Education, skills and wages are barriers to social inclusion. In 2004, it was reported that 84% of households could not afford the average terraced house in the AONB. A plan to provide affordable homes was established, together with the Office of the **Rural Housing Enabler**, who monitors housing needs and negotiates with developers to ensure that appropriate housing is built.

The Local Plan for North Norfolk established growth towns and villages where 30-40% of all new builds were to be social housing. Over 300 properties were completed between 1998 and 2005. These properties were available at below market values and could be shared-ownership projects, sub-rent schemes or subsidised low-cost homes.

Local government acts as **social landlords**, or registers private trusts to enable low wage earners to enter the property market. But now a new problem has emerged – a **dumb-bell market**. The young economically active population are forced elsewhere. The attempt to provide sustainable mixed communities is still not working.

> ● A **dumb-bell market** is a housing market with high-priced houses at one end of the scale and low-priced social housing at the other end, but few properties in-between.

Jaywick and Southwold

Jaywick – an East Anglian outpost

Jaywick developed in the 1930s as a seaside resort for Londoners, just outside Clacton-on-Sea. The houses were built as temporary holiday homes and were not expected to last long. Seventy years later, many have become the permanent homes of families from East London. Most of the homes are timber built and the settlement lacks key services and facilities. It is the most deprived area in Essex and is among the worst-off 1% of areas in England.

The physical environment in Jaywick shows all the characteristics of deprivation:
- unmade roads and footpaths
- no surface drainage
- poor street lighting
- vacant and semi-derelict buildings
- obsolete and substandard accommodation
- abandoned vehicles
- fly-tipping
- little social opportunity
- low skills levels and high unemployment – 59% have no qualifications
- high levels of crime
- poor transport
- poor health – 20% are classified as 'not good'.

Jaywick has become isolated from Clacton, its nearest neighbour, and the residents feel socially excluded because they have limited access to better opportunities. Within Jaywick, the shops, cafes and clubs have been closing down and new private investment is absent.

The Jaywick Regeneration Masterplan was developed in 1999 to tackle these problems. The views and needs of residents were taken into account and funds were acquired from central and local government and the EU.
So far, the plan has achieved:
- a new school and health centre
- Guinness Trust low-rent housing with priority for the disabled
- new access roads
- a major beach renourishment and coastal protection scheme.

Jaywick may still have six of the region's cheapest streets, but the tightly knit community provides its wealth and optimism for the future. Jaywick is community-rich but property-poor.

▲ *Deprivation in Jaywick*

▲ *Easing the inequalities – regeneration in Jaywick*

Southwold and Walberswick – Hampstead-on-Sea

Southwold and its neighbour Walberswick have earned the title 'Hampstead-on-Sea'. Wealthy people from north-west London have moved into these genteel resorts and turned them into expensive dormitory settlements. Southwold has a permanent population of 1500 (out of a total population of 4025) and, of its 1250 properties, 450 are second homes or holiday lets. The local pub – the Sole Bay – has been transformed into a 'fancy eatery' and convenience stores have become antique shops and boutiques.

▲ *Southwold seafront*

Suffolk Coastal District Council has found that areas with high numbers of second homes experience:

- a decline in traditional businesses, but a demand for more decorators
- the closure of local shops as second home-owners make less use of them
- high property prices – so local residents can not afford to live and work in the area
- a decline in the use of local schools and potential closures
- a fluctuating demand on local waste collection and transport services
- the disappearance of traditional support networks as communities change.

A social divide has developed with resentment between locals and incomers, and the sustainability of the community is threatened. Places like Southwold are the opposite of Jaywick – property-rich but community-poor.

> *Sometimes the image of a place is little more than an illusion – behind the façade are serious social and economic problems. Is this true of Southwold?*
>
> **What do you think?**

There is a smell of money in Southwold, with exquisite Regency homes, manicured greens, delis, greengrocers and exclusive little clothes shops. But there are problems: a lack of jobs; the emigration of young people; and prohibitively high house prices. Its position is perilous.

Local estate agents declare: 'Southwold looks wealthy because of the type of people living here, but there are so many second homes that aren't being lived in … you need people to live there and spend money locally.

The influx of Londoners spending little time here has corroded the heart of the community. Increasingly, general stores are being driven out by specialist boutiques. We have an ageing population and there just aren't the people to support the older people in the community. We have lost the broad mix that there used to be.

It would be nice if the local community stayed as it is, rather than outsiders wanting to turn it into some kind of heritage Disneyland.'

Adapted from an article in *The Guardian* by Esther Addley, 2007

● Over to you

1 Consider the causes of the rural housing crisis and explain why they have occurred in East Anglia.
2 Look at the data for Southwold and Jaywick in the tables on pages 218-19. Use an appropriate technique to present these graphically.
3 Read the newspaper extract above. In pairs, discuss how local service provision might change as the community gentrifies.
4 What are the economic and social gains and losses of becoming a popular second home location? Present your answer as a table.

● On your own

5 Define the following terms from the text: rural housing crisis; second homes; affordable homes; dumb-bell market; rural housing enabler; social landlords.
6 Examine the diagram of positive and negative multipliers and explain how they apply to each of the areas covered in this unit.
7 Government policy says that local authorities should try to develop vibrant, mixed and sustainable rural communities. How successful have North Norfolk, Southwold and Jaywick been?

In this unit you'll explore some of the ways in which communities can change, and see whether they can be sustainable.

Meeting present needs

Sustainability is about meeting the needs of the present generation without compromising the ability of future generations to fulfil their own needs. Are rural communities in East Anglia sustainable? If a village cannot provide homes for the young adult population, employment and services for existing residents, or care for the elderly, it may not be a community at all. The diagram below shows criteria used for assessing how well communities cope with current changes.

Criteria for economic sustainability

- The range of economic benefits (jobs/incomes) extends to the maximum number of people.
- It is likely to make people better off in the medium- to long-term, as well as the short-term.
- It focuses on changes that bring economic benefits through social welfare improvements. For example, education and health care will benefit society as well as individuals.

Criteria for environmental sustainability

- Pollution levels are not increased.
- Resources are not exploited faster than they can be renewed or replaced, without considering the alternatives.
- It contributes to the conservation of special environments by raising awareness of natural systems and processes.
- It conserves and protects rare species and special environments.

Assessing sustainability in East Anglia

Criteria for social/cultural sustainability

- The benefits of change extend to the maximum number of people.
- Local people are able to afford land, homes, and basic commodities in the area.
- Jobs and the social benefits of change are long lasting.
- Cultural traditions have been retained.
- The local community is involved in the planning and development of changes.

Clearly, the needs of the present generation are not being met when:

- centralisation of healthcare makes access difficult without transport
- overall skill and qualification levels remain low
- only 8% of graduates from the region's six universities find their first job in East Anglia
- out-migration of the young creates a 'brain drain'
- access to urban-based training centres is difficult because of poor transport
- falling employment in agriculture and food processing is not being replaced
- village shops, post offices and pubs continue to close
- public transport is limited, making car ownership essential
- 365 000 people commute daily out of the region for work (two thirds of these to London).

UNEQUAL SPACES

Creating sustainable communities

Wealth brought into East Anglia by people who commute to London contrasts strongly with the loss of jobs and poor access to education, health and housing experienced by the region's long-term inhabitants. How can rural initiatives create more sustainable futures?

Providing homes

Central government has set out plans for new housing in East Anglia. The region is to accommodate 505 000 new homes by 2021. Over 70 000 are targeted for Norfolk, 74 000 for the Haven Gateway across Suffolk and Essex coastal areas and 123 000 in the London commuter belt.

Local schemes, like those employed by the North Norfolk Partnership (see page 219) and extended by Suffolk ACRE (a community charity working with local government), involve the use of the government's **Exception Policy** and a Rural Housing Enabler. This involves building on land where it would not normally be permitted, in rural villages where housing needs have been proven. New affordable homes are built for local people and can never be sold on the open market. Normally, occupants need to:

- be connected to the village by birth
- have lived there for a number of years
- work in the village.

Suffolk ACRE has provided 8 homes so far.

Providing services

Community transport
Car ownership is vital in rural areas – access to jobs, training and services depend on it. Yet, in rural Suffolk, 47% of adults do not have access to a car and, in North Norfolk, 30% do not have driving licences.

The Bittern Community Rail Partnership is an initiative that serves the needs of North Norfolk for domestic and tourist purposes. The Bittern Line runs between Sheringham and Norwich and links facilities for cyclists, buses and car users. It extends to tourist locations as well as business and work places, and is timed to suit shoppers and day-trippers. Subsidised by government grants and local funding, it is a successful transport project.

The Vital Villages Programme, administered by the East of England Development Agency (EEDA), manages local dial-a-bus systems that can be pre-booked. Fleets of small mini-buses run on flexible routes and can take postal deliveries and shopping. Medi-bus links serve those who need regular health care.

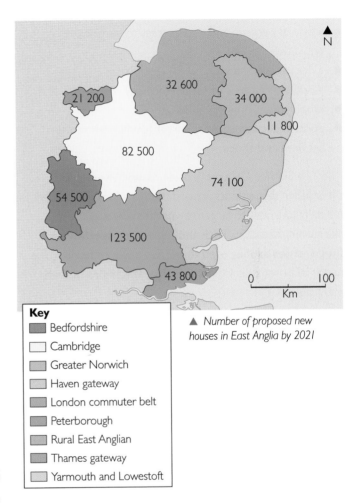

Key
- Bedfordshire
- Cambridge
- Greater Norwich
- Haven gateway
- London commuter belt
- Peterborough
- Rural East Anglian
- Thames gateway
- Yarmouth and Lowestoft

▲ Number of proposed new houses in East Anglia by 2021

Can rural communities be sustainable? – 2

Key Service Centres

The East of England Development Agency (EEDA) is creating a system of Key Service Centres to provide for the local population. Key Service Centres are located in places with over 3000 people, but those in sparsely populated areas have only 1500. Key Service Centres should have:

- a primary school and good access to secondary education
- a doctor's surgery
- a range of shops and services for day-to-day needs
- employment opportunities
- good public transport to larger towns with more services and shops.

The Key Service Centres provide essential services for the smaller settlements around them, reducing the need for daily journeys to larger towns. In many ways, it is like turning the clock back and represents **settlement hierarchies** as they used to be. It is a way of fulfilling the needs of potentially isolated small communities of under 150 homes.

Combined services

Examples of combining vital services are found throughout North Norfolk, Suffolk and Fenlands. As Post Offices close and nearby supermarkets take trade away from the village shop, so communities have grouped together to save their services.

- The Brancaster Community Shop (pictured) remains the hub of the village, combining the services of several outlets that have closed. So far, 23 similar ventures exist across East Anglia, co-ordinated by the Village Retail Services Association.
- The Norwich Network sees churches share travelling priests and runs postal services in churches and village halls.
- Eight rural post offices in North Norfolk have a police desk designed to make the reporting of, and dealing with, rural crimes more accessible.
- Some very small primary schools have merged. Peasenhall with just 18 pupils joined with nearby Middleton to preserve a vital educational resource in Suffolk.

The Brancaster Community Shop ▶

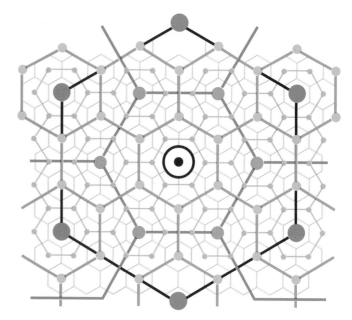

▲ *Fenlands settlement hierarchy*

Key

- ◉ Key service centre – villages with population over 1500
- ● Limited service centre – population 500–900
- ● Small village – population 160–500
- ● Rural hamlet – population under 160
- • Single homes/ribbons – population under 50

● A **settlement hierarchy** is a ranking of settlements (with the largest at the top). Larger settlements have larger populations and more functions and services than smaller ones.

Providing jobs

Many people in the peripheral areas of North Norfolk, Suffolk and Fenlands depend on low-paid agricultural and food-processing jobs. Up to 33% of households have incomes below the EU definition of poverty, and lack the skills to change jobs. Attempts to widen the range of agricultural jobs and broaden skills are vital.

Agricultural jobs

A number of new projects are aimed at farmers:

- Farmers markets – about 100 now regularly happen across the region.
- Online sales – *Fenweb* and *Organics* online receive EU funding and send fresh foods beyond the region.
- The FoxAsh Growers of Essex supply and pack fresh *Anglia Salads* direct to M&S and Sainsbury's.
- 27 farms have turned to viticulture (producing wines).
- Within rural-urban fringe areas, farmers are running their own shops, turning to Pick-Your-Own, road-side sales, and horticultural small holdings.
- Open farms – Easton Farm Park is one of many working farms that also serve as rural educational and recreational facilities.
- Recreational farms – rare breed sanctuaries, clay-pigeon shooting, golf ranges, etc.

Tourism

The leisure industry provides 250 000 jobs throughout East Anglia and earns around £5 billion per year, with Norfolk taking a 28% share and Essex having the highest number of day visitors. Tourism can be all-year-round, like at CenterParcs in Thetford Forest, or weather-dependent, as in the coastal towns. It offers an economic lifeline, but for most workers it has meant swapping low-paid, low-skilled, seasonal work in agriculture for similar low-paid, low-skilled seasonal work in tourism.

Tackling social exclusion

55 km away from Ipswich and Lowestoft is the 'East Suffolk Objective 2 Transition Area'. This is an area where traditional farming and associated industries are in decline. It is a remote district of many dispersed settlements; poorly connected by road, and access to services, jobs and training is limited. Young people with low levels of skills, qualifications and aspirations leave for the towns. The resulting unbalanced population structure and low population density makes providing services difficult. The flow diagram shows how social exclusion is being tackled.

▲ *Diversification on Suffolk farms - from vineyards to owl sanctuaries*

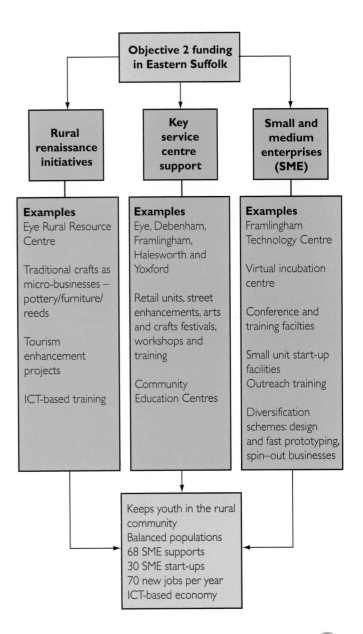

Objective 2 funding in Eastern Suffolk

Rural renaissance initiatives	Key service centre support	Small and medium enterprises (SME)
Examples Eye Rural Resource Centre Traditional crafts as micro-businesses – pottery/furniture/reeds Tourism enhancement projects ICT-based training	**Examples** Eye, Debenham, Framlingham, Halesworth and Yoxford Retail units, street enhancements, arts and crafts festivals, workshops and training Community Education Centres	**Examples** Framlingham Technology Centre Virtual incubation centre Conference and training facilties Small unit start-up facilities Outreach training Diversification schemes: design and fast prototyping, spin-out businesses

Keeps youth in the rural community
Balanced populations
68 SME supports
30 SME start-ups
70 new jobs per year
ICT-based economy

Protecting the landscape

Population growth, increased leisure time, affluence and mobility, along with economic change, threaten rural tranquility. The East Anglian countryside is a working landscape, but it also includes areas protected as AONBs, Heritage Coasts. Meeting the needs of the current population within protected areas presents challenges.

> In August there are queues in the shops and the car parks are full, and you can't walk on the pavement. Then they all disappear like swallows in Autumn, and we return to our dull orderly way of life.
> **A resident of the Suffolk Heritage Coast**

▼ *Responsibilities and aims for protecting the East Anglian landscape*

Norfolk and Suffolk Broadlands

Aims:

To conserve and enhance natural beauty.

- To promote understanding and enjoyment of the area.
- To protect the interests of navigation.
- To respect the needs of agriculture, forestry and the economic and social interests of people who live, or work, in the Broads.

Key
- Environmentally Sensitive Areas
- Area of Outstanding Natural Beauty
- Broads Authority
- National nature reserve

North Norfolk AONB, Suffolk Heritage Coast and Dedham Vale

The same aims as for the Broads, and also to promote sustainable forms of social and economic development. Five key objectives:

- To improve degraded landscapes, manage and protect heritage.
- To monitor and manage natural systems.
- To promote and manage recreational uses.
- To safeguard local interests, promote affordable housing and use local businesses.
- To provide education and awareness of sustainable management techniques.

Since 2005, a Sustainable Development Fund has supported projects designed and managed by local communities and businesses in AONBs – bringing benefits to local populations, economies and the environment. The EU Leader+ scheme offers further support to sensitive rural areas by directing funds to the most vulnerable groups.

> *Is it right that rural areas are changing so unevenly for different places and different social groups?*
>
> **What do you think?**

▼ *Ra, the world's first solar powered passenger boat, on the Broads*

Examples of sustainable development projects in East Anglia

- Covering the start up and promotional costs for a sustainable/renewable fuels business.
- Building composting toilets for a field study educational centre.
- Refurbishing an empty schovol to provide holiday lets.
- The development of an eco-friendly boat design for the Broads.
- Schemes to help with wheelchair access to boats on the Broads.
- Renewable energy schemes.
- Providing: educational programmes about local sustainable lifestyles; educational software for teaching sustainable development; special programmes for special-needs and urban children; skills and training programmes on traditional rural crafts.

Background

Rurality and the rural poor

Many of those moving to rural East Anglia do not depend on the countryside for their livelihood, so the area becomes less rural – its **rurality** declines. In 1971, a geographer, P. Cloke, devised a scale known as the Cloke index to show how rural (or urban) an area was. The map shows the results of Cloke's index of rurality in 1971, using the indicators of rural life below that he used to calculate the index. Will anywhere in East Anglia be classified as 'extreme rural' by 2020?

Indicators of rural life used by Cloke:

- Female population aged 15-45
- Occupational structure
- Population change 1961-71
- Number of people working outside the settlement
- Distance from an urban centre of 50 000+ people
- Population over 65
- Household amenities
- Population density
- Inward migration over the previous 5 years

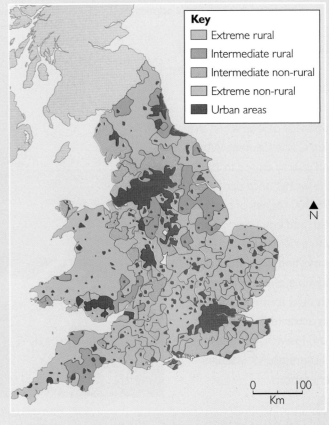

Key
- Extreme rural
- Intermediate rural
- Intermediate non-rural
- Extreme non-rural
- Urban areas

N

0   100
Km

▲ *Rurality in England and Wales, 1971, according to Cloke's index*

● Over to you

1 In pairs, devise a list of those things needed by the rural population.

2 Identify which needs are not being met for the following groups: low wage earners, young children, single parent families, the elderly, disabled, migrant workers.

3 Why should so many people choose to live in rural areas where these needs are not met?

4 **a** Draw a sketch map of East Anglia showing the main towns and villages as a series of colour-coded dots (towns – red, villages – blue).

   b Draw a circle with a radius equivalent to 15 miles around each major town.

   c Identify the smaller settlements within each circle and estimate which ones are most likely to lose key services because people there could travel to the town by car.

   d In pairs, use websites UpMyStreet.co.uk and UKvillage.co.uk to find:

   i population sizes **ii** services of villages inside and outside two of the circles.

e Do villages outside the circles have higher levels of service provision?

f Are the needs of people living there being met?

● On your own

5 Define the term 'sustainability'. What does it mean for rural inhabitants?

6 Should rural settlements fulfil the needs of their populations? Why?

7 Define the following terms from the text: Exception Policy, Key Service Centres, hierarchy, rurality.

8 Draw a table comparing the advantages and disadvantages of strategies being used to create sustainable communities.

9 With the presence of out-of-town supermarkets in the rural-urban fringe and increased car use, how realistic do you think the Key Service Centre policy is?

Exam question: Referring to named examples, evaluate attempts to make rural communities more sustainable. (10 marks)

Planning for the future

In this unit you'll find out that imaginative solutions to rural problems don't always meet with everyone's approval.

Reducing inequalities?

Great Blakenham (population 1083) lies 4 miles north east of Ipswich. The village lost hundreds of jobs when the Blue Circle Cement Works and quarry closed in 1999. The works once produced 195 000 tonnes of cement a year, and employed people with a range of skills. The old works and quarry is a classic example of a brownfield site. It is next to the village that has attracted commuters from Ipswich, but there is little there for younger generations.

▲ *From old quarry to winter sports complex*

In 2004, a company applied to Mid Suffolk District Council for permission to build SnOasis – Europe's largest indoor winter sports facility. It planned to provide 2000 permanent jobs and 1500 related opportunities. Outline planning permission was granted in 2006, but local opposition forced a public enquiry before work could begin.

As well as providing jobs, the scheme planned to:

- double the size of the village
- provide affordable homes
- build a new railway station
- improve road systems.

Efforts were made to secure work, homes and local services for all groups living in Great Blakenham and the surrounding area.

SnOasis anticipates 650 000 visitors a year, bringing £50 million into the region, but this could increase if it successfully taps into the European market. It will cost around £350 million to develop.

▼ *The key features of SnOasis*

What it includes	Year round leisure facility
ski centre	• snow dome slopes; 40-70 m wide piste, 475 m long 100 m drop; nursery slope, ice wall, bobsleigh run, après ski bars
village centre	• leisure rink, bars, restaurants, retail outlets, rollerblade park, convenience store, crèche
academy centre	• 1500-seater dome, museum, conference centre, exhibition hall, sports hall, speed skating track, hostel
entertainment dome	• bowling, bars, nightclub, casino, pool, snooker, food court, cinema, retail outlets
education and kids	• liaison with local schools for training and SSSI education centre, sports academy
resort	• 4-star hotel (350 rooms), 350 chalets, 100 apartments, golf course
lakes	• water sports, amenity sailing, fishing, wind surfing
parkland	• cross-country trail, roller-skating, fitness centre
Infrastructure	
new village	• 500 mixed homes with village green (35% affordable)
new village centre	• new shops and services
railway station	• village railway station
road improvements	• enhanced access to the A12 and A14 roads

Not everyone is in favour

Some people believe that the scheme tackles several local social and economic issues, while putting Suffolk 'on the map'. They believe that it could have the following benefits:

- Acting as a major catalyst for the regeneration of the Ipswich/Cambridge A14 corridor.
- Giving Suffolk its own *Eden Project*-style landmark.
- Easing traffic congestion with the new railway station (but the station could also attract more commuters to the village).
- Helping young people with affordable homes.
- Putting life back into the community with the proposed expansion of services.

Why is it difficult to please all of the people all of the time when planning new developments?

What do you think?

Other people see it as the wrong development in the wrong place. The sNOasis campaigners (against SnOasis) claim that nearly a third of the site is arable land, lakes and reclaimed grasslands – home to wildlife – and surrounded by an area of special landscape with one of the largest bat roosts in Britain, including the endangered pond bat.

Groups in favour	Groups opposing
Suffolk for SnOasis	The 'Alliance' sNOasis
• Institute of Directors (Suffolk)	• Parish Councils of 15 villages
• Suffolk Tourism Partnership	• Suffolk Preservation Society
• Suffolk Chamber of Commerce	• Residents of Great Blakenham
• Ipswich and Suffolk Small Business Association	• SnOasis Concern
• East of England Regional Assembly	
• GO-East	

SnOasis will provide a stimulus to business and create a huge positive impact on the economy, not just in Suffolk and the east of England. **Local businessman**

People are always moaning that there's not enough for young people to do in this area, and they're right, there isn't. But as soon as a development comes along that will change the situation, people try to stop it. It's just the people who live nearby who are stuck in their ways and scared of change who are opposed to the development. **Village teenager**

Stop living in the past and acknowledge that Suffolk has to move with the times – and that means providing jobs (even low-paid ones) and building for the future. **Ipswich teenager**

The village will lose its character when it's doubled in size, but will gain more attractions and encourage young people to stay. **Local parish councillor**

Those in favour

Local roads often become gridlocked already, and it will happen more frequently. **Bramford resident**

Leave Suffolk alone, we don't need this monstrosity, especially here. **Great Blakenham resident**

SnOasis is an energy guzzling site. Politicians of all parties are urging us to be more energy efficient before it's too late. Have any of those in favour of building SnOasis stopped to think about why the natural snow on the ski slopes of the world is melting? **Bramford resident**

The number of trees being planted to help reduce the impact of the CO_2 being created by the development is woefully inadequate. **Retired person living in Baylham**

SnOasis will threaten my farm, which employs up to 120 people. Light pollution from SnOasis might affect my crops. **Hop farmer from Little Blakenham**

83% of the new jobs will be low-skilled, low-paid and seasonal – and will not offer attractive careers for local youngsters – it is more likely to attract temporary workers from outside the region. **sNOasis Alliance publicity leaflet**

Those against

● Over to you

In pairs, complete the following:

1 Summarise the current needs of Great Blakenham.
2 Using all of the information in this unit, draw up a table showing the positives and negatives of the SnOasis plan.
3 Make a list of any extra information that you would need before deciding whether the plan should go ahead.
4 Draw up a list of questions to ask the developers about how they intend to help the local community.
5 How could you assess the environmental impact?
6 What criteria could you use to decide whether the development is sustainable **a** economically, **b** socially, and **c** environmentally?

Botswana: a different kind of inequality

In this unit you'll learn about inequalities in a middle-income country.

Botswana – pula

Botswana is a vast landlocked country. It is a sparsely populated place, a mostly desolate land of diamonds and desert. One word *pula* holds great significance. *Pula* means rain, it is a form of greeting, an expression of good wishes for the future, and the name of the currency. It shows how important rain is in this mostly dry part of south-central Africa.

Botswana has experienced high rates of economic growth since its independence in 1966. However, 47% of the population lives below the UN Poverty Line of $2 per day. The wealth gained from diamond exports is not equally distributed.

Urban yet rural

The rate of **urbanisation** in Botswana has been rapid since 1970. The number of people classified as urban has increased from 4% in 1966, to 55% today. Natural population growth rates reached a high of 4.1% between 1971-91, and migration to the large towns caused a major shift in the country's population. By 1981, Gaborone had become the **primate city**, double the size of Francistown the next largest city. Gaborone's population increased from 17 700 in 1971 to 186 000 in 2001, which is similar to LEDCs that have experienced **hyper-urbanisation**.

Although urbanisation has been rapid, many people retain their rural lifestyles. This traditionally involves regular migrations between three homesteads. Living in the village is combined with spells on the lands and the cattle-post. Half the population now combines periods in the towns for work with rural activities. The main towns, particularly Gaborone, attract hundreds of daily and weekly workers who are unable to afford homes in town. Villages in their **urban shadow** have become dormitory settlements, doubling in size in the last 10 years.

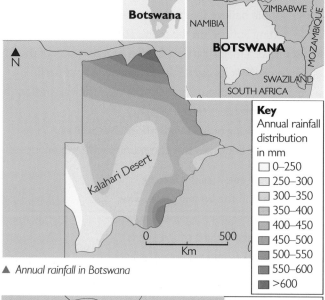

▲ *Annual rainfall in Botswana*

Key
Annual rainfall distribution in mm
- ☐ 0–250
- ☐ 250–300
- ☐ 300–350
- ☐ 350–400
- ☐ 400–450
- ☐ 450–500
- ☐ 500–550
- ☐ 550–600
- ☐ >600

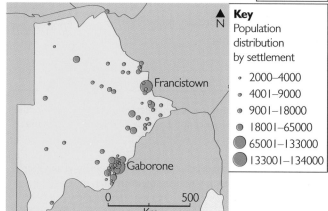

Key
Population distribution by settlement
- · 2000–4000
- · 4001–9000
- • 9001–18000
- ● 18001–65000
- ● 65001–133000
- ● 133001–134000

▲ *Population density and distribution*

Towns and urban villages	1991	2001	% change 1991-01	% growth p.a. 1991-01
Gaborone	133 500	186 000	39.3	3.3
Francistown	65 200	83 000	21.4	2.4
Lobatse	26 000	29 700	14.2	1.3
Selibi-Phikwe	39 800	49 800	25.1	2.2
Orapa	8800	9200	4.5	0.4
Jwaneng	11 200	15 200	35.7	3.1
Palapye	17 300	26 300	52.0	4.2
Tlokweng	12 500	21 100	68.8	5.2
Mogoditshane	14 200	36 300	155.6	9.4
Maun	26 800	43 500	62.3	4.8
Letlhakane	8600	15 000	74.4	5.5
Kasane	4300	6700	55.8	4.4
Ghanzi	5500	9700	76.4	5.7
Bobonong	7700	14 400	87.0	6.3

▲ *Urbanisation in Botswana, 1991-2001*

Background

Botswana fact file

- Botswana has grown economically from one of the 10 poorest countries in 1966 (GNP $360 per capita) to a middle-income country in 2006 (GNP $5360 per capita).
- The government's 49% share of the Debswana (DeBeers/Botswana) diamond mining operation provides over 50% of Botswana's revenue.
- Income disparities exist, with remote areas and female-headed households worst off. 23% of the population lives on less than $1 a day, with high concentrations of poverty in western areas.
- The HIV/AIDS pandemic has had a negative impact on household incomes, putting a strain on health and education and hindering development.
- Unemployment is high at 23.8%.
- Botswana is unevenly and sparsely populated. 80% of the population is in the south east.
- Rapid urban growth and urbanisation is putting strains on the infrastructure, service provision and resources.
- Most women have low socio-economic status; men decide about family size and the use of contraceptives.

Urban in Botswana means any settlement with more than 5000 inhabitants, and in which 75% of people have non-agricultural occupations.

Urbanisation is an increase in the percentage population living in urban areas.

Hyper-urbanisation is where urbanisation happens so fast that the economy cannot support the huge increases in the urban population.

Poverty and inequality in Botswana – the UN report a high degree of income inequality in the country. The most recent estimate of the GINI coefficient (a measure of inequality) gives Botswana a value of 0.54, on a scale of zero to 1 – where zero would represent a situation of complete income equality, and 1 an extreme concentration of income to one person.

Why do rural people never really sever their ties with the countryside, even though they work in the towns?

What do you think?

Poverty in Botswana persists despite growth

The Inter Press Service News Agency reported that 'Despite Botswana's high rate of economic growth over the past two decades, it is unlikely that the country will halve the proportion of people living on less than a dollar a day by 2015'.

'Progress is slow,' says David Morwang of a children's orphanage. 'The authorities still lack strategies for integrating poor rural communities into the mainstream economy.'

The chairperson of Molepolole Village Development Community argues that political commitment is vital to change levels of inequality in Botswana. 'These levels are striking given Botswana's status as an upper middle-income country.'

According to the United Nations Development Programme, the poorest 20% of the population get a measly 4% of the total national income. The richest 20% of Botswana receive almost 60% of the total national income.

Four out of five households in rural areas are still dependent on income from a family member in the urban areas. About one-fifth of rural households do not have any income source.

Over to you

1 On an outline map of Botswana, use the data in the table and on the map of population density to show where urbanisation occurred from 1991 to 2001. Explain why it happened there.

2 Suggest how the natural environment shapes human development in Botswana.

3 Why do you think that
   a the 'poorest 20% of the population get a measly 4% of the total national income'?
   b the 'richest 20% receive almost 60% of the total national income'?

On your own

4 Write about 300 words with the heading 'How developed is Botswana?'

5 Read 'Poverty in Botswana persists despite growth'. Suggest three ways in which Botswana could attempt to reduce poverty. Justify your suggestions.

The marginalised Batswana

In this unit you'll find out about marginalised groups in rural and urban areas in Botswana, and about what is being done to reduce the inequalities.

Common threads

After thirty years of economic growth, 55% of Botswana's population is urban, although in Botswana urban means more like rural. You are as likely to see a small cattle herd in the street as you are groups of teenagers. Rural-to-urban migrants perceive better lifestyles in towns, but traditional land rights mean that they keep their rural homestead and cattle post. Sharing time between three locations is common and offers security if urban life does not meet expectations.

Botswana data file	
Area (km²) (equal to France)	582 000
Capital	Gaborone
Population in 2003 (million)	1.8
Expected population in 2015 (million)	1.7
Population growth per annum (%)	-0.4
Per capita GDP ($ ppp 2004)	9945
Per capita GDP/annual growth 1990-2004 (%)	4.7
Doctors per 100 000 population	40
Adult literacy (%)	82.4
Life expectancy at birth (years)	34.9
Access to adequate sanitation (% of population)	42
Access to clean water (% of population)	95
People with HIV/AIDS in the 15-49 age group (%)	24.1

Land rights in Botswana
- Botswana has three types of land tenure: tribal/customary 72%, state 24%, and freehold 4%. These apply in both rural and urban areas.
- Traditionally, all tribal lands were common lands administered by chiefs, who ensured that:
  - everyone had a homestead and arable and grazing lands (only the homestead was privately owned)
  - land was distributed according to need
  - land rights were inherited by sons.
- Today, Land Boards administer tribal lands for the benefit of citizens.

What is happening?
- **The poor** are losing access to land as it is privatised and fenced off as part of modernisation and the intensification of farming.
- **Women** traditionally never acquired land. Changes to the law are addressing this issue.
- **Minority groups**, known as the Remote Area Dwellers, make up 6% of the population of Botswana but their 'invisibility', and the development of key resources, have eroded their rights to land.
- **HIV/AIDS** is undermining communities by depriving them of people, worsening gender inequalities and levels of poverty, and creating an orphan problem.

Many Batswana (the majority population of Botswana) who migrate, intend either to send earnings back to their families or eventually return home. Often they cannot find a home in the city. Men living within 100 km of Gaborone make the daily or weekly journey to a low-paid job in overloaded pickup trucks. Some never return – the death toll on the unlit roads is extremely high.

The urban and rural populations in Botswana have the same needs: a homestead, land for grazing and crops, energy, food, health, education, and transport. Access to these determines people's well-being, and there are inequalities in their distribution.

▲ Remote Area Dwellers, a San Bushman family

Rural Botswana

Only 5% of the land in Botswana is suitable for arable farming, and this is mostly in the east of the country. 20% of people are employed in agriculture, and **subsistence farming** is frequently threatened by drought and disease. 60% of Botswana has thin desert soils with poor water retention. Beef exports have traditionally been strong, but increasing cattle numbers have caused serious over-grazing in communal areas.

Botswana's rural poor

Surveys by Botswana's government and the EU show that:
- 48% of the rural population is poor; 40% of these are very poor.
- The Central and North Eastern districts contain 33% of the population and the highest density of poor people.

- In the western districts of Ghanzi and Kgalagadi, 71% live below the poverty line and 59% are very poor. This is where the Basarwa (San) live (see pages 234–235).
- 50% of households are headed by women, and half of these are very poor. They have limited assets, the burden of child-raising, and low earning capacity (because of poor education).

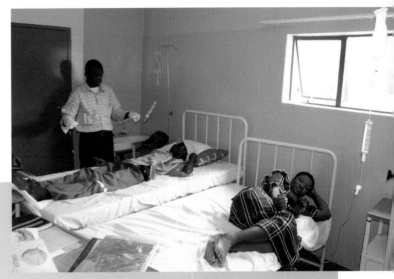

▲ A new village health centre in Botswana

Managing inequality – nationwide initiatives

Botswana's government produces a National Development Plan (NDP) every 6 years to try to cut poverty. By targeting rural areas, they are attempting to stem the flow of people to the towns. Strategies include:
- increasing commercial farming, with new research and training centres and farming for export (e.g. ostrich, fruit, vegetables, sunflowers, cotton). This improves incomes but takes land out of local use.
- taking water from the Letsibogo Dam in the north to Gaborone and Lobatse in the south, where demand is high, in the North-South Water Carrier Project. 88% of people now have access to clean water.
- completing a national ring road connecting all of the towns and main villages as part of NDP8. Establishing feeder roads to all settlements within 10 km of any major road as part of NDP9.
- providing loans and materials for upgrading homes in towns and rural areas through Self-Help Housing Associations.
- electrifying rural areas – 72 villages every 2 years.
- providing education and health posts for all villages, and permanent facilities in district centres to teach family planning, safe sex and education on HIV/AIDS.
- easing the problems of the poorest Batswana with the Remote Area Development Programme.

▲ HIV/AIDs education (see page 238 for its importance)

Remote Area Dwellers

Remote Area Dwellers are the poorest of Botswana's poor. The San Bushmen are the best known, but there are other groups who remain remote from most of Botswana, living in small, dispersed nomadic communities far from basic services. They lack access to formal education and training, and so remain poverty-stricken and marginalised.

Recent Remote Area Development Programmes allocated $14 million to fund poverty reduction. The main aims are to develop projects in health, education, training and skills which are designed to boost incomes. The tables show the achievements of the Programmes, and the number of people benefiting from the projects.

The plight of the San Bushmen

The San Bushmen are hunter-gatherers who lived in the Central Kalahari Game Reserve. They were perceived as primitive, frequently exploited as farm labourers, and plagued with alcoholism. As part of the Remote Area Development Programme, they were moved to the three new settlements of Xere, Kgoesakeni (New Xade) and Kaudwane in 1997, and their villages were destroyed. They were offered income-generating and training schemes. The government declared that the land they had occupied was set aside for wildlife and tourism development and that the relocation of 2500 San Bushmen was part of the poverty alleviation programme. But human rights groups claimed they were forcibly removed to make way for diamond exploration in the reserve.

◄ A San welder in one of the new settlements

	Advantages	Disadvantages
Water supplies	Clean water provided in 64 settlements	Nomadic lifestyle replaced by permanent settlement
Education	Primary schools set up in 63 settlements; 16 492 enrolled pupils; 3786 secondary and tertiary students; boarding schools in 17 settlements	Children from the Central Kalahari Game Reserve have to be bussed to boarding schools
Health	Health posts in 58 settlements; a maternity clinic in the largest	None
Settlement	3 permanent new settlements Kgoesakeni (New Xade), Kaudwane and Xere	Nomadic lifestyle replaced by permanent settlement
Economic empowerment	Small-scale and micro-enterprise schemes have been established	Traditional skills are lost to commercialization; they are dependent on market success

▲ Achievements of the Remote Area Development Programmes, 1978-2007

Activity	Kgoesakeni (New Xade)	Kaudwane	Xere
Basket weaving	1		
Candle making	1		
Brick moulding	3	1	
Knitting	6	4	
Sewing	6	7	5
Poultry	4		
Carpentry/woodwork	4	4	2
Bakery	2		2
Welding and Fabrication	2	1	
Leatherworks	2	1	
Tannery		18	4
Backyard garden	60	30	
Backyard poultry units	60	30	
Self-help houses	60	60	
Toilets	50	30	
Crafts	30	20	
Veld products	30	20	
Community based resource management	1	1	

▲ Number of people benefiting from small-scale schemes

Water supplies, food rations and social services in the old settlements were cut off in 2002 to ensure that the San could not return. New Xade and Kaudwane are said to be depressing villages where the San live on handouts. The free education is inaccessible because the lessons are taught in languages that the children do not speak. In January 2007, the High Court ruled that the San were wrongfully evicted.

> Driving through New Xade you see people queuing for drought-relief and food-for-work programmes. The government offers free education and health care, old-age pensions, drought aid, free food for AIDS orphans and free antiretrovirals for people with HIV/AIDS.
> *A government spokesman*

> Government relief is another way of killing a person – in the Reserve we knew how to provide for ourselves.
> *A San bushman living in one of the new settlements*

> I am proud of my culture and I love the land. I'd like to live in the bush and wear clothes like my ancestors, but the government won't let me. What are my goals? To help my community understand the importance of education and to fight alcoholism. And my dreams? That the San community will be respected like all others.
> *A young San bushman*

> Do they really want to plant tomatoes, build bricks and weave baskets with reeds trucked in from the Okavango delta because there is no bush around New Xade?
> *An NGO worker*

> Displacement of the landless has created an underclass of rural poor who are dependent on the state. 'Brick-and-mortar' solutions (schools and hospitals) and relief food do not solve social exclusion, low self-esteem and discrimination. The government tried so hard to be culture-neutral that it became culture-blind.
> *A researcher*

Managing inequality – rural initiatives

Veld Products

Farming in Botswana faces a 30% chance of crop failure due to drought. In the arid western district, the chances of crop failure are higher. Veld Products Research and Development is a Swedish government-aided project devoted to making rural communities sustainable. By demonstrating the market value of indigenous products, such as cochineal dyes and marula beer, they have helped to reduce the dependency on arable agriculture, and also women's workloads and environmental degradation. Starting in 1996 in the villages of Motokwe, Khekhenye and Tshwaane, the project has improved the living standard of rural communities.

Three assumptions underline the project:

- People will get involved if they can see sustainable rewards.
- Veld products exist in communal lands, so the Remote Area Dwellers are the best groups to manage them.
- Veld products are known, but their market potential has not yet been realized.

The Oodi Weavers and Thamaga Pottery

In 1973, Oodi and Thamaga were subsistence-based villages. Co-operatives were set up in the two villages as home-based ways of increasing incomes. The women of Oodi were taught how to weave tapestries, rugs and blankets, and pottery was made in Thamaga. The Oodi Weavers employ 32 people (just 2 men); in Thamaga, similar numbers are employed, and both produce for the local tourist and international markets. All proceeds are ploughed back into the villages and surrounding areas. Such small-scale enterprises have helped both women and the development of the rural communities.

Urban Botswana

Beer is flowing, Chibuku cartoons change hands, there is laughter and animated talk ranging from this and that. The makeshift houses and one-roomed mud houses in Peleng belong to the labourers and their unemployed dependents. In many homesteads, they are tending to some frail life bundled in a blanket.

It is the rampant unemployment that gives many Lobatse residents sleepless nights. 'We educate our children thinking that they will help us to live a better life. They finish school and cannot get jobs.

How are we to feed them when we are not working ourselves?' asked Paetonyana Mudende.

In most households in Peleng, particularly those headed by women, people are no longer concerned about unemployment. They have almost accepted it as a way of life. They have instead fixed their thin hope on the government's food rations, which are dispensed to those who live on the edge.

Gaborone's poor

Gabarone's rapid urbanisation (see page 230) has put a strain on the provision of water, electricity, sewerage and housing services. Education, health and employment prospects have also suffered.

In Gaborone, most rural migrants are drawn to run-down areas such as Old Naledi, a deprived community south of the city centre, where they continue to run their farms while earning an extra income from **informal** and **formal** jobs in the city. 40 000 people in Gaborone live below the poverty line (a third of these are very poor).

Old Naledi

Old Naledi grew as a temporary settlement for builders in 1964. By 1971 it was dirty, dangerous, and:
- housed 6000 people illegally on a 24-hectare site
- lacked water, waste disposal, drainage or lighting
- was squeezed in-between the main road and railway line on unused land.

Plots were self-allocated and varied from 49-155 people per hectare. Shacks with mud, cardboard or plywood walls, and loose thatch roofs, contrasted with brick-built homes with tin or asbestos roofing.

In 1973, the City Council wanted to bulldoze the site and resettle residents in New Naledi and Broadhurst, but this proved too expensive. Instead, the Self Help Housing Association was set up to provide cheap homes for poor people. Starting in 1975, Old Naledi was legalized and upgraded under this scheme. The text box opposite shows how the scheme works.

▲ Not everyone in Gabarone lives in poverty

▲ Old Naledi today

The Self Help Housing Association scheme

Phase 1: Basic site and service provision

- The SHHA identified the land area and marked out the primary services – main roads (unsurfaced), water supply, standpipes (1 per 20 homes), main drains and some street lighting.
- Building plots of 450 sq m each were allocated to families and they earned the right to build a 'core dwelling and toilet' there within 1 year of the allocation to gain a Certificate of Rights (the right to live there) for 99 years – on completion of the building and payment for water and waste collection.
- The SHHA provided cheap building materials and loans repayable within 15 years.
- Residents were able to choose their own building materials and styles.

Phase 2: Enhanced site and service provision

- A second level of water supply, drains and pathways was developed between each plot.
- Standpipes and refuse collection points were put in for every 5 homes.
- Schools, shops and community facilities were developed.
- Main roads were surfaced.

Phase 3: Ongoing improvements up to 2009

- Self-determined and SHHA-provided improvements are taking place, such as the upgrading of houses.
- 1700 Certificates of Rights have been issued.
- Standpipes for every home and sanitation by water-borne sewerage are being provided.
- Electricity is being provided to every home for a few hours a day.
- Local small business enterprises are being set up.
- Plots have been reduced to 375 sq m in size.
- The possible demolition of 800 homes has been considered in order to widen roads and expand schools, play areas and clinics.

By 2007, Old Naledi had deteriorated again and was overcrowded with 6307 buildings on 1704 residential plots. Many households only spend time in Naledi when they are not required on the cattle post, and they send urban earnings back to their rural homes to enlarge their herds. In the eyes of the residents, the upgrading works because it meets their needs.

● Over to you

1 In pairs, create a table to show how land rights lead to inequalities.
2 Why do you think the National Development Plans have targeted rural areas in particular?
3 Have the rural initiatives made the lives of the Remote Area Dwellers better or worse? Why?
4 Draw up a table to compare the conditions that rural people endure with those endured by urban dwellers. Whose life seems better? Justify your answer.

Botswana is a middle-income country. Why hasn't its wealth prevented the marginalisation of vulnerable people?

What do you think?

● On your own

5 Define the following terms:
marginalised groups, minority groups, subsistence farming, remote rural dweller, co-operatives, micro-enterprises, Self Help Housing Association.
6 How well do you think Self Help Housing schemes have worked in Gaborone?
7 Write 600 words comparing the impacts of the National Development Plans, the rural initiatives designed to help the Remote Area Dwellers, and the Self Help schemes in Gaborone. Which seems most effective, and why?

Exam question: Referring to an urban area, evaluate the attempts to improve the quality of life for deprived communities. (10 marks)

In this unit you'll investigate two of the contemporary issues that divide Botswana – HIV/AIDS and the move towards sustainable tourism.

Botswana today

Botswana suffers from the highest rate of **HIV/AIDS prevalence** in the world, at 39% for the 15-49 age group, and has an increasing number of AIDS orphans. The impacts of the pandemic threaten Botswana's future. Tourism has been seen as an important rural development strategy but it merely increases localised disparities.

HIV/AIDS

Migration to South Africa for work, and internal migration between urban and rural areas, have spread HIV/AIDS across Botswana. Life expectancy rates have fallen from over 60 years in 1996 to 35 in 2007. With many people dying young, the size of the work force is declining and the number of widows and orphans is increasing.

Palapye in Central Eastern Botswana (population 26 293) is a large village built around a colliery and power station. Thatched, one-room mud huts lacking running water and sewerage exist on the fringes, and smart multi-roomed well-serviced houses are found near the centre. 54% of the population is female, HIV/AIDS prevalence is 43% and there are 1743 registered orphans. So many people are infected that it is referred to as a **pandemic**, not an epidemic.

In Palapye, many women are heads of households and unemployed. The village is a major truck stop on the route between Gaborone and Francistown, and is a place where poverty-stricken women resort to prostitution.

Accepting government benefits and food rations to support orphans is seen as humiliating, and many carers will not register the orphans in their homes. When parents die, it is normal for grandparents, aunts and friends to take over childcare. But many households are too poor to look after extra children, and too proud to accept benefits. Consequently, more children are living in terrible conditions, underfed and increasingly vulnerable. The HIV/AIDS pandemic has made the children's lives worse. Despite policies designed to deal with the problem, the situation is not getting better.

AIDS poster campaign in Gaborone ▶

● **HIV prevalence** is the proportion of the population infected with HIV.

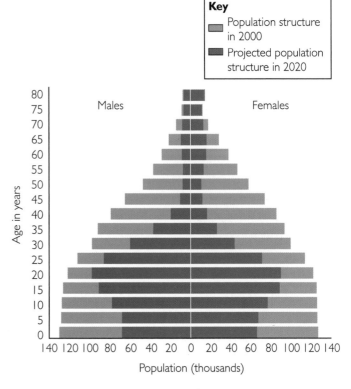

Key
- Population structure in 2000
- Projected population structure in 2020

▲ *Botswana's population structure in 2000, and projected population structure in 2020*

Botswana – HIV/AIDS statistics:
- HIV infected – 350 000
- AIDS deaths – 26 000
- AIDS orphans – 69 000
- The prevalence of HIV is highest in towns, lower in cities, and lowest in villages
- HIV is most prevalent among people aged 25-34
- The prevalence among women aged 15-19 is 9.8%, and 3.1% for men aged 15-19

Sustainable tourism

Tourism is the world's fastest growing industry and Botswana has cashed in. Income from tourism has increased by 26% in recent years, and the number of tourists has increased by 18%, although the overall numbers are still relatively low. As part of the government's attempts to reduce poverty, tourism is seen as a good source of employment.

The relatively low number of tourists, and the high costs of their holidays, has reduced the risks of over-saturation and land degradation in sensitive areas, whilst still providing a good income for the country. Since 2001, Botswana's National Ecotourism Strategy has been designed to spread the benefits of tourism in a sustainable way. Involving local communities (particularly women and young people) is a vital aspect of the strategy.

Hot spots like Chobe National Park and the Okavango Delta have attracted large numbers of tourists because of their wildlife. Government licensing and the issuing of visitor permits have controlled the number of visitors. The licence fees are shared between local communities and the government, and local people are employed to manage the visitors.

However, conflicts have arisen between farmers and tourists, because they see wildlife differently, and 'veterinary fences' have been used to segregate domestic and wild animals.

▲ *Tourists travelling by dugout canoe in the Okavango Delta with a local guide*

Tourism can undermine the local culture and should only ever be used as a last resort when trying to develop rural areas.

What do you think?

Ecotourism

The International Ecotourism Society defines ecotourism as: 'Responsible travel to natural areas that conserves the environment and improves the well-being of local people'. Ecotourism should follow certain principles including:

- minimising impact
- building environmental and cultural awareness and respect
- providing positive experiences for visitors and hosts
- providing financial benefits for conservation and local people.

It's not all good news

Tourism can create additional inequalities by changing local customs and values.

- For example, fake San Bushmen can be hired to ambush unsuspecting tourists on game drives and perform traditional dances for cash.
- Traditional crafts are modified, e.g. it is common practice for 'woodworkers' in Gaborone's Mall to sell plaster cast elephants instead of carved animals.
- Local people's demand for western foods, clothes and possessions can erode their own identity and values and replace them with those of the visitors.

● Over to you

1. Using the information in this unit, create a table to classify the impacts of HIV/AIDs into economic and social, and short-, medium- and long-term.
2. How significant has HIV/AIDS been in terms of Botswana's economic prospects? Explain your answer.
3. Why do you think that people are unwilling to admit that
   a members of their family have HIV?
   b they cannot cope with the increased burden of caring for orphans?
4. Why is tourism important in a country affected by HIV/AIDS, such as Botswana?
5. Create a table to assess tourism as a way of bringing help to the poorest. Think about the benefits and problems of tourism and the economic, social and environmental aspects.
6. Can tourism be a sustainable rural initiative if it increases regional inequalities?

5.9 Capital inequalities: London

This unit looks at some of the inequalities across London, and the reasons for them.

Wealth and poverty

> London is the most dynamic, cosmopolitan and diverse of our major cities, and is one of a handful of truly 'world cities'. It is also a city that is characterised by huge inequalities in income, employment, and quality of life.
> **The Mayor of London, 2004**

The National Census provides data on the following ways of showing poverty:

- income and employment deprivation
- deprivation in health and education
- barriers to housing and services
- environmental deprivation and crime.

These are known as Indicators of Multiple Deprivation (IMD). Data are given for every ward; scores are then ranked from 1 (worst) to 8414 (best) to give each ward a national ranking. The data shows that there is some serious poverty in London.

- The **spatial distribution** of income shows that the boroughs with the worst deprivation are all inner-city locations: Tower Hamlets, Hackney, Newham, and Islington. Towards London's outskirts, income rises; Richmond-upon-Thames has the lowest deprivation.
- **Minority groups** are concentrated in specific areas and suffer high levels of child poverty and unemployment.

▲ The City and Spitalfields in London

Over time, should ethnic enclaves become more, or less, distinctive?

What do you think?

London – Index of Multiple Deprivation, 2000. Some of the most deprived wards in England can be found in six London boroughs ▶

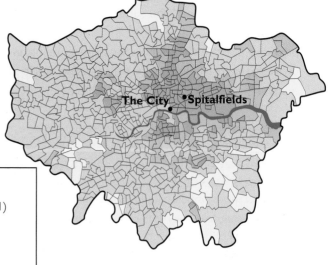

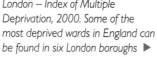

Key

☐ In the top 10% most deprived wards in England (ranks 1 to 841)

☐ In the next 10% most deprived wards (ranks 842 to 1683)

☐ In the next 70% most deprived wards (ranks 1684 to 7573)

☐ In the 10% least deprived wards (ranks 7574 to 8414)

Urban land use and change

In theory, the most accessible part of a city is the Central Business District (CBD). Demand for land is highest here and, therefore, it's the most expensive area. Zones appear across the city based on what people can afford. Each has its own socio-economic character – inequalities are born, and **social segregation** occurs. But changes in the urban economy may mean that places change character and use. Once, Fulham in London was run-down – now it is one of London's most desirable places to live.

The area around the CBD is a zone where land use often changes. Old and new users sit uneasily side by side. Within Tower Hamlets is an **inner-city zone of transition**. Where the CBD expands into old derelict areas, shining new office blocks stand next to derelict industrial sites, and wealthy business people rub shoulders with local residents on income support. Environmental quality swings from ultra-new to ultra-decaying.

The transition zone contains low-quality, cheaper houses that attract low-income groups. Those who rent property there suffer **social exclusion**, unqualified to play a part in society, often having a different language and culture. The Spitalfields area in East London has long been associated with migrant populations living close to places of work in the CBD. Once a Jewish area, it is now the heart of Banglatown, with the largest Bangladeshi community outside Bangladesh. With a 63% Asian/Asian British population, it has become an **ethnic enclave**, known locally as 'Spitalfields and Banglatown'.

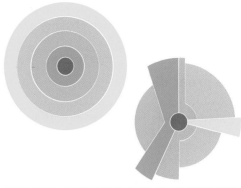

Key

■	Central Business District (CBD)
■	Wholesale light manufacturing (transitional)
■	Low-class residential (old inner-city areas)
■	Medium-class residential (inter-war areas)
■	High-class residential (modern suburbia)

▲ Urban land use models, Burgess (left) and Hoyt

2003 data	Spitalfields & Banglatown
Overall rank	46th
Income	15th
Child poverty index	41st
Employment	68th
Health deprivation and disability	1316th
Education, skills and training	320th
Housing	1st
Geographical access to services	8233rd

▲ Banglatown, IMD rankings out of 8414 wards in England (1st = worst, 8414th = best)

Why ethnic enclaves like this form

The pull of the East End
- **Employment opportunities** – it is close to unskilled jobs in the CBD.
- **Cheap housing stock** – being low-paid, people are forced into cheap housing. The area around Brick Lane is part of the old inner city and has been vacated by wealthier people.

The need to stay together
It is easier for migrants:
- they face **discrimination** in housing and jobs, even though this is illegal in the UK.
- when there is **security in numbers**, with support from relatives and friends.
- when people speak the **same language**, share other **cultural characteristics** and practise the same beliefs.
- if people gain political power.

● Over to you

1 Study the map showing the Index of Multiple Deprivation in London, and the two urban land use models. What links are there between them?
2 Explain why transition zones and ethnic enclaves develop around the CBD.
3 How and why might environmental quality vary across a large city?
4 What are the advantages and disadvantages of having areas of cities which are culturally distinctive? Draw a table to list them. Which is the longer list?

● On your own

5 Use the website and www.nationalstatistics.gov.uk look up the IMD rankings for your own area. Do similar patterns of inequality exist?

In this unit you'll find out about the huge inequalities between Hackney and Hampstead in London.

The great divide

> If you stand where the kites fly on the summit of Parliament Hill, in Hampstead, on a clear day you will see... the immense tower blocks of the Nightingale Estate, almost in the centre of Hackney. It is less than 4 miles from Hampstead, but if social distances were measured in miles it would be half-way round the globe. The gulf between inner city and desirable neighbourhood is a measure of the wide gap between wealth and poverty in Britain.
>
> Paul Harrison, journalist, 1983

Government strategies for deprivation in cities

Stark contrasts still exist between Hampstead and Hackney (as you can see from the panel on the right), but a number of attempts have been made to alleviate the problems associated with urban decay.

Government Strategies since the Second World War

Post Second World War to the 1960s – concentrating on housing improvements
- Urban slum clearance and building outer-city council estates
- Green belts and New Towns
- Redevelopment with tower blocks

1967-1977 – adding education and employment initiatives (Urban Aid Programme)
- Urban renewal with community projects
- Housing Action Areas
- Small-scale schemes focusing on local needs

1977-1990s – inner-city focus
- Urban regeneration by Urban Development Corporations and Enterprise Zones
- Inner City Initiatives, City Challenges
- Single Regeneration Budgets

1990s-21st century
- Urban New Deals
- Rebranding cities

Hampstead and Hackney – two decades on

IMD ranking out of 8414 wards in England, 2001 (1 = worst, 8414 = best)	Hampstead (Camden)	Westdown (Hackney)	Eastdown (Hackney)
Overall rank	7279	225	238
Income	6510	154	185
Child poverty index	6658	303	349
Employment	5253	136	203
Health deprivation and disability	6927	1239	896
Education, skills and training	8129	1126	1044
Housing	2706	80	69
Geographical access to services	8186	8038	7964

Key
- Hackney
- Hampstead

◀ Hackney in 2004

▶ Hampstead in 2006

The Nightingale Estate

The Nightingale Estate in Hackney (home to 2000 people) demonstrates the strengths and weaknesses of the entire sequence of government strategies. Tower blocks were built between 1967-72 to replace the poorly constructed, squalid rows of terraced houses. There were landscaped open spaces, wider roads, and improved access – and the tower blocks provided modern, well-equipped homes.

However, the old neighbourliness of the terraces was gone, the elderly felt trapped, children's play areas were lacking, and poor building design seemed to encourage vandalism and crime in the dimly lit, hidden passageways and lifts. Social problems increased and work places were not located on, or close to, the estate. Those who were able to move out did so, leaving a residual population of those on low incomes, the unemployed, or the elderly. It was perceived as a problem area.

▲ *Demolishing old tower blocks on the Nightingale Estate*

The 1990s – The Age of Renewal

By the 1990s, Hackney Borough Council began to plan for change with their Comprehensive Estates Initiative. An EU/central government funded Urban Regeneration Project saw the estate completely rebuilt.

The key characteristics of the regeneration of the Nightingale Estate are that it has:

- been a ten-year programme (1997-2006), costing £60 million.
- incorporated traditional homes and gardens, safe play areas, new schools, a community centre, job training and social support agencies.
- provided homes for local people, with a mixture of homes for sale, rent and joint ownership.
- involved residents alongside the architects.
- demolished and replaced five tower blocks with low-rise, high-density homes.
- refurbished completely a 22-storey tower block with new toilets, bathrooms and kitchens, new balconies, double-glazed windows, insulation and new roofing.
- made security a priority – a concierge unit has been added with a state-of-the-art security system linked to cameras and alarms covering three nearby shops and a community square where there is a children's play area.
- dramatically improved security further by introducing CCTV, entry-phone systems and high-security front doors.
- designed Olympus Square as the heart of the estate, to include a children's play area, nursery, back-to-work club, pensioners' luncheon club, café and local station.

▲ *New low-rise housing to replace the tower blocks*

A New Deal for Hackney

The government's New Deal initiative, established in 1999, targets the country's most deprived communities. It is about solving long-standing problems, by finding bottom-up solutions (working at community level). Shoreditch, in south-west Hackney, is ethnically diverse and, in 1999: 81% of residents lived on council estates; unemployment was 23%; and high crime rates meant that 38% of people were afraid to go out at night.

The New Deal targets five key issues over ten years in a coordinated way:

1. Tackling unemployment and lack of jobs through training schemes, career workshops, and attracting private finance for small-scale business ventures.
2. Improving health through education and community care facilities.
3. Tackling crime through 'diversionary projects' for young offenders. The Crib is a youth inclusion project that offers skills training, clubs and police-based programmes. SKY partnership provides information services and community awareness programmes.
4. Improving educational achievement through business-school partnerships, work-based learning, and Outreach projects.
5. Improving housing and the physical environment through the use of Housing Associations and Trusts

Shoreditch's successes include:

- 413 supported businesses; 29 new start-ups surviving
- the creation of 511 new jobs and the saving of 273 others
- the training and use of local workers by construction companies
- falling youth crime
- affordable homes and an expanded range of shops
- helping those who are dependent on public transport, and who suffered from the old limited North-South services, with the new Hoppa Bus East-West service

The New Deal is encouraging new ways of working to tackle social exclusion in deprived neighbourhoods, and has put the community in the driving seat. Organisations such as the council, the police and the health services work across traditional boundaries to provide lasting and imaginative solutions.

There is a large amount of untapped strength in Shoreditch – as well as some very innovative ideas … the New Deal will harness that energy for the first time.

The New Deal has captured the hearts and minds of local people.

Hundreds of residents, voluntary groups, local businesses and agencies have worked together to assess the problems of the community and come up with solutions. We will now be able to put our plans into action.

▲ *Local people's views about the New Deal*

Gentrification in Hackney and Tower Hamlets

Gentrification occurs when higher-income groups move into lower-cost housing areas. The traditional population is displaced and property prices rise.

A BBC news article about Hackney reported that 'real deprivation and conspicuous consumption exist side by side'. Gentrification might improve buildings and the environment but, as the old homes are transformed, their former inhabitants are forced elsewhere.

▼ *To gentrify or not?*

Benefits of gentrification	Drawbacks of gentrification
• Housing quality improves	• Homes are priced out of the reach of local, lower-income groups
• Physical environment improves	
• Wealth of the area increases and attracts new businesses and services	• Inappropriate services develop in the area to serve the minority
	• Car owners congest the streets
• Council gains higher tax revenues	• Schools and local services are not used so they close down
• The costs are born by individuals and not councils	• There can be social friction

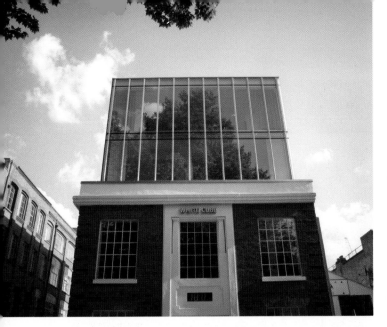

It is only a matter of time before gated communities appear in Hackney, as higher-income groups return to the inner city. The communities inside and outside gated communities do not mix and there is no social cohesion. The gates separate the haves from the have-nots.

Hampstead Garden Suburb

Hampstead remains one of the wealthiest areas in the UK and the Garden Suburb, designed in 1909 to 'house a broad cross section of people', has become an exclusive enclave. The original plans contained most of the ideas now included in twenty first century regeneration and community projects: a mix of housing styles; local services; close to places of work, and open spaces for leisure. Managed by a private trust, it has become highly sought after, boasting excellent education facilities, security conscious neighbourhood watch committees and strong environmental management rules.

The original plans for Hampstead Garden Suburb ▶

- Persons of all classes of society and standards of income should be accommodated and the handicapped should be welcomed.
- Cottages and houses should be limited on average to eight to an acre.
- Roads should be 40 feet wide, and the fronts of the houses should be at least 50 feet apart, gardens occupying the intervening space.
- Every road should be lined with trees.
- The woods and public gardens should be free to all the tenants.
- Noise should be avoided, even to the prohibition of Church, Chapel or Institute bells.
- Lower ground rents should be charged in certain areas to enable weekly wage-earners to live on the Estate.

● Over to you

1 Construct a table and complete it to show how well the approaches of the Nightingale Estate, Shoreditch and Hampstead Garden Suburb have dealt with urban problems. Include issues such as housing, services, employment and crime.

2 Do you think these schemes have increased social inclusion?

3 Which scheme achieved most improvement?

● On your own

4 a Select one Inner and one Outer London borough.

   b Research data that you think are good indicators of deprivation and affluence using the website www.nationalstatistics.gov.uk. Use the 'focus on London' and 'Neighbourhood' sections to look up your chosen areas. Justify your choice of data.

   c Present the data for your two boroughs to show how quality of life varies across London.

Exam question: Explain using examples how and why quality of life varies across cities. (10 marks)

5.11 Sustainable urban communities

This unit considers whether sustainable urban living is achievable.

Unsustainable spaces?

Population mobility, economic change and technological advances have altered the ways in which cities work. Manufacturing has moved out of UK cities, providing opportunities for regeneration.

Closely-knit communities and affordable homes within walking distance of work, shops and services are the characteristics that today's planners are after when creating sustainable communities. These features were typical of inner-city areas in the nineteenth century. Workers' homes were built in long rows of high-density terraces next to factories, with limited facilities (no inside toilet, bathroom or electricity), few open spaces and no gardens.

Sustainable Leicester

De-industrialisation, the relocation of Leicester's textile industry, and changes in transport, have all undermined the reasons for the existence of areas of dense old-style housing. A residual, predominantly Asian, population has been left in this old inner-city housing – stranded in a deprived zone.

As part of a major inner-city regeneration programme, and its status as Britain's 'First Environmental City', Leicester city has designated five new zones that aim to be sustainable:

- **The new business quarter** – close to the railway station; with offices, restaurants, leisure facilities, shops, apartments; pedestrian links to the city centre; new public spaces; 4000 jobs.
- **The retail circuit** – the refurbishment of shopping streets; pedestrian-friendly with the completion of the £250 million New Shires Mall, establishing Leicester as a leading retail destination.
- **The new science and technology park** – providing accommodation for hi-tech businesses to capitalise on the city's two universities; 1800 new jobs and new riverside housing.
- **Waterside** – 11 miles of canal and riverside; new high quality housing; shops, restaurants and facilities for the local community; open spaces and waterside walks.

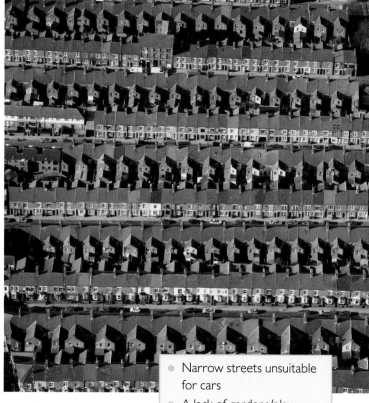

▲ *Leicester inner-city*

- Narrow streets unsuitable for cars
- A lack of gardens/play areas; doors open directly on to the street
- A mix of derelict industrial sites and housing
- Old decaying buildings
- The Latimer and Spinney Hills wards are 60-75% Asian
- 21% of inner-city residents are on income support

- **A new community** – 1800 homes close to the city centre; to appeal to families with appropriate housing, shops, open spaces and access to healthcare and education.

Bringing residential populations back into central Leicester, close to workplaces and within walking distance of services is seen as more sustainable than counter-urbanisation. Providing a mixture of employment opportunities will suit lower-income ethnic groups who live in central Leicester, and retain some of the university graduates who currently leave the city.

Small-scale sustainability

The Upton Project

This scheme was granted planning permission in 2003. Upton, to the south east of Northampton, is being developed as a sustainable extension to the city. The project comprises:

- 1200 low-energy eco-homes
- a primary school
- neighbourhood shops
- open spaces, a playing field and a country park.

English Partnerships and the Prince's Foundation are regenerating the land, together with a range of local stakeholders. Neighbourhoods include a variety of house types: townhouses, apartments, semi-detached and detached homes – with 22% affordable homes.

Other key features of the site are:

- **renewable technologies** – solar power, water recycling
- **drainage systems** which direct surface water run-off into reed beds, and create wildlife habitats.
- **safety** – homes have been designed to provide good views on to the street from inside.
- **accessibility** for wheelchair users in all areas.

Pathfinder Projects

The negative effects of economic decline make it difficult for inner-city residents to sell their properties. In 2003, the government set up Pathfinder Projects in nine northern cities to challenge developers to 'tackle the problems of communities … where housing markets have failed'. These projects aim to upgrade housing quality for people who own their own homes and wish to stay where they are. Its critics refer to it as 'government-sponsored gentrification' because, as the homes become worth more, they are sold to higher-income earners, thus destroying the community.

The Rochdale Pathfinder Renewal Scheme aims to make East Central Rochdale a thriving, safe and desirable place to live, by offering people a wider choice of property, improving community safety and the environment, and making the most of the area's natural selling points, e.g. the River Roch and the proximity to the town centre and the planned Kingsway Business Park.

Over £5.8 million has already been spent refurbishing homes, introducing community safety measures, and demolishing property to make way for new housing.

▲ The Upton Project

Are the Pathfinder Projects little more than 'government-sponsored gentrification'?

What do you think ?

● Over to you

1 In pairs, list the characteristics of a sustainable urban community.
2 Use your list to judge how sustainable the following are:
   - Nineteenth-century industrial communities
   - Out-of-town suburbs
   - Planned modern housing communities
3 In groups of three, examine the proposals for Leicester, Upton and Rochdale and complete a SWOT analysis (Strengths, Weaknesses, Opportunities, Threats) for people, the economy and the environment.

● On your own

4 Use the websites: www.englishpartnerships.co.uk/ pathfinders.htm and www.oldhamrochdalehmr.co.uk index.htm to review the Pathfinder projects. Is urban deprivation being successfully tackled this way?

Unit summary

What key words do I have to know?

There is no set list of words in the specification that you must know. However, examiners will use some or all of the following words in the examinations, and would expect you to know them, and use them in your answers. These words and phrases are explained either in the glossary on pages 297–301, or in key word boxes throughout this chapter.

affordable homes
counter-urbanisation
deprivation
dumb-bell market
environmental
  magnetism
exception policy
minority group
negative externalities
pandemic
positive and negative
  multipliers

positive externalities
ripple effect
rural housing enablers
rurality
rural-urban continuum
settlement hierarchy
social exclusion
social landlords
subsistence farming
sustainability
urbanisation and
  hyper-urbanisation

Try these questions

1 With reference to **one or more** named **urban** areas, examine the impacts of deprivation on people's lives. (10 marks)

2 Describe the **results** of your **fieldwork and research** into how to reduce inequality, and explain how these help you to judge the success of **either** the urban **or** rural schemes involved. (15 marks)

3 Explain how you would decide whether schemes designed to reduce deprivation had been a success or not. (10 marks)

4 Using named examples, explain how inequalities have arisen in **either** urban **or** rural areas. (10 marks)

6 Rebranding places

What do I have to know?

This topic is about rebranding places. Rebranding of places is the way in which a place is re-developed and marketed so that it gains a new identity.

1 Time to rebrand: What is rebranding and why is it needed in some places?

What you need to learn
- Why rebranding is needed in some places
- How places re-invent and market themselves
- The social, economic and environmental factors involved in rebranding

This is a general question and does not need specific knowledge. In-depth case studies come later.

2 Rebranding strategies: Who are the 'rebranding players' and what strategies exist for places to improve themselves?

What you need to learn
- Who the different players are, and the part they play in the rebranding process
- Examples of how urban places rebrand, e.g. commercial and residential areas, gentrification, culture and identity, sport as a catalyst for change, resorts
- Whether rebranding is sustainable for the future

Here, detailed case study knowledge is needed, e.g. on the London Olympics and regeneration in East London.

3 Managing rural rebranding: How successful has rebranding been in the countryside?

What you need to learn
- Examples of rural strategies
  - Rural tourism
  - Adding value locally
  - Rural diversification
  - Rural economy in LEDCs

Again, detailed examples are needed, e.g. on Cornwall, rural regeneration and the strategies used in Objective One funding. Most cases are from MEDCs, but two examples of projects are used from Uganda to provide answers from LEDCs.

4 Managing urban rebranding: How successful have urban areas been in rebranding themselves?

What you need to learn
- Knowing how to evaluate and judge the success of schemes

This section is less about knowledge and more about knowing how to use different ways to work out whether or not rebranding schemes have been successful.

6.1 Rebranding places

In this unit you'll find out what 'rebranding places' is all about.

Why rebrand Doncaster?

Doncaster, in South Yorkshire, has been hit hard by industrial change. Its traditional coal mining and engineering industries have declined, and high unemployment and derelict land characterise the area. Outlying mining villages, such as Rossington, have traditionally had little to offer residents except coal mining and, therefore, when Rossington pit closed in 2006, there were few other jobs available.

Rossington's community newsletter commented that '*job-seekers from Rossington pit are being offered alternative employment ... as beauty and nail technicians!*'. The lack of suitable alternative jobs has resulted in social and economic **deprivation**, affecting both people's health and wealth. A big problem in Rossington is the high percentage of people without qualifications.

Rossington and Doncaster need more full-time jobs. How can this be achieved? Rebranding is one way. Doncaster and its industrial past need to be re-shaped so that new companies will invest there.

How can Doncaster be rebranded?

More than 11 000 people signed a petition opposing the name of South Yorkshire's new airport. The former RAF Finningley is being redeveloped into a commercial airport and has been named Robin Hood Airport by its owners. But Councillor John Mounsey is leading the campaign to get the name changed to Doncaster Finningley Airport.

Councillor Mounsey said: 'We are fully behind the airport, but they should have consulted the people of Doncaster before they came up with the name. We're the ones who are going to use this airport, it's our town.'

Sharon Bennett, the airport's marketing and community manager, said: 'We're very happy with the effect of rebranding. The name Robin Hood is welcomed by the public. The name has to raise awareness of the airport, and of the whole area. If Robin Hood Airport is to bring jobs, it has to have some commercial impact'.

▲ *Can Robin Hood save South Yorkshire?*

● **Rebranding** is the way or ways in which a place is re-developed and marketed so that it gains a new identity. It can then attract new investors and visitors.

Rossington compared with Doncaster	
Children in households with no wage earner	+46%
People with 'not good' general health	+26%
Single parents	+30%
People with no qualifications	+30%
People not employed	+30%
Households with no car	+17%

▲ *Socio-economic indicators – Rossington compared with the Doncaster average. These figures are from the 2001 Census. Doncaster as a whole is one of the most deprived places in the UK, but Rossington is even worse off – these figures mean that, for instance, there are 30% more unemployed people, single parents and people with no qualifications in Rossington than in Doncaster.*

▼ *Robin Hood Airport, Doncaster*

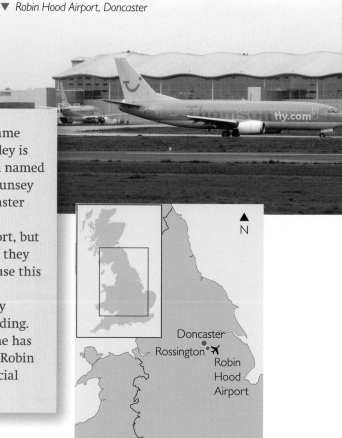

▲ Liverpool docks – rebranding former wharves and warehouses as luxury apartments, shops, cafes/restaurants, and a marina for private vessels

Background

Why do places need to rebrand?

Places need to rebrand because of economic change. Two factors have caused this:

- The decline in primary and secondary industries.
- The growth of tertiary industries, particularly tourism.

Industry in decline …

In the 1980s, the British government planned changes to the British economy. Goods produced by the UK's primary industries (e.g. mining) and secondary (manufacturing) companies were more expensive than those produced overseas. For instance:

- British coal is deeper underground and more expensive to mine than elsewhere.
- The UK's wages were also higher than overseas, making British products more expensive. The growth of manufacturing in Asia, with its cheaper workforce, led to cheaper imported goods.

Large numbers of mines and secondary industries closed during the 1980s. These changes caused huge unemployment in the North and Midlands.

Relic feature	Rebranded as
Industrial past	'Industrial Heritage' environments, e.g. Ironbridge, Shropshire
Derelict land and docks	Brownfield sites for re-building and expansion of office space, e.g. London docklands Waterside locations for new housing, e.g. Leeds canalside
Old factories and warehouses	Loft apartments, e.g. Birmingham canalside Shopping centres, e.g. Fort Dunlop, West Midlands

▲ Different ways of rebranding industrial sites

… the growth of tertiary industries

To replace the lost jobs, the government encouraged growth in a new **post-industrial** economy. This includes:

- **'knowledge' industries** – using expertise in design, process, IT, and consultancy
- **banking and finance** – international banks using the UK as a base to finance global investment
- **tourism** – the world's fastest-growing industry.

These sectors can locate anywhere; they are '**footloose**'. Decisions about where to locate are often made on preference, the quality of the environment, and financial incentives. **Image** is important.

The growth in tourism has resulted from greater prosperity in MEDCs, cheaper air travel and increased car ownership. Many areas of the UK and Europe have been cashing in, creating tourist attractions. Some attractions are new, like the London Eye, while other places have sought to **rebrand** their past – creating a new image. Former industrial areas hope that this will bring economic growth.

● Over to you

1 Draw a spider diagram to outline why Rossington has suffered deprivation.
2 What is wrong with offering employment as 'beauty and nail technicians' in Rossington?
3 a How might Doncaster's new airport help in 'rebranding' the area?
  b What risks are there in this attempt to rebrand?
4 Look at the table above. Use the same headings and complete a table for your area, or a locality that you know.

● On your own

5 Define these words:
  rebrand, post-industrial economy, knowledge industries, footloose industries
6 Research a local area that has suffered decline, and outline **how** you would rebrand it.

Should traditional industries be kept alive, even if they are losing money?

What do you think?

In this unit you'll learn about the need for regeneration and rebranding in East London

London – global city

London is a true global city – financially, in its diversity of people, and in its links with the rest of the world. Yet parts of it are amongst the poorest areas of the UK. This unit is about East London and attempts to regenerate and rebrand it.

Why was regeneration needed?

In 1981, the last of London's docks closed. Until the mid-1970s, they were the biggest docks in the UK. But, as ships increased in size, they needed deeper water, so Tilbury, 20 miles downstream, provided a better site. Container ships also replaced the need for people to work in the docks. Felixstowe, 70 miles away, is now the largest container port in the UK, and has replaced London as the UK's biggest port.

Closing London's docks had devastating effects:

- Between 1978 and 1983, over 12 000 jobs were lost. In some parts of East London, over 60% of adult men were unemployed in 1981.
- The docks were left derelict. Dereliction in some parts was so severe that the costs of regeneration were considered too high and, therefore, the area was unattractive to investors.

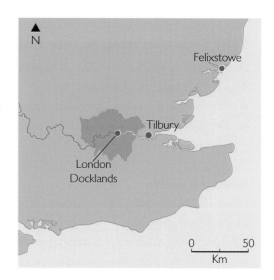

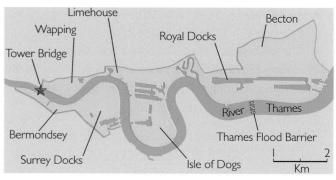

▲ *The former Port of London, now London Docklands, and its two main replacements*

▲ *London's busy docks as they used to be*

▲ *Dereliction in London Docklands in the early 1980s*

The rapid succession of industrial and dock closures in East London in the 1970s and early 1980s, led to huge unemployment. Economic decline was matched by population decline, as people left to find work elsewhere. The population declined by 18.5% between 1971 and 1981. Half of the remaining population worked in lower-paid manual occupations.

Background

Regeneration 1980s style

Similar levels of high unemployment in cities such as Manchester, Leeds, Newcastle and Liverpool, gave inner-city areas a poor image – and one in which there seemed little economic future. Deprivation and lack of opportunity, and ethnic tension, led to inner-city riots in Liverpool (Toxteth), Leeds (Chapeltown) and London (Brixton) in 1981.

The government (led by Margaret Thatcher) reacted by attempting to rebrand inner cities. During the 1980s and early 1990s, a number of 'Garden Festivals' were held in five cities. They tried to provide a 'green image' of inner cities.

Over the past 20 years, a variety of government regeneration programmes have been launched. Some, like Garden Festivals, have been dropped. Other programmes have tried to boost cities by attracting tourists. The European 'Capital of Culture' initiative (Glasgow in 1990 and Liverpool in 2008) attempts to focus on cultural regeneration in European cities.

Garden Festival City	Year	Eventual outcome
Liverpool	1984	Now a mixture of housing and derelict sites.
Stoke-on-Trent	1986	Now mostly maturing garden parkland, with some retail and offices.
Glasgow	1988	Now Glasgow Science Centre and a digital media village beside the River Clyde.
Gateshead	1990	Now a housing estate.
Ebbw Vale (South Wales)	1992	Now mostly housing, in parkland and woodland.

◀ Garden Festivals in the UK

Regenerating Canary Wharf

A huge plus for London was that Docklands offered 21 km² of land available for redevelopment – and very close to the City, the financial heart of London. Regeneration has been spectacular. Key features include:

- the Canary Wharf building itself – tenants of which include global investment banks (e.g. Bear Sterns), newspapers (Trinity Mirror), and professional services firms which carry out business consultancy and accounting (e.g. KPMG)
- huge skyscrapers in which some of the world's major financial institutions have their global headquarters – such as Citigroup, and HSBC.

100 000 commuters now travel to work in these office towers each day, adding to the 325 000 who work in the City of London.

Transport developments (known as **infrastructure**) have been completed to ensure that commuters can get to work. These include:

- extending the London Underground network, e.g. the Jubilee Line
- developing the Docklands Light Railway, a surface rail network which covers most of the Docklands area
- new road links, such as the Limehouse Road Link, taking traffic through Docklands
- London City Airport (located about 6 miles from Central London), which provides easy access into the City and Canary Wharf from European cities such as Paris and Geneva.

▲ Canary Wharf as it is today, a financial and business hub

Regeneration in East London – 2

Background

How does regeneration take place?

Land regeneration involves many different **key players**, such as:

- landowners
- designers and developers, with ideas about how the land could be redeveloped
- investors, who provide finance
- local authorities which give planning permission
- local people, who can object to redevelopment plans
- central government, which creates policies about where land should be developed and how.

The London Docklands Development Corporation was set up in 1981 to manage the development of the docks. As long as planning permission was granted by 1991,

companies could obtain huge tax breaks on any new building. These financial incentives were designed to attract companies to invest, and arre still in place.

This kind of development, where private companies make the decisions and are given benefits (such as tax breaks), is known as **market-led regeneration**. In the Docklands area, economic regeneration was a priority and the government felt that property developers would know how to develop land in ways that would attract business. Jobs would be created, and wealth, it was argued, would 'trickle down' to poorer communities.

Many cities in the UK have followed the Docklands model, e.g. Leeds, Manchester, Newcastle, and Bristol.

People and the environment

The Docklands regeneration has been criticised by many people, including geographer Sir Peter Hall, because it failed to make an impact on local people and the environment. Local people argue that East London still suffers deprivation, poor health and poor environmental quality. In fact, geographers who study health statistics, found that on a journey along the Jubilee Line between Westminster in Central London and Stratford, its terminus in East London, life expectancy drops by nine years – one year for every station. Wealth and poverty sit almost side by side.

The table on the right shows data comparing Canning Town and Millwall (the location of Canary Wharf). Canning Town is just a short distance from the office blocks of Canary Wharf, but has low levels of employment and few of its residents play any part in the growing economy of Canary Wharf. In 2002, Queen Mary College University of London (QMUL), surveyed people in Canning Town in order to assess their needs and quality of life. The results are shown on the next page.

	Canning Town	Millwall
Ethnicity (%)		
White	61.0	66.6
Asian or Asian British	6.4	18.4
Black or Black British	26.9	5.5
Other	5.7	9.5
General health (%)		
General health: good	65.0	75.7
General health: not good or with a limiting long-term illness	20.7	11.4
Employment status: aged 16-74 (%)		
Employed full-time	37.6	52.7
Employed part-time	8.1	5.0
Permanently sick / disabled	9.1	4.3
Highest qualification level reached (%)		
No qualifications	43.1	20.8
Qualified with a university degree	16.5	45.2

▲ 2001 Census data comparing Canning Town and Millwall (which contains Canary Wharf)

You can never regenerate areas and help everyone. As long as the majority of people win, that's ok.

What do you think?

The 2002 Canning Town survey from QMUL

A Is there a strong sense of community here?

Yes	33%
No	64%

B How many of your neighbours do you know by name?

< 4	44%
4-9	33%
10 +	19%

Canning Town, one of London's most deprived communities

C What do you like about living here?

Nothing	20%
Family / friends / community	17%
It is a quiet area	12%

D What do you dislike about living here?

Crime	32%
Dirty streets	10%
Lack of facilities	9%
Everything	9%
Noise	8%

E If you could do one thing to improve the area, what would it be?

More facilities, including those for children, and community schemes	30%
Reduce crime / more police	18%
Cleaner streets	16%
More green areas / parks	6%

F What people said about living in Canning Town:

> Before, people used to stay and you got to know them … the trouble is now I hardly know anybody any more down this street.

> In cases where the properties are let and the turnover of residents is quite high, you lack that close community – because people tend not to stay in the same place for very long.

> Front gardens are full of rubbish and they don't care where it goes or anything.

> Those who succeed move out and those who aren't able to get work and don't succeed economically stay where they are.

> Improvements begin when kids get good results in school and improve the prosperity … If I had a magic wand and could improve one thing, it would be the schools.

● **Over to you**

1 Using a website, such as Google Maps, produce a map of East London which shows the features mentioned in this unit.

2 Now produce a map which shows Tilbury and Felixstowe docks. Annotate it to show how these caused London's docks to close.

3 Using the comparison table opposite, construct graphs on a spreadsheet to test the following hypotheses:
   - Canning Town residents show evidence of deprivation.
   - Millwall / Canary Wharf residents are much better off.

4 Suggest reasons why the people in Canning Town expressed the opinions they did in the QMUL survey.

5 Using the results of the QMUL survey, draw up a list of Canning Town's greatest needs.

● **On your own**

6 Define the following terms from the text:
   global city, container port, derelict land, regenerate, redevelop, manual occupations, deprivation, infrastructure

7 Briefly outline East London's recent history, using each of the above terms.

8 What do you think of the Garden Festival City idea to rebrand Britain's inner cities?

9 Why might deprivation lead to urban riots, as it did in the 1980s?

10 List the advantages and disadvantages brought by the Canary Wharf development. Which is the longer list, and why?

11 Research an example of regeneration in another city, such as Leeds, Cardiff, or Newcastle. Write a 500-word case study and include illustrations.

REBRANDING PLACES

In this unit you'll find out about the impact that a major sports event can have on regeneration.

London wins Olympic bid

6 July 2005 is a day that many Londoners remember. Expecting that Paris might win, but hoping it wouldn't, people cheered long and loud when the International Olympic Committee (IOC) announced that London was to host the 2012 Olympics. The competition had been keen. London was said to have won because it plans to use the 2012 Olympics as a way of regenerating East London. This unit looks at how successful this is likely to be; the area, the economy and the people affected.

What do the Olympics involve?

The Olympics involve host cities in massive expense. The hosts have seven years to prepare, with tight schedules. If the Olympics are to be successful, they have to have a **legacy** of sustainable impacts, lasting beyond the period of the Games themselves.

▲ *Winning gold – London wins the right to host the 2012 Olympics!*

The Olympics are a huge commitment. The 16 days of the Games will involve:

- 11 000 athletes competing in 300 events, supported by 5000-6000 coaches and officials, and attended by 4000 5000 other members of the Olympic community
- over 7000 sponsors
- 4000 athletes and 2500 officials in the Paralympics (equivalent to Manchester's 2002 Commonwealth Games)
- 20 000 of the world's newspaper, radio, television, and Internet journalists
- the sale of over 9 million tickets, with half a million spectators a day travelling to events in and around London
- 63 000 operational personnel (47 000 will be volunteers, many as stewards, marshals, and drivers).

Providing Olympic facilities is expensive, and costs need to be weighed against benefits. This is done using a cost-benefit analysis. This assesses:

- costs – including **tangible** costs (things that cost money, e.g. building a stadium), and **intangible** costs (things that cannot be measured but are important, e.g. the effects of building disruption on people)
- benefits – again, these are either tangible (e.g. receipts from tickets), or intangible (e.g. the gains provided by having a new park).

The Olympic bid was made on the basis that benefits would exceed costs, although many of these benefits depend on the eventual value of the regenerated land.

▼ *The locations of key Olympic facilities*

The Olympic Park	Near Stratford, East London. The planned facilities are labelled on the photograph opposite.
What's there?	The main stadia and facilities, including ExCel (the Exhibition Centre for East London), where contact sports (boxing, judo, etc.) will be held.
And outside the Olympic Park?	Other London-based events include: • Wembley stadium (football) • Wimbledon (tennis) • Lord's cricket ground (archery) • Greenwich Park (Equestrian and modern pentathlon) • Hyde Park (Triathlon and road cycling) • Horse Guards Parade (beach volleyball)
Elsewhere in the UK?	• Bisley in Surrey (shooting) • Eton/Dorney rowing lake in Berkshire (rowing) • Weymouth in Dorset (sailing) • Various football grounds, e.g. Old Trafford (football)

Who are the key players?

The Olympics are very complex, with a number of organisations involved, as outlined in the table. The impact of the Olympics on the local economy, people, and environment is affected by the decisions which these organisations make.

Impact on the local economy

The site of the main Olympic facilities is in Lea Valley, Stratford, along an industrial estate known as Marshgate Lane. There were 207 companies here, employing over 4800 people. By December 2006, all the companies had agreed compensation with the ODA. By July 2007, all had moved out. One company moved about 150 metres away to a new location just outside the new Olympic Park; most others have stayed within the East London area. For some workers, this has meant an increased commute but others are now closer to work.

Key player	What they do
International organisation:	
The International Olympic Committee (IOC), with officials from most countries in the world	They decide who will host the Olympics.
UK central government, which has set up:	
The London Development Agency (LDA)	They are responsible for economic and urban development in London.
The London Organising Committee of the Olympic Games (LOCOG)	They are responsible for organising the Games, e.g. events, athletes, drug-testing, etc.
The Olympic Delivery Authority (ODA)	They are responsible for building the facilities, from planning and design, to land purchase, construction and completion of the stadia and other facilities.
Regional government – the London Assembly,	
Ken Livingstone, Mayor of London and leader of the London Assembly	The London Assembly has control over transport in London, and also influences policy and planning decisions across the city.
Transport for London (TfL); part of the Mayor's Office	They are responsible for transport across London, ensuring that half a million people a day can get to and from the Olympic sites effectively.
Local government – the Borough Councils involved	
Four London Borough Councils are affected by the Olympics (Tower Hamlets, Newham, Hackney and Waltham Forest)	They consider the Olympic Planning Application for approval.

▲ *Marshgate Lane industrial estate will be the site of the new Olympic Stadium*

The planned Olympic Park along the Lea Valley in East London ▶

London and the 2012 Olympics – 2

Impact on the local environment

The main feature of the Olympics is the Olympic Park, which lies along the Lea Valley in East London. In the 1960s, this was the UK's biggest manufacturing area; nearly a third of people here worked in manufacturing, ranging from food processing to railway engineering. The closure of London's last docks in 1981, and the relocation of many industries overseas (or to locations outside London), meant the collapse of manufacturing, so now only 7.5% of local people work in this sector.

This manufacturing collapse has led to widespread dereliction in the area. The Olympic Park will re-landscape the area to create a huge new London park. All the main Olympic facilities will be located here. Having the main stadia on one site will help to prevent the traffic congestion which brought Barcelona and Atlanta to a standstill in 1992 and 1996 – when athletes and officials joined spectators trying to travel to the main Olympic venues.

However, it is the legacy of the new park which will transform an area where environmental quality is poor. It is this legacy that helped London to win the Olympic bid. Electricity pylons have long blighted the environmental quality of the area – and these were the first things to be demolished as construction began and the power lines were put underground.

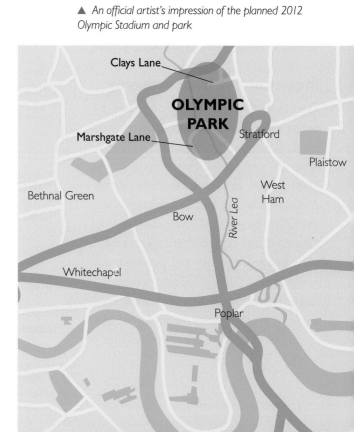

▲ An official artist's impression of the planned 2012 Olympic Stadium and park

▲ The location of the Lea Valley

▼ Dereliction along the Lea Valley

Impact on local people

The biggest social legacy of the Olympics will be housing. Housing needs in London are severe. Large price rises since the 1990s, and particularly since 2006 when the Olympics were won, have made housing in the area unaffordable for many trying to purchase or rent property. There is a serious need for **affordable housing** which low wage earners can either buy, rent, or part-purchase.

The Olympic Village is going to be built along Clays Lane in Stratford. The original site consisted of three types of housing:

- Halls of residence for the University of East London. The university has accepted compensation from the ODA for relocating.
- Clays Lane Housing Co-operative. This was a housing development built in the 1970s for single and homeless people, and those who normally found it difficult to get housing. The 450 residents here were eventually relocated to Housing Association housing throughout East London, thereby breaking up a community. In spite of financial compensation, many of the residents felt badly about this loss of community.
- A travellers' site. Newham Borough Council has had considerable problems finding a new site for the 150 travellers who had to be moved.

▲ *Clays Lane housing estate – all of which had to be demolished*

The plan is to re-model the Olympic Village after 2012 into affordable housing for 3000 people. However, existing houses have had to be lost in the building process.

There are also plans for at least another 9000 houses to be built post-Olympics. The aim is for over 50% of these to be affordable. Some will be for rent, others will be available for part-purchase and shared ownership with Housing Associations.

● Over to you

1 Copy the table below. In pairs, read through this unit and list the costs and benefits of staging the London Olympics in 2012. Compare your results with the rest of the class.

	Costs of the Olympics	Income and benefits from the Olympics
Tangible		
Intangible		

2 Using the 3 photos of the Lea Valley area before the Olympic builders moved in, describe the quality of the environment before development.

3 Now study the two images of the planned Olympic facilities. How far do you agree that the Olympics will leave a legacy of environmental gain?

4 a What are the economic and social gains and losses of the Olympics for the local area? Copy the table and summarise them.

	Gains from the Olympics	Losses from the Olympics
Economic		
Social		

  b Overall, would you say that the Olympics provide more gains than losses? Explain your answer.

● On your own

5 Define the following terms from the text: cost-benefit analysis, legacy, affordable housing

6 How powerful are each of the seven key players shown in the table on page 257? Reach a rank order from most to least powerful, and give your reasons.

7 In 500 words, consider this statement about the Olympics: 'Good for the country, good for London, not so good for local people.'

Exam question: Using examples, assess how well regeneration projects can rebrand urban areas. (10 marks)

The 2012 Olympics – economically, socially and environmentally successful?

What do you think?

In this unit you'll learn about the importance of transport in regeneration.

Stratford – the new hub of London

What made the UK government select East London as the location for the 2012 Olympics? For a long time, the area has been neglected and isolated from the rest of London because of poor east-west transport links across the city. Transport times to West London (e.g. Heathrow) have been slow, and have prevented developers from investing in East London.

This isolation has led to a **negative multiplier** in East London. Because the area has suffered low development and high unemployment for 30 years, low local spending power has kept retail investors away. Low spending power also means that local councils have lower incomes from business taxes, and this means there has been little money for transport development.

● A **Negative multiplier** means a downward spiral, or cycle, where economic conditions prevent growth.

However, things are now changing. To win the Olympic bid, London's Mayor, Ken Livingstone, had to show that transport links would be good enough to handle the number of people expected at the Games. Stratford was chosen because:

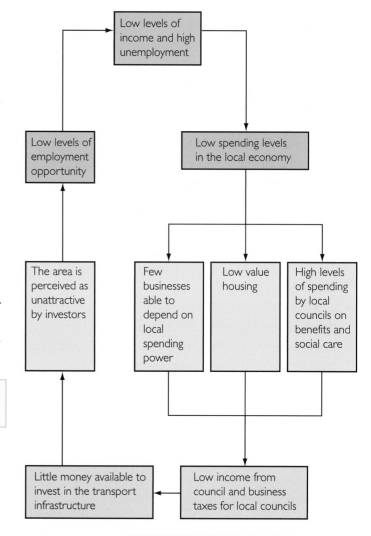

● the completion of the Docklands Light Railway now links Stratford to almost all parts of East London, and South London as far as Lewisham

● there is a fast commuter rail link to Essex and East Anglia, and additional rail links to North, North West and South West London

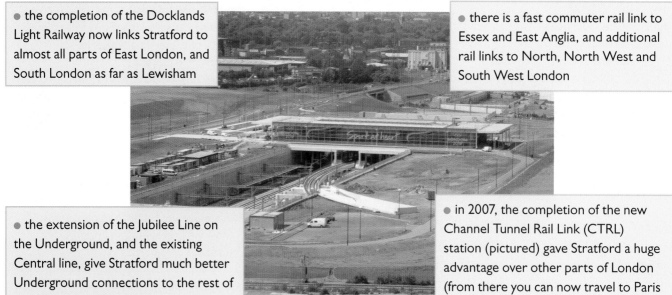

● the extension of the Jubilee Line on the Underground, and the existing Central line, give Stratford much better Underground connections to the rest of London and up into Essex

● in 2007, the completion of the new Channel Tunnel Rail Link (CTRL) station (pictured) gave Stratford a huge advantage over other parts of London (from there you can now travel to Paris or Brussels within 2 hours).

Now that it is at the junction of many rail links, Stratford has many geographical advantages over the rest of London. By 2012, Stratford will be able to handle 500 000 people per day.

Future links – go west!

Olympic investment in East London is part of a bigger regeneration programme. When completed, Stratford's new transport links will make it the best-connected location in the UK. By 2015, a further proposal, CrossRail, will link Stratford to West London and Heathrow. This line alone will be able to handle a 40% increase in passengers across London, and will be a major benefit to commuters. Investors now see Stratford as one of the major economic growth areas of London and the UK. It has become a true **transport hub**.

The biggest planned project is Stratford City – the largest commercial development since Canary Wharf. Costing £4 billion it will consist of 1.2 million square metres of space and be built over 15-20 years. Its architects claim that this space is greater than Bluewater Shopping Centre in Kent, the Broadgate Centre in London's City, and 4500 homes combined. It will have some major impacts:

- The retail development will become a new focus for East London.
- People shopping at Stratford City will increase consumer spending in London, instead of **economic leakage**.
- Commercial space will increase, giving companies the choice of locating their offices close to the CTRL station, so they will be just 2 hours from Paris.

▲ Stratford City as it will look

- A **transport hub** is where several transport links converge in one location.
- **Economic leakage** occurs when money earned in one location 'leaks' or is spent in another location.

Will European banks choose Stratford instead of Canary Wharf as 'the place to be'?

What do you think?

Over to you

1 Explain how East London's original transport links led to its deprivation.
2 Use the example provided of a negative multiplier diagram to devise a **positive** multiplier diagram if the plans for East London are successful. Draw a cycle or spiral.
3 How, and why, should the new CTRL station be so important in East London's regeneration?
4 Work in small groups. Devise a storyboard or sequence of advertisements to advertise Stratford's advantages as a location for European companies to invest in.

On your own

5 Define the following terms from the text: geographical isolation, negative multiplier, investor, transport hub, commuter, economic leakage, CTRL
6 Outline how these terms explain why:
   a East London is poorer than the rest of London
   b and how transport improvements should improve this.

Sydney's Olympic Story – 1

In this unit you'll learn how successful sports-led regeneration was in Sydney, and what we mean by sustainability.

Evaluating rebranding in Sydney

For host cities, the Olympics are about more than just sport. They offer cities the chance to use investment in sport as an opportunity for regeneration. This unit is about Sydney's Olympics in 2000, and an attempt to regenerate and rebrand an area of contaminated land.

Sydney's Olympics

When competing to host the Olympics, a city's proposal has to be unique. Sydney's bid was based on a proposal to hold the world's first 'Green' Olympics. The plan began in early 1992 with a design competition for the Olympic Village, organised by local architects, planners, housing associations and the local Council. One of five winning entries was from Greenpeace Australia, called the 'Greenpeace Australia Guidelines for the Olympic Village'. Their plan was to create an Olympic Village which could then be used for affordable housing.

It was to provide an example of **sustainable development**. The plan would have to provide:
- an effective public transport system
- affordable housing and rents
- a mixture of employment opportunities and housing, thus avoiding long commutes
- local services for people, such as medical services, shops, schools
- other quality of life features, such as parks.

Greenpeace set out to avoid:
- creating anything which was environmentally destructive, e.g. leading to traffic pollution
- using any construction materials which were high energy consuming or polluting, or where supplies were limited.

Having planned the Olympic Village, Greenpeace extended their design to the whole Olympic Park. On the right are their criteria for a sustainable, or 'Green' Olympic Games.

Criteria used by Greenpeace for a 'Green Olympics' ▶

▲ *Sydney's Olympic Park*

A Green Olympics should:
- involve the use or adaptation of existing facilities, rather than building from scratch
- be constructed on **brownfield sites** (previously used industrial and commercial land), so that **greenfield sites** (parks or agricultural land) remain untouched
- be financially solvent
- use building designs which are environmentally friendly, e.g. using solar power and recycled water where possible
- undergo an environmental and social impact assessment, involving local people
- minimise any adverse impacts of the Olympic sites and events on nearby residents
- protect native bushland, forests, wetlands, fauna or threatened ecosystems
- be accessible by public transport.

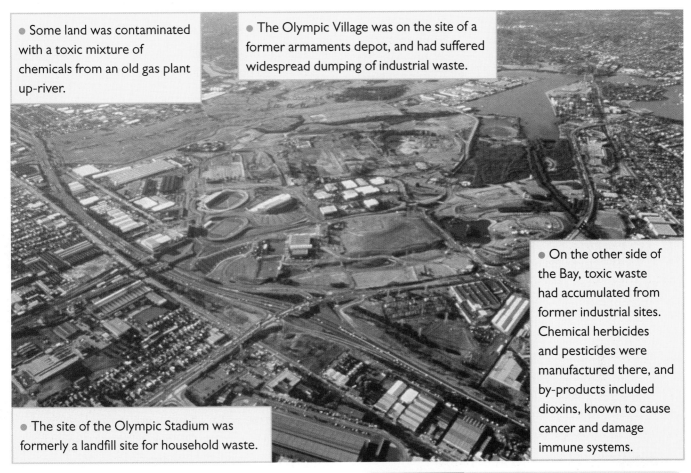

- Some land was contaminated with a toxic mixture of chemicals from an old gas plant up-river.

- The Olympic Village was on the site of a former armaments depot, and had suffered widespread dumping of industrial waste.

- On the other side of the Bay, toxic waste had accumulated from former industrial sites. Chemical herbicides and pesticides were manufactured there, and by-products included dioxins, known to cause cancer and damage immune systems.

- The site of the Olympic Stadium was formerly a landfill site for household waste.

▲ Homebush Bay from the air, showing the former uses of the Olympic sites

The plan was to build the Olympic Park at Homebush Bay, about 14 km west of Sydney's CBD. However, from the start it was difficult to achieve the aims of a 'Green Games'. Homebush Bay, and the land surrounding it, was badly polluted, as the photograph shows.

Before the Olympic sites were redeveloped, consultants from the USA found that pollution levels there were among the worst they had ever seen.

Cleaning up the site

From 1993 until the Games took place in 2000, the organisers went a long way to cleaning up Homebush Bay. A number of different ways were found to reduce the effects of past contamination using new, 'green' technology which reduced dioxins to a mixture of nothing more harmful than table salt and vegetable oil! Soil microbes were also found to be able to process dioxins safely in ways that would cut contamination. Although the site is still far from clean, it is better than it was in 1993.

Background

What is sustainability?

Sustainability is 'the ability to provide for the needs of the world's current population without damaging the ability of future generations to provide for themselves.' It has three main aspects:

- Economic sustainability asks 'Will benefits match costs? Will the facilities last for years?'
- Social sustainability asks 'Does everyone benefit, including the poor and the elderly? Or is the project likely to benefit just those who are well off?'
- Environmental sustainability asks 'How much does a project damage the environment? Are buildings made from environmentally sound materials that do not pollute when they are manufactured or disposed of? Is it efficient in its use of energy and water?'

The rest of this unit considers whether Sydney's Olympics were sustainable economically, socially and environmentally.

Sydney's Olympic Story – 2

A – Judging economic sustainability

In 2001, consultants PriceWaterhouseCoopers published the economic impacts of Sydney's 2000 Olympics and Paralympics (figures are in Australian dollars):

- $3 billion investment came to Sydney, including $2 billion worth of permanent sports facilities.
- 20% of Olympic spending went to businesses in New South Wales (NSW), the state in which Sydney is located.
- Over $6 billion was spent on infrastructure, e.g. redesigning Sydney's airport.
- Over $1.2 billion in conference business came to NSW between 1993 and 2007.
- Tourism received a huge boost: 1.6 million additional tourists spent $6 billion in 2001, turning the 2000 Olympics into a profit-making event.
- The Olympics helped Australians to compete for projects overseas. The Australian firm Multiplex built the new Wembley Stadium in London.
- Other income included broadcasting rights ($1.1 billion), sponsorship ($680 million) and ticket sales ($600 million).

B – Judging social sustainability

- Sydney's main aboriginal suburb, Redfern, suffered increased police raids in the month before the Games.
- There were protests in Bondi at the loss of access to the beach during construction of the Beach Volleyball Stadium.
- Local residents were exposed to toxic dust from the Olympic clean up.
- Between 1995 and 2000, housing rents in the Olympic area rose 7% above inflation each year, affecting low-income families most.
- 76% of Sydney's boarding houses – for those on social security, or with psychiatric disabilities – were converted to backpacker hostels for tourists.
- Homelessness increased in Sydney during the Olympic period.
- The cost of the Games diverted government funding from health, education, and transport.
- Many rural police stations closed during the Games, because the police officers were sent to Sydney.
- The plan to turn the Olympic Village into low-cost housing suffered, because the clean up of Homebush Bay was so expensive. Everything was eventually sold at market value.

Will London 2012 repeat Sydney's successes – and its failings?

What do you think?

◄ The former Olympic Village, now Newington suburb, which advertises itself as 'Australia's first solar suburb'

C – Judging environmental sustainability

'Green' rules	How well does Sydney match up?
Use brownfield, not greenfield sites	Homebush Bay was an industrial area. Much of the land was derelict in 1993 when Sydney won the 2000 Olympics.
Use existing materials, rather than build from scratch	Building foundations were recycled concrete and masonry from the demolition of an old abattoir on the site. During the construction of Sydney showground, 95% of waste was recycled.
Design and use 'green' buildings and materials	Where possible, non-toxic materials were used, e.g. natural fibres, non-toxic paints, etc. CFCs, and PVC were banned. Buildings were designed for insulation; no air-conditioning was used.
Minimise the impacts of Olympic events on residents	Most stadia and the Olympic Village were located on one site, avoiding traffic jams during the Olympics.
Minimise waste, and recycle wherever possible	Renewable energy sources were used. Compostable paper plates, packaging, cutlery and bin liners, and recyclable plastic beakers, wine 'glasses' and food packaging were used.
Protect native ecosystems	Mangrove and salt marsh in Homebush Bay, near the Olympic Park, were protected and extended, and habitats were restored.
Make Olympic sites accessible by public transport.	A new rail link was built to the Olympic Park. Admission tickets included public transport costs to the Games, from up to 200 km away.
Manage water sustainably	Sewage and storm water from the Olympic Park and Olympic Village were filtered and then recycled for irrigation and toilets. Pool water was ozone-filtered to reduce chlorine.
Use energy efficiently	All 665 houses at the Olympic Village had solar energy cells and solar hot water heaters.
Create an amenity for people	450 hectares were restored near the Olympic Park to form a public recreational and ecological area.
Reduce past contamination	Australian scientists developed a way of destroying dioxin, turning toxic chemicals into vegetable oil and table salt!
Cut the use of CFCs in cooling drinks	Coca-Cola was asked to provide 'green' refrigeration during the Olympics. Although this was not achieved in time for Sydney's Games, it was for Athens in 2004.

● Over to you

1 In pairs, study the economic panel (A) and classify each statement according to whether:

   a it benefited Sydney, New South Wales (NSW), or Australia as a whole.

   b the impacts were short-, medium-, or long-term. Create a table to classify the statements. You can split them up if you want, or place them in more than one box.

2 Using this table, explain in 300 words whether Sydney, NSW or Australia as a whole benefited most from the 2000 Olympics, and whether the benefits were short-, medium- or long-term.

3 Take each statement in the social panel (B). Copy the table below, and decide whether each issue is a price worth paying for the Olympics. Give your reasons.

Statement	Yes, this is a price worth paying because …	No, this is not a price worth paying because …
Police raids in Redfern.		

4 Take each statement in the environmental panel (C). Award each one up to 5 points for how well you think Sydney provided a 'green' Olympics to get a score out of 60. Compare your score with other people.

5 'The Sydney Olympics – economic and environmental benefits, but at a social cost'. Debate in class how much you agree with this statement.

● On your own

6 Define these terms from the text: sustainable development, brownfield sites, greenfield sites

7 'Economic and environmental benefits, but at a social cost'. Follow up the class debate and write about 750 words to explain how far you think this description is true of Sydney's Olympics.

Exam question: Referring to examples, evaluate attempts to improve the environmental image of urban areas. (10 marks)

Re-imaging Manchester

In this unit you'll find out how music and leisure can be used to regenerate and re-image cities.

In need of a new image

Manchester suffered economic collapse in the 1980s, as its textile industry declined. It had a poor image, with derelict industrial land, poor housing, and a declining economy. However, Manchester City Council developed a policy of re-imaging the city as a 'place to do business with', involving music, television, sport and culture.

Manchester – music and television

Manchester has had a vibrant music scene for decades; its bands have enjoyed national and international success. These bands include:
- 1960s – 1970s pop acts (e.g. Herman's Hermits, 10CC)
- 1970s punk – 1980s indie (e.g. Buzzcocks, The Smiths, Joy Division and New Order)
- late 1980s 'Madchester scene' (e.g. Stone Roses, Happy Mondays)
- early 1990s mainstream pop and 'Britpop' (e.g. Take That, Oasis)
- late 1990s (the 'new Manchester dance underground').

Television also helped Manchester's regeneration and re-imaging. In the 1990s, Granada Studio Tours developed a successful tourist attraction based on Coronation Street. Other TV series – such as *Shameless*, *Clocking Off*, *Queer as Folk*, *Cold Feet*, and *Cutting It* – have created the image of a city with a strong business and youth culture. The BBC has moved several of its operations to Manchester.

Manchester – sport and culture

Aside from its famous football clubs, since the late 1980s the city's image has focused on sport.
- The National Cycling Centre (now the Manchester Velodrome) opened in 1994.
- The city bid unsuccessfully for the 1996 and 2000 Olympics, but hosted the World Table Tennis Championships in 1997 and the Commonwealth Games in 2002.

Non-sporting events hosted by Manchester include music conventions, the Arts Council City of Drama, Digital Summer 98, and major concerts at The Nynex Arena (Europe's largest indoor concert venue) and Bridgewater Hall (home of Manchester's Hallé Orchestra). The redevelopment of Central Station into G-Mex has provided a venue for major exhibitions and conferences.

▲ *Derelict factory site in East Manchester*

▲ *Granada's Coronation Street set, one of the city's major tourist attractions*

▲ *G-Mex, the former Manchester Central Station, now a conference and exhibition centre*

How did Manchester's success happen?

Re-imaging has given Manchester a reputation as 'the most successful city' outside London. The City Council's strategy to re-image Manchester, its music and its talent was key to this. However, a good transport infrastructure (the Metrolink tram system, developed in the early 1990s, provides a superb transport network) and the availability of cheap property were also crucial.

What are the issues for the future?

Re-imaging Manchester's old warehouses and factories has brought problems. The redevelopment of Castlefield canal basin occurred because land there was cheap. Cheap, short-term leases on office and workshop space are attractive to the music industry – but cheap locations are now hard to find. The redevelopment of canalside areas with bars and nightclubs, such as Manchester's Gay Village, has made city centre property more expensive.

Property cost is not the only problem. Several clubs have been forced out because of:

- pressure from organised crime
- the inability of magistrates and police to manage the sector effectively
- problems in the provision of safe, late-night transport.

Manchester's new image will take as much effort to maintain as it took to create.

▲ *Castlefield, one of the city's major regeneration sites*

Is music a good way of regenerating cities? Or do tastes change too often?

What do you think?

● Over to you

1. Using Google Maps and Images, develop an annotated photo to locate Manchester's features mentioned in this unit.
2. In pairs, decide which of these factors has played the greatest part in re-imaging Manchester:
   - local people with musical talent
   - talent within Granada TV and other media companies
   - decisions by the council to bid for events such as the Commonwealth Games
   - the availability of cheap buildings
   - the vision of developers in turning derelict areas into attractions

3. How serious do you think the four future issues (including property cost) will be for Manchester?

● On your own

4. Research another city with a music, sport or cultural industry that has helped to develop its image, e.g. Liverpool, Nashville. How has it used its image? How far has the image of the city changed?

6.7 Can Walton-on-the Naze be resuscitated?

In this unit you'll learn how rebranding is needed in the traditional British seaside resort.

The traditional seaside resort

▲ Albion Beach, Walton-on-the-Naze, Essex, in its heyday in the 1960s

▲ The resort today – locked beach huts and ageing holidaymakers

In the 1950s and 1960s, Walton-on-the-Naze in Essex was in its prime. Located about 75 miles from London, it had the longest beach in Essex and the second longest pier in the UK. Its amusement arcades resounded with the sound of slot machines, and stalls selling candyfloss and kiss-me-quick hats, and fairground rides, were found on the pier. Day-trippers travelled either by train from East London, or by bus from other towns within Essex, such as Chelmsford and Colchester. Many people spent their annual holiday there. At this time, few people travelled far on holiday – most had only two weeks holiday a year, wages were low, and only better-off people had cars.

Now, a combination of longer holidays, higher incomes, increased car ownership, and cheaper travel overseas have meant that resorts such as Walton-on-the-Naze have declined. Day-trippers still visit the town, but hotels and guesthouses have lost business – often closing or left dilapidated. If there was ever a case for rebranding, Walton-on-the-Naze is surely it.

Walton-on-the-Naze

▼ The High Street in Walton-on-the-Naze

Walton-on-the-Naze in the 21st century

Walton-on-the-Naze is an ageing town, inhabited by the people who retired there after happy holidays earlier in their lives. Its population structure makes it one of the UK's most 'top-heavy' towns. Many people live on state pensions, with little spare money for luxuries. While the pubs cater for tourists, there are few restaurants. The town centre has a mixture of convenience shops, charity shops – and two funeral parlours. Many young families leave, not just because there is little well-paid work, but because there is little for them to do.

How should Walton-on-the-Naze rebrand itself?

Walton-on-the-Naze has several options. See the list below and other examples that you have studied in this chapter. Consider which rebranding projects you think have been successful, and why. Then make a reasoned assessment about which might work best for Walton-on-the-Naze.

Options	What?	Why?
Option 1	**New commuter housing** Expand the town and renovate older property	Housing in Essex is much cheaper than in London, although London is only 70 minutes away by train. This would attract younger, professional people, and boost spending power in the town.
Option 2	**Gentrification** A rebrand to attract a different socio-economic group	A rebrand appealing to London's middle classes, with smart restaurants and hotels. Southwold and Aldeburgh in Suffolk have branded themselves in this way, and their economies benefit from 'London money'.
Option 3	**The sunshine coast** Upgrade Walton-on-the-Naze's tourist image and encourage more visitors	This builds on Walton-on-the-Naze's existing image as a holiday resort. The weather is drier and sunnier here than in most parts of the UK – only the Sussex coast is sunnier.
Option 4	**Environmental value** Rebrand Walton-on-the-Naze on the basis of environmental quality	The river estuaries to the north of Walton-on-the-Naze teem with bird life. The Naze itself, a coastal spit around a river estuary, is a Site of Special Scientific Interest (SSSI).
Option 5	**Flagship development** Attract the casino market and create a new gambling hotspot	Gambling is a huge new growth industry, offering both jobs and future expansion. Casinos elsewhere offer conference facilities as well as their gambling business.
Option 6	**Retirement centre** Rebrand the retirement market and focus on the over-60s as the core population	Rebrand Walton-on-the-Naze as a 'retirement capital' and upgrade the facilities and amenities for that age group, e.g. secure housing units and shopping and entertainment focused on older age groups. Similar 'resorts' exist in the USA and Australia.
Option 7	**Festival town** Attract the festival and conference market, with special events through the year.	The conference market is large and company spending power is huge. The Walton-on-the-Naze Festival is already established in September, and there are a number of different festivals during the summer.

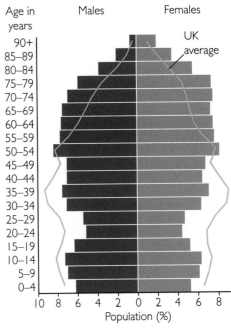

▲ The population structure of Walton-on-the-Naze

> *Have British seaside resorts had their day?*
>
> **What do you think?**

● Over to you

1 Is there a 'cycle of poverty' for UK seaside resorts? Explain your answer.
2 Create a table to list the advantages and disadvantages for Walton-on-the-Naze of each of the 7 options.
3 Which two options would you recommend Walton-on-the-Naze develops in the future? Write a 600-word report with your reasons for your choices, and explain why you rejected the other options.
4 Now research Walton-on-the-Naze's own website – www.Walton-on-the-Naze.com – and see how it is trying to rebrand.

● On your own

5 Research another resort, e.g. Blackpool, Skegness, Southend-on-Sea, Scarborough. How far is it trying to rebrand by selling its image to attract tourists? How does it do it? Use the resorts' own websites.
6 Seaside resorts often suffer serious problems of homelessness and poverty, particularly during the winter months. Find out why this is.

Exam question: Using examples, explain why some seaside resorts are difficult to regenerate and rebrand. (10 marks)

In this unit you'll investigate deprivation in a rural area.

Cornwall and deprivation

> Cornwall remains one of Europe's most deprived areas. Many coastal communities are still struggling to find ways to face social, economic and environmental challenges.
> Andrew Mitchell, Cornwall County Council

Every year, over 4 million tourists visit Cornwall for its spectacular coastal scenery and beaches. The photo above is typical of the image that many of these visitors have of Cornwall – it is a beautiful area and the UK's top destination for family holidays and short breaks. Yet many of the people who live there feel that it is in crisis and in need of regeneration. Low wages, the decline in the rural economy, the lack of rural services and employment opportunities have all contributed to this feeling.

Deprivation caused by low wages

Deprivation means a lack of something. Deprivation can be measured using average wages, and, in rural areas, they tend to be lower than in urban areas. Cornwall has the lowest weekly wages in Britain (£329.30 in 2005 – 25% below the UK average), and the gap between it and the rest of Britain is getting wider. Within Cornwall, the poorest borough is North Cornwall (average weekly wage £307.60 in 2005).

Decline in the rural economy

Why are rural areas in crisis? Mostly, it is because of the decline in traditional employment. In Cornwall, **primary employment** dominated until the 1970s. This included farming, fishing, tin mining, and quarrying china clay. Each sector declined for the reasons shown in the table. Rural areas are now producing less than they used to – so geographers talk of the **post-production countryside**, just like 'post-industrial' is used for towns and cities.

The decline in Cornwall's primary employment ▶

▲ *The Cornish countryside – how can this be a place in crisis?*

- **Primary employment** involves working in the production of raw materials or natural products, e.g. farming, fishing, forestry, mining, quarrying.
- **Post-production countryside** means how the countryside should be used if farming declines even more.

Sector	Reasons for its decline
Farming	• Falling farm prices, as supermarkets seek the lowest prices from their suppliers. • Importing food from overseas, where wages and costs are lower. For example, milk from the new EU countries costs 14p per litre to produce, while UK farmers can only produce milk at 21p per litre. • Withdrawing EU subsidies, which has led to a rapid and accelerating decline.
Fishing	• EU quotas have allocated fish supplies to other European countries. • The decline in overall fish stocks caused by previous over-fishing.
Mining	• The exhaustion of the tin reserves in Cornwall. • A collapse in tin prices caused by overseas competition. • The strength of the pound has made UK tin more expensive to buy overseas.
Quarrying	• The St Austell area has some of the world's best china clay reserves. However, fewer and larger quarries, using technology rather than people to extract the clay, has resulted in cutbacks in the workforce.

The tourist economy

Tourism is an established industry in Cornwall, and has helped to offset job losses in primary industries. It directly employs 25% of people in some areas; indirectly it employs more – including shopkeepers, decorators and builders. Cornwall's main tourist resource is its environment, especially the coast. But tourism brings problems:

- The jobs are mainly seasonal, part-time and poorly paid.
- The visitor numbers vary between seasons and depend a lot on the weather.
- Only 33% of the profits from tourism stays in Cornwall – the rest 'leaks' out of the county, e.g. going to national hotel and pub chains.

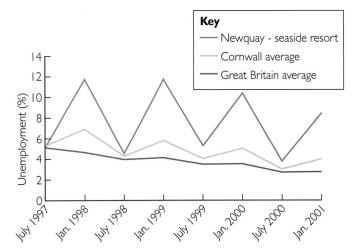

▲ *Seasonal unemployment (Newquay is Cornwall's biggest resort and suffers the worst seasonal unemployment)*

Background

Core and periphery theory

The main reason for deprivation in rural areas is their remoteness. Geographers use a theory called **core and periphery** to explain this.

The UK's **core** is where most wealth is produced. It extends from London to Leeds and Manchester. This core:

- produces 75% of the UK's goods and services
- has the highest wages and levels of investment
- has the densest transport **infrastructure** – most motorways, the busiest airports and the densest rail networks are there.

Look at the map on the right. Notice how far Cornwall is from the economic core.

Rural areas within the core are better off than more remote areas. The UK's most **affluent** rural county is Surrey; average wages were £500 per week in 2006. 35% of Surrey's residents work in London in professional employment – their average wages were even higher in 2006 (£800 per week).

Remote rural areas are described as the **periphery**, and are poorly served by transport infrastructure. They are a long way from core markets, and job opportunities are fewer. There are some benefits – peripheral areas can be cheaper to live in – but they often suffer depopulation as young people leave to find better jobs.

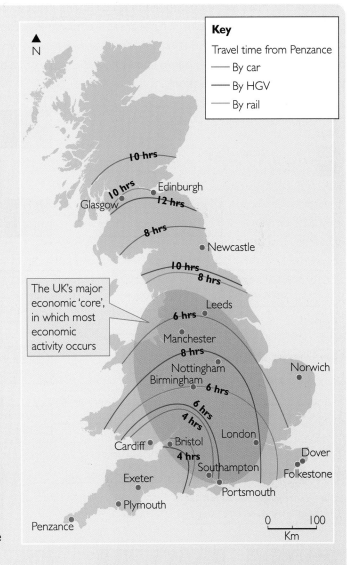

Cornwall in crisis? – 2

Lack of rural services

In the UK, 99% of areas which suffer from a lack of services are rural.

- Visiting a Post Office is now more than a 2 km trip for 1 out of 15 rural households.
- 72% of UK villages don't have a village shop.
- 39% of households in rural areas live over 2 km from a cash point.
- Only 14% of rural parishes have a doctor's surgery, and 29% of all rural settlements have no bus service.

Some services remain open because of tourism, but they remain fragile.

▲ *The village store and garage in Gorran, south Cornwall, closed in 2002 and has been redeveloped as a small housing complex.*

Lack of employment opportunity

For many years, at least half of Cornwall's young people have created a 'brain drain' leaving the county to look for better opportunities elsewhere. In a survey of St Austell, half the people aged 18 did not expect to still be living there in five years. Most felt that there were no opportunities for them. Most employment in Cornwall is low paid. Low-income households in rural areas are more likely to remain on a low income than urban households. This results in low spending power, which in turn limits business opportunities. The diagram below shows how this is part of a bigger chain of events, known as a 'cycle of deprivation'.

▼ *Cornwall's cycle of deprivation*

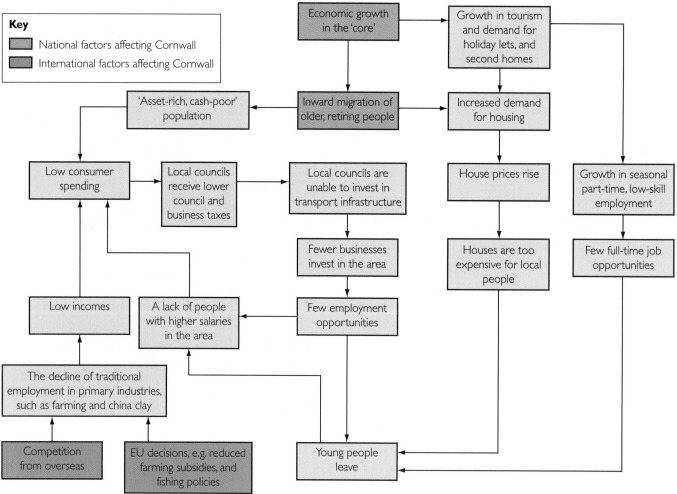

In 1979, a geographer named Shaw said that deprivation was a trap, because people in rural areas had poorer:

- resources – low incomes and a low share of national wealth
- opportunities – poor services, e.g. education, health
- mobility – poor transport and high costs, making jobs inaccessible.

The UK government produces a **multiple index of deprivation** to illustrate this. It uses six ways of measuring deprivation: income, employment, health and disability, education, housing, and services. These measures are compiled into a single figure and ranked to describe how deprived an area is compared to other areas in the UK. Cornwall is considered deprived in terms of income and employment, but education and health services are good.

Oddly, while Cornwall is losing its young people, it has the fastest growing population of any UK county! High environmental quality has led to **inward migration** of older or retired people. However, many of them live on low, fixed pensions, which adds to the large number of Cornish residents on low incomes. Without additional funds from central government, local Cornish councils have low incomes from council or business taxes.

Can high environmental quality make up for deprivation in areas like Cornwall?

What do you think ?

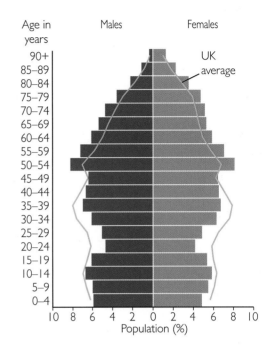

▲ *Cornwall's population structure*

● Over to you

1 Study the table on page 270. Classify the causes of Cornwall's fall in primary employment – are they
   a local or regional
   b national
   c international or global?

2 How far do the causes of Cornwall's deprivation lie outside the county?

3 a In pairs, consider reasons why rural shops are forced to close, in spite of tourism.
   b Using these reasons, construct a cycle of deprivation to explain why villages have few services.

4 a Make a list of the advantages and disadvantages of living in peripheral areas.
   b Which would appeal more as a place to live – the core or the periphery – to i 16-18 year olds, ii retired people? Explain why.

● On your own

5 Define the following terms from the text:
   rural deprivation, primary employment, post-production countryside, core and periphery, infrastructure, affluent, multiple index of deprivation, inward migration

6 Describe and explain the pattern of unemployment in Cornwall.

7 Explain how and why deprivation can be a spiral.

8 Using the cycle of deprivation and information on core and periphery, explain:
   a why many people leave Cornwall
   b why many jobs in Cornwall are low paid
   c why low-income rural households are more likely to remain poor than urban ones
   d why central government assists deprived areas.

9 Research one other area of the UK that you consider part of the 'periphery'. Using the census website www.statistics.gov.uk find out whether this area is also
   a deprived b in need of regeneration.

6.9 # Cornwall's Garden of Eden?

In this unit you'll find out how flagship projects for re-imaging rural areas can bring major benefits.

A new landscape

In March 2001, the Eden Project opened to visitors. It consists of two large conservatories – known at Eden as 'biomes' – exhibiting the world's major plant types. Between them, they include 100 000 plants representing 5000 species. Eden consists of:

● a humid tropics biome, housing humid tropical plants, and;

● a warm temperate biome with Mediterranean plants and plants from places such as South Africa.

In 2007, a third biome was planned to explore the theme of sustainability.

The Eden Project has huge potential for tourism, but how successful has it been in regenerating and re-imaging the area?

Located 4 miles north-east of St Austell, in the borough of Restormel, the Eden Project has transformed the local landscape. The area around St Austell is quarried for china clay, which leaves huge white waste heaps and open quarries on the landscape. In contrast to its image as a holiday destination, Cornwall has the UK's highest percentage of derelict land. So building the Eden Project in the bottom of a china clay pit was an opportunity to re-image the environment.

Eden project – the problems

There have been criticisms of the Eden Project, as summarised in the table below.

Traffic levels	• Visitor levels have produced huge traffic congestion in the area, with traffic sometimes at a standstill.
	• Local lorry drivers have estimated that journeys have increased by 30 minutes because of the congestion.
Pollution and air quality	• The Eden Project is ironically the major source of pollution in the area.
	• Almost all visitors arrive by road, and few by cycle, bus, or rail.
	• Although there is a bus link from St Austell, few use it.
	• The 3500 cars that fill Eden's parking spaces, generate more CO_2 emissions than all other sources in St Austell combined.

▲ *The biomes at the Eden Project*

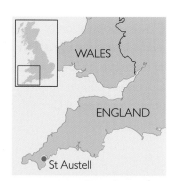

▼ *The location of the Eden Project in Cornwall*

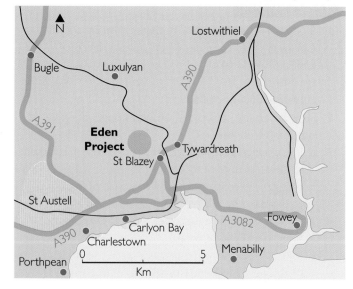

◄ *Problems created by the Eden Project*

Eden project – the benefits

How successful has the Eden Project been? The table below summarises its benefits to the local economy.

Visitor numbers	• Initial estimates were for 750 000 visitors in its first year. In fact, 1.9 million visitors came! Numbers have now stabilized at 750 000 per year. • Within six months, it was the UK's third most-visited attraction. • Over 6 million people visited in the first 4 years.
Visitor spending	• Each visitor to the Eden Project spends on average £150 in Cornwall. • Tourists visiting Eden in its first three years spent £600 million.
Accommodation	• Since 2001, demand for holiday cottage accommodation has doubled. • Holiday cottage owners normally get 16-20 weeks' booking per property per year. Since 2001, many now get 36 weeks. • Most cottages are owned by local people and so benefit them.
Employment	• The Eden Project employs 400 full-time staff. • Including part-time and seasonal staff, 600 equivalent full-time jobs were created. • Eden claims that all staff were recruited locally. • 75% of staff were previously unemployed, and 40% are over 50 years old. These do not include construction staff. • Overall, Eden has reduced Cornwall's unemployment by 6%.
The impact on local producers	• The company sources all food and drink locally, where possible. This has boosted local farmers and food-processing companies.
Impact on other attractions	• The Eden Project has created spin-off interest for other attractions, e.g. the National Maritime Museum at Falmouth, Tate of the West at St Ives.
Impact on the economy	• In 2003, an average of 80% of Cornwall's businesses said they felt that Eden had brought very positive impacts for them and the Cornish economy.

◀ *The benefits of the Eden Project*

Does Cornwall need re-imaging?

What do you think?

A **multiplier effect**, increases the local economy:
- Visitor spending on entry tickets, food and gifts at the Eden Project increases turnover, and therefore employment – not just at the Project but for local suppliers as well.
- Visitors also spend money on accommodation, food, other leisure activities and travel, which creates a second ripple of benefits for employment in the area.
- Finally, the spending power of those employed by Eden and local suppliers increases as a result of their employment.

● The **multiplier effect** is the increased economic activity which occurs when one business creates demand and benefits others.

● Over to you

1 Why do you think the majority of local businesses feel that Eden has been very positive?
2 Study the information provided in this unit. Classify the problems and benefits created by the Eden Project into **a** economic, **b** social and **c** environmental.
3 Which of these classifications seems to offer **a** the greatest benefits, **b** the greatest problems?
4 Are the benefits worth the problems? Explain your answer.

● On your own

5 Define the term 'multiplier effect'.
6 'Economic benefits outweigh the environmental impacts'. Explain in 400 words how far you agree with this statement about the Eden Project.
7 Research progress into Eden's planned third biome. For example, when will it be completed? What benefits are expected? What impact will it have on visitor numbers, extra employment and so on?

In this unit you'll find out how investment in rural areas can help regeneration.

The need to rebrand

Together with the Eden Project, a lot has been done in recent years to regenerate Cornwall and rebrand the county. But why should Cornwall need to rebrand itself?

Cornwall's rural economy is suffering from a collapse in farm prices. It needs to **diversify** by creating a year-round economy, not just one that thrives in summer. Cornwall has for years been a favourite holiday destination for families and the retired, but what about young adults who might normally holiday overseas? As Units 6.8 and 6.9 have shown, problems of low economic growth and seasonal, part-time, and low-paid jobs, have led to a 'brain drain' of young people out of the county. Many of them think Cornwall has too little entertainment, especially in winter.

▲ Family holidays in Cornwall – but what opportunities are there for young people?

Now there are serious attempts to rebrand Cornwall and attract new tourists. The main concept is **destination tourism**, where people visit a place simply because of a single attraction. The hope is that they will then visit others. The Eden Project is an excellent example of this.

Two factors have helped to rebrand Cornwall and promote 'destination tourism':

A **individuals**, such as TV chef Rick Stein, whose seafood restaurant at Padstow in north Cornwall has led to large numbers of tourists who are interested in his restaurants and food shops. Now local people refer to Padstow as 'Padstein'!

B the **expansion of Newquay Airport** by the local council to accept flights from UK cities such as Leeds, Manchester and Edinburgh, as well as London's Gatwick and Stansted airports. This has helped to reduce Cornwall's geographical isolation.

▼ Watergate Bay, near Newquay Airport, now advertises to London's City workers to 'leave the office at 4, be on the beach by 7.30'. In this way, weekend groups are attracted to come to Newquay for short breaks or special events, such as stag and hen weekends.

Background

Objective One funding – turning £20 000 into £320 000!

In 1999, Cornwall was successful in gaining EU Objective One funding, designed to boost the local economy. Objective One aims to reduce social and economic differences within the European Union (EU). The funding comes from the EU and is granted to areas where the gross domestic product is 75% of the EU average, or less. It aims to encourage investment to boost local economies.

▼ *Who are the key players in Objective One?*

- Private individuals who invest their own capital
- Local councils (e.g. Cornwall and Isles of Scilly Council)
- Development agencies (e.g. the South West Development Agency – who use government money to stimulate and regenerate the economy)
- Business interests (e.g. Cornwall Farmers Ltd, Eden Project, Cornwall Tourist Board)
- **Objective One key players**
- Central government and its agencies (e.g. Arts Council, Countryside Agency, Heritage Lottery Fund, Jobcentre Plus)
- Education interests (e.g. Combined Universities in Cornwall, University of Exeter, Cornwall FE Colleges)
- Environmental groups (e.g. Cornwall Heritage Trust, English Heritage, English Nature)

How does Objective One work?

Investors have to start the process, and then apply for equal amounts of money from other sources, called match funding. This might come from public sources, e.g. local councils, or from private sources, e.g. banks. So, investors can:

- set up £20 000 of their own money
- get a bank loan for another £20 000, making £40 000
- ask local councils to match it to make £80 000
- bid for the South West Development Agency to match it to £160 000
- finally, bid for Objective One funds to match it, creating £320 000.

How successful has it been?

Helped by Objective One, the Cornish economy has grown faster than the UK average. It is still the weakest economy in England, but from 1994-2004 it grew at 5.8% per year, ahead of the UK average of 5.4%. In 2003, the Cornish economy showed the greatest improvement of any region in the EU. However, its schemes have had varying success – from the greater-than-expected success of the Eden Project, to failing schemes like South West Film Studios (see page 279).

But Professor Peter Gripaios of Plymouth Business School criticises it. He claims that it was wrong to give Objective One funding to some projects. By 2007, Objective One had backed 580 projects in Cornwall with £230m, but Mr Gripaios claims that 'the jam has been spread too thinly on too many projects'.

Rebranding 'cool' Cornwall – 2

The projects included here were all funded by Objective One. Some are focused on places along Cornwall's coast, perhaps its greatest asset.

The Extreme Sports Academy at Watergate Bay

The Extreme Sports Academy at Watergate Bay, near Newquay Airport, is targeting a younger age group. It offers courses in surfing, wave skiing, and kite surfing. The owners also run the Watergate Bay Hotel, overlooking the Academy and the beach. The hotel has a new restaurant, bar and accommodation. They are open all year, employing 50-60 people year-round in 2006, compared to 15-20 in 2003, as a result of increased trade.

Surfing at Watergate Bay ▶

Jamie Oliver's Fifteen restaurant

In 2006, TV chef Jamie Oliver opened the Fifteen restaurant, overlooking Watergate Bay. The 100-seater restaurant trains local young people in catering skills. Thirty 16-24 year olds from disadvantaged backgrounds are selected each year. Fifteen of them work in the restaurant, training at Cornwall College from January and in the kitchens from May, supported by professional chefs. Profits fund further training and development.

Jamie Oliver's 'Fifteen' restaurant, overlooking Watergate Bay ▶

CUC – stopping the brain drain?

To increase university courses and develop a 'knowledge economy', University College Falmouth and Exeter University joined forces to create the Combined Universities in Cornwall (CUC). CUC helps graduates set up businesses or secure jobs in knowledge-based companies in Cornwall, trying to cut the 'brain drain' of graduates leaving Cornwall. Examples include:

- Sixixis – which shapes Cornish timber into handmade furniture
- Neutralize, a computer company helping top clients to improve their placing on Internet search engine results, e.g. London's Stock Exchange.

> It is exciting to set up your own business, and we all love it in Cornwall. The quality of life down here is excellent.
> **Chris Jaratt, Sixixis**

> It's no disadvantage to be here, most of our communication with clients, for example, is done by e-mail anyway. We were in London before, but it was too expensive; I grew up in Cornwall and wanted to come back.
> **Edward Cowell (co-founder of Neutralize)**

▲ *What new businesses say about Cornwall*

Using Arts and Culture

To attract Arts tourists, Fowey holds an annual Du Maurier Festival in May, named after local author Daphne Du Maurier. It hosts authors, musicians, and broadcasters for 11 days. Investment in Arts and Culture in Cornwall has grown from the museum dedicated to sculptress Barbara Hepworth in the 1970s, to the opening of Tate St Ives in 1993. Theatre, music and dance are now available at Hall for Cornwall in Truro, which opened in 1997.

Fowey, home of the Du Maurier Festival ▶

But some fail ... South West Film Studios

In 2002, South West Film Studios proposed Cornwall's first film business at St Agnes on Cornwall's north coast. Many films have been made in Cornwall, e.g. Die Another Day, but no studio facilities existed. The £5.7m complex received nearly £2m from Objective One. Two film studios were built. It was expected that the company would create 200 permanent jobs and would bring millions of pounds into the economy.

In 2004, the studios went bankrupt. Building work was never completed and re-financing was sought, caused by increased building costs. Investigators claim that they found no accounts concerning the studio's finances. The owner, Alex Swann, was arrested for fraudulently obtaining money from Objective One and jailed in 2007.

● Over to you

1   In groups of four, select one of the following projects each – Extreme Academy, Fifteen, Combined Universities of Cornwall, and the various Arts festivals. Copy the table, and assess how successful you think your project has been.

How well does it rebrand Cornwall?	
How well does it attract new tourists?	
How well does it bring:	
a) economic benefits?	
b) social benefits?	
c) environmental benefits?	
Does it:	
a) provide year-round employment?	
b) improve incomes?	
c) make good use of Cornwall's environment?	
d) help to stop the 'brain drain'?	
e) improve Cornwall's lack of entertainment?	

2   Now reach an agreed order – which are the best projects? Present your order with reasons to the rest of your class.

3   Discuss whether you think there was ever a market for a film studios in Cornwall.

● On your own

4   Define these terms from the text: diversify, destination tourism, Objective One, match funding

5   How well do you think Objective One benefits Cornwall? Write about 400 words, using examples from this study.

A Cornish MP said: 'We could have taken all the Objective One money and given £20 000 a year, for 5 years, to the poorest 5% of the Cornish population. Would that have been better?'

What do you think❓

Rebranding farming

In this unit you'll learn how farmers can survive by rebranding their farm businesses.

Farming in crisis

The UK farming industry is one that few people know about and understand, and it's in crisis. Imagine a job where you have to work 7 days a week, 12 hours a day; perhaps getting up at 5.00 a.m. to milk cows in all weathers. Then imagine you are paid about half the national minimum wage. Small wonder that, in 2006, record numbers of farmers sold up.

The problem for some years has been that prices paid to farmers are not enough to earn a decent living. Farm incomes fell sharply between 1973 and 2003. The biggest fall occurred between 1996 and 2001/2, when average farm incomes (after costs) fell from £80 000 in 1995/6 to £2500 during 2001; recovering to £12 500 in 2002/3. But as an hourly rate, this is still less than the national minimum wage. Livestock farmers have been badly affected.

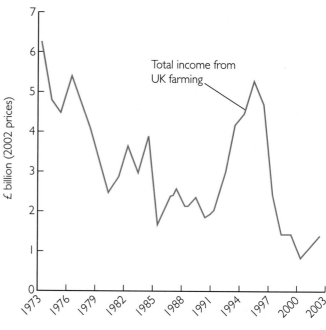

Total income from UK farming

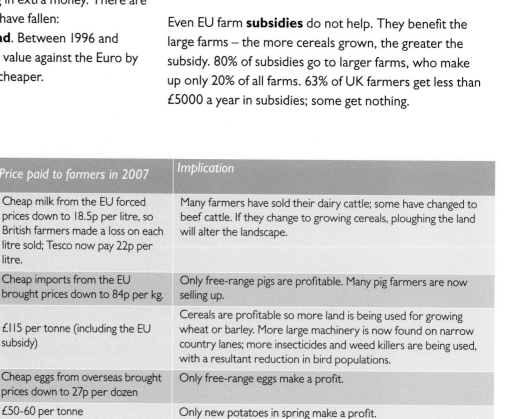

▲ *Farm incomes, 1973-2003*

Why has farming collapsed?

For small farmers, especially, incomes remain low. 69% of farmers rely on farming for their entire income, but many also have part-time jobs to bring in extra money. There are two main reasons why incomes have fallen:

- **The strength of the pound**. Between 1996 and 2007, the pound increased in value against the Euro by 33%, making imported food cheaper.

- **The power of supermarkets**. Competition between big supermarkets is intense. Prices get forced down to attract customers, and, as a result, farmers are paid less.

Even EU farm **subsidies** do not help. They benefit the large farms – the more cereals grown, the greater the subsidy. 80% of subsidies go to larger farms, who make up only 20% of all farms. 63% of UK farmers get less than £5000 a year in subsidies; some get nothing.

▼ *Comparing costs of production and farm prices*

	Cost of production	Price paid to farmers in 2007	Implication
Milk	21p per litre	Cheap milk from the EU forced prices down to 18.5p per litre, so British farmers made a loss on each litre sold; Tesco now pay 22p per litre.	Many farmers have sold their dairy cattle; some have changed to beef cattle. If they change to growing cereals, ploughing the land will alter the landscape.
Pork	95p per kg	Cheap imports from the EU brought prices down to 84p per kg.	Only free-range pigs are profitable. Many pig farmers are now selling up.
Bread wheat	£95 per tonne	£115 per tonne (including the EU subsidy)	Cereals are profitable so more land is being used for growing wheat or barley. More large machinery is now found on narrow country lanes; more insecticides and weed killers are being used, with a resultant reduction in bird populations.
Eggs	45.3p per dozen	Cheap eggs from overseas brought prices down to 27p per dozen	Only free-range eggs make a profit.
Potatoes	£77 per tonne	£50-60 per tonne	Only new potatoes in spring make a profit.

New life on the land?

In an attempt to increase incomes, some farms are now rebranding what they do. There are various options:

- Embracing tourism – tourists have a chance to see and understand how a farm works; buildings are converted for farm holidays.
- Focusing on higher-value products, such as vegetables, cheese, and wine.
- Adopting **eco-approaches** to managing the environment, such as planting mixed woodland.
- Adopting welfare approaches towards animal rearing which focus on the quality of care.

> *Cheap food or a healthy and profitable farming industry – which would you rather have?*
>
> **What do you think ?**

Lobb's Farm Shop

Until 2003, three brothers from the Lobb family, in south Cornwall, were making just £30 000 from their 800 acres, shared between three families. Their farm is near the Lost Gardens of Heligan, south Cornwall's second biggest tourist attraction after the Eden Project. They created a shop to sell their beef and lamb to a potential market of 463 000 visitors a year visiting the Gardens. The project, Lobb's Farm Shop, was financed using £200 000 funding from Objective One (see page 277) and central government.

The shop's focus is meat and vegetables produced on the farm, and other locally sourced products such as Cornish wine, cheeses, etc. It has created 14 new jobs, with more in the summer months. These include:

- 5 butchers (2 full-time, 2 part-time and a trainee)
- 1 Full-time and 6 part-time shop assistants
- 2 administrative staff.

It has generated over £600 000 in additional sales in three years. However, it is more than just a shop. It includes:

- a visitor centre informing visitors about farming, and tours to show them what happens on a farm
- ways of improving environmental quality on the farm by attracting wildlife – planting in ways that will attract birds and insects
- managing and raising beef cattle in a welfare-conscious way.

▲ *Lobb's Farm Shop*

▲ *Environmental quality on Lobb's Farm*

● Over to you

1. In pairs, make a list of all the negative features of working on a farm.
2. Now consider the positive features. Which list is longer? Why?
3. In what ways does Lobb's Farm Shop rebrand farming?
4. List all the benefits brought by the farm shop. Then classify these into economic, social and environmental.

● On your own

5. Define the following terms from the text:
   the strength of the pound, the power of supermarkets, farm subsidies, eco-approaches to farming
6. Explain how the strength of the pound and the power of supermarkets have led to many farmers giving up. Show this in a series of stages as a flow chart.

Exam question: Evaluate the success of a range of projects designed to rebrand a named rural area. (10 marks)

In this unit you'll learn how rebranding rural areas in LEDCs is just as essential as in MEDCs, and can have major impacts.

Rebranding and Uganda

Rebranding plays an important role in regenerating both cities and rural areas in the UK. But does it have the same possibilities in poorer countries of the world? This unit focuses on rural Uganda, and looks at two projects which try to improve people's image of rural areas, and the rural poverty trap. Can regenerating rural areas actually help people's lives in one of Africa's poorest countries?

UGANDA

Uganda factfile

Uganda lies across the Equator, but because much of it is highland, it is much cooler than you might expect. Unlike people's usual views of rural Africa, Uganda is mostly green and lush – apart from the much drier north.

Indicator	2005	Compared to UK
Surface area (sq km)	241 000	250 000
Population (million)	28.8	60.2
Annual population growth rate (%)	3.5	0.7%
Population living below the poverty line (%)	34	No data
Life expectancy at birth (years)	50	78.9
Fertility rate (births per woman)	7.1	1.8
Infant mortality per 1000 live births	79	5
Under-5s mortality rate per 1000 births	136	6
Under-5s malnourished (%)	23	No data
15-49 year olds infected with HIV (%)	6.4	.01
Primary school enrolment (% of age group)	117.5*	100
Secondary school enrolment (% of age group)	16	99
Literacy rate (% of over-15s who completed primary school)	67	99
GDP total (US$)	8.7 billion	2.2 trillion
GDP per person per year (US$)	302	36 544
GDP growth per year (%)	6.6	1.8
Fixed line & mobile phone subscribers per 1000 people	56	1615
Internet users per 1000 people	17	473

*Note: primary school education has attracted many who missed out before it became free. Many classes have older students learning with young children – hence the figure is 117.5% of the age group.

The rural cycle of poverty in Uganda

Rural incomes are low in Uganda. Over time, farms have become divided amongst sons when a farmer dies, making them smaller and smaller. Families with larger plots might earn £200 a year from coffee, but small plots only produce small incomes. Much farming is **subsistence**, and farming families can therefore get trapped in a cycle of poverty like that below.

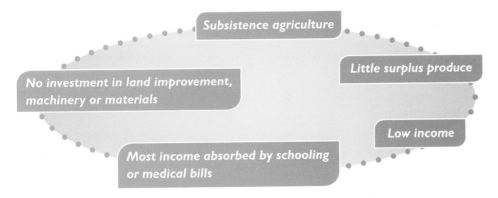

● **Subsistence farming** is when most of the produce farmers grow is used to feed themselves and their families, with little surplus to sell for cash.

◄ *The cycle of poverty for subsistence farmers*

Communities become locked in a **poverty trap**:

- Rural places offer few opportunities for people. For those who can afford the fare, migration to cities is a good opportunity.
- Few children receive secondary education because it costs about £50 per year – for many families this would be their total income from coffee. If a family has six children, who should be chosen to go to school?
- University places are expensive, with few government grants.
- Girls are less likely to receive secondary education. In primary schools, the ratio of girls to boys is 1:1; by age 11 it is 1:2, and by 16 it may become 1:6 or even 1:10.
- Girls may marry as young as 13 or 14 in rural communities, and have their first child soon after. The **fertility rate** is high – 7 children per woman.
- Women are poorest and have few rights; local custom means that they rarely own land, and are most likely to work as landless labourers as and when jobs are available. They have little control over how much they earn, and even less over their careers.
- Men often leave rural villages to go to cities – e.g. Kampala, the capital – to look for work and send money home (called **remittance payments**) to their families. Family income increases significantly with remittance payments.
- The rate of HIV/AIDS infection is high, even though Uganda has made huge strides in education and HIV prevention programmes. Deaths from AIDS have reduced Uganda's life expectancy from 65 to 49 for men and 51 for women.
- Families affected by AIDS suffer; costs of treatment are high, people who are ill have less energy and can do less work so family incomes suffer. Death means the loss of a wage earner and the debt of funeral costs.
- Health care systems are poorly developed in remote rural areas. In one rural district, there are two Government doctors and two health centres for 300 000 people. Poor sanitation means that malaria and cholera are big killers.

● **Fertility rate** is the number of children, on average, that a woman will have in her lifetime.

One person makes a difference

Equatorial College School (ECS) is in Kyarutanga, a village near Ibanda, 120 miles west of Kampala. The village is small, and dispersed over several square miles; access is by dirt road and the journey to Kampala can take 8 hours by car or 12 by bus. There is little secondary education; only 5% of primary school leavers attend secondary school, and fees are too expensive for most parents. There are few opportunities beyond working within the family. Those who do leave the village go to one of Uganda's main cities for employment, as do students who go to university.

In 2003, one man – Robert Kamasaka – set up ECS on a hillside covered in scrub. There is no state funding of secondary education, so ECS relies on:

- **Fees**. Each term's fees are £20, or £50 a year. Boarding fees are double.
- **UK Fund-raising**. As a boy, Robert studied in the UK, and his former school in North London now raises money for ECS. Between 2003-7, it raised over £30 000.

ECS is a mixed school with 470 students, aged 13-19. It teaches traditional school subjects, such as maths, English, geography.

- Some children walk 10 miles to school each way each day.
- Food is provided with the school fees; it means a daily meal is guaranteed. Many families pay for their school fees with food.
- The biggest costs are the teachers' salaries (a teacher earns £100 a month) and generators for lighting. A solar panel powers computers, TV and video, and provides lighting for homework.
- In 2007, the school opened its first science labs. ICT was provided in 2007 with 20 computers from the UK.
- Girls have equal opportunities with boys. But some girls drop out of school for marriage and because the family gets a dowry.
- ECS has raised money for 50 AIDS orphans who would otherwise drop out of school; few enter secondary education.
- The school's first exam results have been among the best in the region. Teachers hope that the first students will go to university in 2008-9.

▲ Equatorial College School

▲ The village of Kyarutanga where ECS is based

UGANDA

Kabujogera

Lake Victoria

0   200
Km

The Village Phone Initiative

Few people have phones in Uganda because they are expensive. Yet not having one leaves people isolated. In 2007, Nokia began an initiative with a micro-loan organisation to bring mobile phone technology to rural villages. Its purpose was to stimulate local business and reduce rural isolation. The key players are:

- MTN, an African mobile service provider
- the Nokia phone company, using old phone models
- the Grameem Bank, a bank which specialises in small micro-loans, offering small loans which are easily repaid (normally 6-12 months).

The scheme works like this:

- Someone setting up a mobile phone business buys a village phone kit, using a micro-loan. It costs about £100.

- The kit consists of an adapted Nokia phone, a 10-metre cable connected to an external aerial, and a battery to keep it charged; mains electricity is rarely available in rural areas.
- The aerial extends, picking up signals 30 km away, putting most parts of Uganda within reach.
- The phone is used by anyone wanting to make a call, paying the owner for however much time they use.
- The profit helps the phone's owner to earn an income and repay the loan.

Most uses so far have been for family reasons, but business use is increasing, as are remittance payment transfers, and phone calls to TV and radio talk shows!

▼ A phone kiosk in a rural village

Ugandan phone box

Over to you

1 Using the cycle of poverty diagram, show how AIDS has made the poverty trap worse in Uganda. How does it change the cycle of poverty?
2 What would happen to a child's education in families affected by AIDS? How is this worse for girls?
3 Draw a table comparing life choices faced by girls **with** and **without** secondary schooling. How does each affect Uganda's fertility rate?
4 In pairs, draw and explain what will happen to the cycle of poverty if **a** ECS, **b** the Village Phone Initiative are successful.

On your own

5 Define the following terms from the text: fertility rate, GDP, subsistence, poverty trap, remittance payments
6 In 500 words, assess how **a** ECS and **b** the Village Phone Initiative can help to rebrand Uganda's rural areas.

Mobile phones in Uganda's villages – good business opportunity or just expanding consumerism?

What do you think?

Unit summary

What key words do I have to know?

There is no set list of words in the specification that you must know. However, examiners will use some or all of the following words in the examinations, and would expect you to know them, and use them in your answers. These words and phrases are explained either in the glossary on pages 297–301, or in key word boxes throughout this chapter.

affluent
affordable housing
brownfield sites
container port
core and periphery
cost-benefit analysis
deprivation
destination tourism
diversify
eco-approaches to
  farming
economic leakage

fertility rate
footloose industry
greenfield sites
index of deprivation
infrastructure
inward migration
key players
legacy
market-led regeneration
multiplier effect
multiple index of deprivation
negative multiplier

post-production
  countryside
poverty trap
primary employment
remittance payments
subsidies
subsistence farming
sustainable development
tangible and intangible
  costs
transport hub

Try these questions

1 Using named examples, explain why rebranding is considered necessary in **either** some urban **or** some rural areas. (10 marks)

2 Evaluate the success of **either** named urban **or** named rural developments that you have studied, which are designed to rebrand the areas in which they are located. (15 marks)

Exams: how to be successful

AS Geography – your teacher has chosen the Edexcel specification, but you have chosen to study geography. If you want to be successful, you need to know and understand the geography that you have been studying, but you also need to know – how you will be examined, what kinds of questions you will come up against in the exam, how to use what you know, and what you will get marks for. That is where this chapter can help.

What is the Edexcel AS Geography specification all about?

It is a brand new specification written for the twenty-first century. It includes new ideas, e.g. 'Rebranding Places' and offers a new twist to old geographical favourites, e.g. 'Crowded Coasts'. Along with the A2 it provides a balance between your own physical, human and/or environmental interests, and key geographical topics that provide you with the knowledge, understanding and skills to ready you for further study in higher education – or for employment.

What does the AS specification include?

The AS specification consists of two units broken down into topics.

- **Unit 1 Global Challenges** consists of two compulsory topics looking at some of today's big issues.
  - *Topic 1 World at Risk* looks at a range of natural hazards which threaten some areas of the world more than others, and then focuses on climate change and global warming.
  - *Topic 2 Going Global* looks at today's rapid economic changes which impact unequally on people around the world, and the related issues of population change and migration.

- **Unit 2 Geographical Investigations** takes a closer look at how physical and human issues influence lives and can be managed. Fieldwork and research skills are a key feature of this unit. You can choose two topics from four in this unit, but you must do one physical and one human topic.

Physical topics:

- *Topic 1 Extreme Weather* seems to be more frequent and intense, and many people are at risk from extreme weather events.
- *Topic 2 Crowded Coasts* looks at how increasing development is testing our ability to manage coastal environments.

Human topics

- *Topic 3 Unequal Spaces* investigates the causes and consequences of rural and urban inequalities and how to manage them.
- *Topic 4 Rebranding Places* focuses on how we need to re-image and regenerate rural and urban places, using appropriate strategies.

Extreme weather – the Boscastle flood of August 2004. See Units 3.4-3.6 for more information ▼

Cornwall – a county in need of rebranding. See Units 6.8-6.10 for more information ▼

How will you be assessed?

There are two exams for AS Geography (and no coursework). The table below shows what the exams are like.

Unit	Assessment information	Marks and method of marking
Unit 1 Global Challenges	This exam is 90 minutes long and consists of two sections. You will have a resource booklet. You will have to answer all the questions in Section A, and one question in Section B. Section A uses the resource booklet and consists of multiple choice – or choosing one of several (fixed right answers, no interpretation), data response and short answer questions. Section B also uses the resource booklet, plus your own ideas, and consists of a choice of Going Global or World at Risk longer/guided essay questions.	There is a total of 90 marks for this exam Section A is point marked. Section B is level marked.
Unit 2 Geographical Investigations	This exam is 60 minutes long and consists of two sections. You will have a resource booklet. You will choose one physical question from Section A and one human question from Section B, based on the topics you have studied for Unit 2. The questions have three parts and need longer responses. They are designed to include data response, investigation and evaluation skills and related impacts/management issues. You will be expected to use the resource booklet and your own ideas from fieldwork and research that you have done. You cannot take any materials into the exam.	There is a total of 70 marks for this exam. Sections A and B are level marked.

Know your exam papers

What will the questions be like?

The table explains what type of questions you will get in the exams. Here are 2 examples.

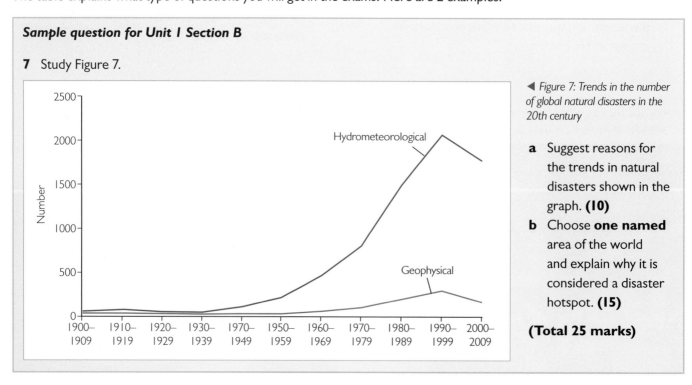

Sample question for Unit 1 Section B

7 Study Figure 7.

◀ Figure 7: Trends in the number of global natural disasters in the 20th century

a Suggest reasons for the trends in natural disasters shown in the graph. **(10)**

b Choose **one named** area of the world and explain why it is considered a disaster hotspot. **(15)**

(Total 25 marks)

Sample question for Unit 2

Sections A and B have the same type of question. This one is from Section B.

3 Study Figure 3.

▼ *Figure 3: Best and worst places to live in the UK*

	WINCHESTER	HACKNEY
Lifestyle		
Average house price	£280 000	£430 000
Average income per year	£22 120	£25 000
Income support claimants	1540 people	18 920 people
Crime figures 2005/6		
Violence against persons	1310	7471
Theft of a motor vehicle	115	1749
Education		
15-year-olds with 5 or more GCSE passes A* - C	77.2%	17.2%
Environment		
Total area of gardens	23 970 sq m	3496 sq m
Total area of green space	609 738 sq m	62 city parks
Population		
Total numbers	110 000	205 000
Type	White, middle-class	–

Winchester comment:
"It's very nice here, being so near the countryside and with so few problems. The schools are really good. But finding work isn't easy and the night life is a bit limited for younger people."

Hackney comment:
"We manage to rub along and help each other but we live under pressure all the time, especially financially. Hackney is a grubby, overcrowded place that needs big money spending on it, and right now!"

a Comment on how well this information shows the inequalities of living in Winchester and Hackney. **(10)**

b With reference to **one or more** named **urban** areas, examine the impacts of deprivation on people's lives. **(10)**

c Describe the **results** of your **fieldwork and research** into how to reduce inequality, and explain how these help you to judge the success of **either** the urban **or** rural schemes involved. **(15)**

(Total 35 marks)

Command words

One of the first things to do when you look at an exam question is check out the command word. This tells you what the examiner wants you to do. This table gives you some of the most commonly used command words.

Command word	What it means	Example
Account for	Explain the reasons for – you get marks for explanation rather than description.	Account for the changes in Cornwall's employment structure.
Analyse	Identify the main characteristics and rate the factors with respect to importance.	Analyse the economic effects of Objective One funding in Cornwall.
Assess	Examine closely and 'weigh up' a particular situation, e.g. strengths and weaknesses, for and against.	Assess the success of coastal management schemes, using named examples.
Comment on	This is asking you to assess a statement. You need to put both sides of the argument.	Comment on the view that desertification is caused by human rather than physical factors.
Compare	Identify similarities *and* differences between two or more things.	Compare the causes and impacts of hazards in California and the Philippines.
Contrast	Identify the differences between two or more things.	Contrast the impacts of hazards in California and the Philippines.
Define	Give a clear meaning.	Define river discharge.
Describe	Say what something is like; identify trends.	Describe the trends shown in the graph.
Discuss	Similar to assess.	'Global warming is a myth.' Discuss.
Evaluate	The same as assess.	Evaluate the success of flood management schemes.
Examine	You need to describe *and* explain.	Examine the attempts by governments to control population growth.
Explain	Give reasons why something happens.	Explain, using examples, whether you think hazards are really 'natural'.
How far?	You need to put both sides of an argument.	How far is Kiribati a nation facing multiple hazards?
Illustrate	Use specific examples to support a statement.	Illustrate the ways that debt makes development difficult in some developing countries.
Justify	Give evidence to support your statements.	Suggest three ways in which Botswana could attempt to reduce poverty. Justify your suggestions.
List	Just state the factors, nothing else is needed.	List the different methods of coastal protection.
Outline	You need to describe *and* explain, but more description than explanation.	Use the graphs to outline the trends in global disasters.
To what extent?	The same as 'How far?'	'The Sydney Olympics – economic and environmental benefits, but at a social cost.' To what extent is this true?

Interpreting the questions

You have probably already been told by your teacher to read the exam question carefully and answer the question set – not the one you think it might be. That means you need to interpret it to work out exactly what it is asking. So, look at the key words. These include:

Command words – these have distinct meanings which are listed opposite.

Theme or topic – this is what the question is about. The examiner who wrote the question will have tried to narrow the theme down, and you need to spot how so that you do not write everything you know about the theme.

Focus – this shows how the theme has been narrowed down.

Case studies – look to see if you are asked for specific examples.

Here is an example of a question that has been interpreted using the key words above.

Command words
- *'Suggest reasons' You must give more than one reason for the trends.*
- *'Choose one named area' You are being asked for a case study.*
- *'Explain' Give reasons why your named area is considered a disaster hotspot.*

Theme or topic
- *This question is for World at Risk, looking specifically at natural disasters and disaster hotspots.*

Study Figure 7.
a Suggest reasons for the trends in natural disasters shown in the graph. **(10)**
b Choose one named area of the world and explain why it is considered a disaster hotspot. **(15)**
(Total 25 marks)

Case studies
- *Choose one named area only.*

Focus
- *Part (a) asks you to talk about trends. You need to suggest reasons for the trends, not everything you know about natural disasters.*
- *Part (b) asks you to explain why your chosen area is a disaster hotspot. Give examples to support your answer.*

How to use case studies in the exam

Case studies are in-depth examples. There are lots of case studies in this book, some may be 2 pages long while others may be 6 or 8 pages. You will need to use these to answer some of the exam questions. This is an example where you would need to use a case study to answer a question on World at Risk.

> **b** Choose **one named** area of the world and explain why it is considered a disaster hotspot.

Here is another example from a question on Rebranding Places.

> **b** With reference to **named examples**, explain how rebranding strategies such as those shown in Figure 4 might regenerate **rural** areas.

- This question involves using more than one case study – you will limit your marks if you restrict yourself to just one example.
- For each case study, you need to say how it might help to regenerate rural areas – so don't just describe it. Focus on regeneration. For example, 'The Eden Project helped to regenerate Cornwall because …'.

There are some things you can do to help yourself learn and use case studies.

A portfolio of case studies

Build up a 'portfolio' of case studies, so that you know which part of the specification and content the case studies fit. A grid like this might help.

Topic name

Subject content	Relevant theory	Case studies

The grid works like an index – or pigeon hole. If you fill it in you will know exactly what case study and theory is appropriate for each part of a topic. It also helps to make links between topics.

How to remember your case study

Look again at the question from World at Risk. There are two case studies in this book that you could use to answer this question, California and the Philippines. Here is one way of remembering your case study – draw a spider diagram with all the key information on it. This diagram is only a start, you would need to fill in more details – for example about the Pinatubo eruption in 1991, the Guinsaugon landslide and so on.

Using your case study

Once you have revised and learned your case study, you can use it to answer exam questions. Look again at the question on World at Risk. It asks you for one named area – you can use the Philippines. It asks you to explain (give reasons) why it is considered a disaster hotspot. The reasons are its location in terms of:

- tectonic hazards
- threat from tsunami
- threat from typhoons
- topography (mountainous landscape)

Tectonic hazards
- Lies on destructive boundary between two plates.
- Risk of earthquakes and volcanoes.

Threat from tsunami
- Northern and eastern coasts face the Pacific – the world's most tsunami-prone ocean.

Philippines hazard hotspot

Landslides
- Landscape is mostly mountainous and landslides are common.

Typhoons
- Lies within South-East Asia's major typhoon belt.

How to use fieldwork and research in the exam

Fieldwork and research skills are a key feature of Unit 2 Geographical Investigations. Whichever two topics you have studied, you will be expected to use fieldwork and research in the exam. Part of the question you have to answer will be something like this:

> **c** Describe and explain **a programme of fieldwork and research** that you would use to investigate the impacts of **either** coastal erosion **or** coastal flooding, along a stretch of coastline. **(15)**

Step 1: Understand the question

This is a complex question and it would be easy to write a lot and gain few marks. To avoid this, two skills are essential:

- Read the question. It is not 'all you know about a stretch of coast', but about how you would **investigate the impacts** of coastal flooding or coastal erosion.
- Underline the command words so that you do what the question asks. So – describe and then give reasons.

Use your pen or pencil to highlight the key words.

Step 2: Focus the question

Focus on either coastal erosion or coastal flooding along a stretch of coast.

- You **must** mention a named stretch of coast – examiners only award about half marks for generalised answers that do not refer to places.
- If you choose flooding, it must be a coastal flood – inland will not do.

Step 3: Answer the question

To answer the question, you need not have visited the coast – simply researched it. Fieldwork gets credit, but so too does your own research. You might never have visited Holderness, but instead seen videos, used material in this book, and carried out some research.

- Research could include photos about impacts of erosion, news articles about a coastal flood, or results of a survey about which methods of flood protection proved most effective in a recent coastal flood.

- Remember – it is the **programme** of research that you should describe – what you would do first, then second, etc.
- Focus the programme on **impacts** – which can be classified as economic, social and environmental.

Example 1	**economic** impacts might include damage caused, or transport disruption, using news websites (with examples)
Example 2	**social** impacts are about people – perhaps using a questionnaire in the field. Think of questions that you could ask, and why.
Example 3	**environmental** impacts might include visual impacts of coastal erosion, e.g. using Environmental Quality Surveys.

You will get credit for:

- naming a stretch of coast that you have studied
- describing your outline programme in stages, and explaining why you would do it in this way
- focusing each stage on **what** you would do and then **why**. For example *'I would select a local newspaper such as the Western Morning News to find out how local emergency services coped with the flood in Torquay, because this would show how they rescued people, together with photos of the flood.'*

Use these same approaches for questions in other units. Examples include:

Extreme Weather	Describe and explain **a programme of fieldwork and research** you would use to investigate the links between precipitation and flooding along a stretch of river.
Unequal Spaces	Describe the **results** of your **fieldwork and research** into how to reduce inequality, and explain how these help you to judge the success of **either** the urban **or** rural schemes involved.
Rebranding Places	Describe the **results** of your **fieldwork and research** into **urban** rebranding, and explain how these help you to judge the success of the schemes involved

How are the exam papers marked?

An examiner will mark your exam papers and will have clear guidance about how to do it. They must mark fairly so that they will mark the first candidate's exam paper in exactly the same way as the last one. You will be rewarded for what you know and can do, rather than being penalised for what you have left out. If your answer matches the best qualities in the examiner's mark scheme, you will get full marks.

If you look at the table on page 288 it tells you which parts of the exam are point marked and which parts are level marked. Are you clear about what that means?

Point marking

Look at this question.

> **b** Suggest **two** ways of investigating medium and longer term climate change, before global temperature records were available (from 1861). **(2)**

There are two marks for this question and you have to suggest two ways of investigating climate change. So, you get one mark (or point) for each method (way) that you give.

The examiner will have a mark scheme which tells them which methods to accept as correct answers.

Level marking

Look at this question.

> Study Figure 7.
> **a** Suggest reasons for the trends in natural disasters shown in the graph. **(10)**

There are ten marks for this question and the examiner will award these using the following mark scheme.

Level	Mark	Descriptor
Level 1	1-4	Little structure in answer. Describes trends with little explanation. May refer to types of disaster. Geographical terminology is rarely used. Frequent written language errors.
Level 2	5-7	Some structure. Attempts to interpret the trends in the graph. Shows understanding of disasters as opposed to hazards and why they have increased. Geographical terminology is used with some accuracy. Lower end descriptive. There are some written language errors.
Level 3	8-10	Structured detailed account of disaster status and vulnerability of population/property/infrastructure, using appropriate geographical terms and exemplification to show understanding. Written language errors are rare.

So, to gain the most marks, your answer needs to meet the criteria for level 3.

Questions which have more marks might have 4 levels in the mark scheme.

How to gain marks and not lose them

Gaining marks instead of losing them is just a matter of technique. You might get a lower mark than you really ought to, not because you do not understand the material but because you do not know how to use it. Here are some helpful hints based on choosing the right question, case studies, the type of question you are answering and communication.

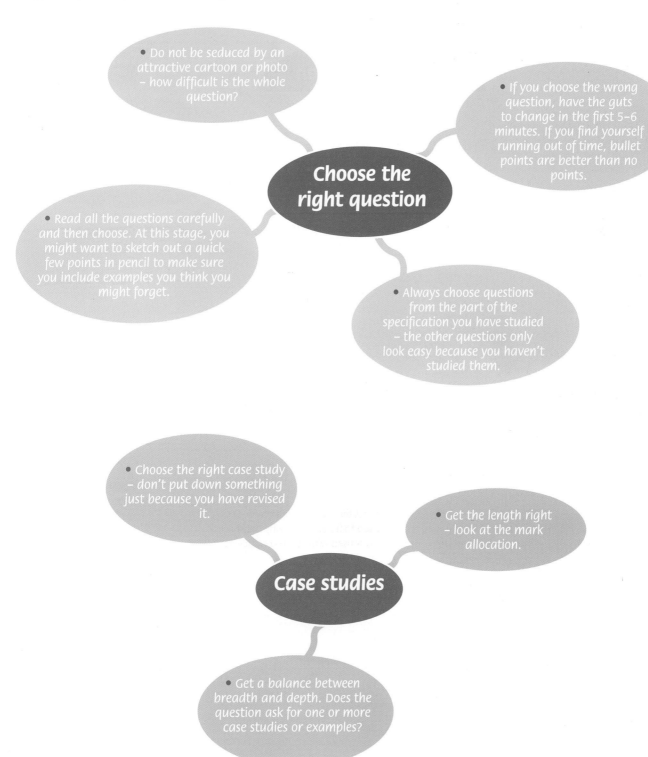

- Do not be seduced by an attractive cartoon or photo – how difficult is the whole question?

- If you choose the wrong question, have the guts to change in the first 5–6 minutes. If you find yourself running out of time, bullet points are better than no points.

Choose the right question

- Read all the questions carefully and then choose. At this stage, you might want to sketch out a quick few points in pencil to make sure you include examples you think you might forget.

- Always choose questions from the part of the specification you have studied – the other questions only look easy because you haven't studied them.

- Choose the right case study – don't put down something just because you have revised it.

- Get the length right – look at the mark allocation.

Case studies

- Get a balance between breadth and depth. Does the question ask for one or more case studies or examples?

Type of question

For longer questions

For other questions

- Read to the end of each question and underline the command words.

- Answers must contain facts, data and information. Do not be vague or include trivial information. Do not waffle.

- Read the whole question and underline the command words. Take a highlighter into the exam with you.

- Include some keywords from the question in every paragraph.

- Use all the resources provided.

- Structure your answer into paragraphs.

- Be prepared to use geographical skills, e.g. in looking for evidence on photos, in reading tables of data to discover trends, in interpreting map evidence.

- Use maps and diagrams – a well labelled diagram takes less time than writing a paragraph of text.

- Be specific and precise. Do not waffle.

- Be accurate when describing trends in graphs or distributions on maps, e.g. 'The number of migrants remains stable until 1998, then rises sharply to 2006.' 'Earthquakes cluster most along the Pacific Rim.'

- Sketch out a quick 1-minute plan for your answer and note the examples you will use, and then make a note that links each example to the question. This will keep you on track and stop you going off the point.

Communication

- Use precise geographical terminology, e.g. 'migration of people' is better than 'movement of people'.

- Make sure your writing is legible.

- Use a suitable style of writing – you are writing for an intelligent adult.

- Use correct spelling, punctuation and grammar. Only use abbreviations where these are commonly understood, e.g. UK, USA, UN.

- Write in logical well-planned paragraphs.

- Avoid sweeping generalisations, e.g. the South of England is rich and the North is poor.

Glossary

*= a cross-reference

A

abrasion – the grinding away of bedrock by fragments of rock which may be incorporated in ice. Also known as corrasion

adaptation – any change in the structure or functioning of an organism or system that makes it better suited to the environment

affluent – having a lot of money and a good standard of living

affordable homes – homes which can be afforded by young adults on or below the average wage. The US guideline is that they should not cost more than 30% of a household's gross income

anoxic – without oxygen

at-risk zones – inhabited areas most at risk from natural disasters

B

baby boom – any period of greatly increased birth rate during a certain period, and usually within certain geographical bounds. Persons born during such a period are often called baby boomers

boreal forest – evergreen forests which occur naturally between 55° and 66° North where winters are long and very cold. Also known as coniferous forest

boulder clay – the unsorted sediment deposited directly below a glacier, which has a range of particle size from fine clay to rock fragments and boulders. Also known as glacial till

burakumin – a Japanese social group associated with 'dirty' occupations and discriminated against by other people

C

carbon credits – a way to reduce greenhouse gas emissions in industry by putting a value and a limit on a company's emissions and allowing the company to sell its emissions shortfall or to buy carbon credits if it exceeds its limit. Carbon credits can be used to finance carbon reduction schemes

carbon offsetting – the act of *mitigating greenhouse gas emissions, e.g. by asking people or companies to pay extra for air travel

carbon sinks – reservoirs of carbon dioxide. The main natural sinks are the oceans and plants that use photosynthesis to remove carbon from the atmosphere

centrally planned economy – a country where (nearly) all business and industry is controlled by the state

chalk – a pure form of limestone consisting of the shells of tiny marine organisms together with egg-shaped particles of calcium carbonate

chemical fertilisers – compounds given to plants to promote growth. Fertilizers can be organic (composed of organic matter), or inorganic (made of simple, inorganic chemicals or minerals)

coastal squeeze – a) where a coastal settlement is prevented from expanding due to the sea on one side and rural areas inland b) an environmental situation where the coastal margin is squeezed between the fixed landward boundary (artificial or otherwise) and the rising sea level

coastalisation – the movement of people to coastal areas. It can be used in the negative sense of intensive housing development along coastlines, leading to environmental degradation

cold front – a *front where the cold air pushes in behind the warm air

congestion charging – making road users pay to drive in a particular city

conservative plate boundary – where the movement of tectonic plates is parallel to the margin. It is also known as a transform fault

container – a metal box of standard size, used for the transport of cargo by road, rail or water

convergence – in meteorology, when airstreams flow to meet each other, leading to an increase in the height of the atmosphere, which often causes weather events; in plate tectonics, when two plates come together

core and periphery – a model of development which tries to represent the emergence of an urban system in four major stages. In between the affluent core and the deprived periphery are two 'transition' regions, one 'upward' and the other 'downward'. In international terms, core areas include North America, Europe and Japan

Coriolis force – when air streams are deflected through the rotation of the Earth on its axis. It can be the cause of storms, including hurricanes

corrasion – see *abrasion

corrosion – the breaking down of something slowly, especially by chemical action

cost-benefit analysis – a technique where projected public schemes are evaluated in terms of social outcomes as well as in terms of profit and loss

crop and pasture management – using management techniques to enhance the quantity and quality of forage supply to grazing animals and to improve the productivity of crops

D

debt – a sum of money that somebody owes

deep sea core sample – a cylindrical section of the Earth's crust removed from the ocean floor. The layers in the core sample tell us about the geologic history of the area

deforestation – the complete clearance of forests by cutting and/or burning

delta – a low-lying area found at the mouth of a river and formed of deposits of alluvium

dependency theory – this explains how poorer and less developed nations depend on wealthier nations for their trade and income

deprivation – lacking in the provision of desired objects or aims

deprivation, index of multiple – used by the UK government as a way of measuring deprivation using six 'domains': health, employment, income, education, the living environment, and barriers to housing and services

destination tourism – when people visit a place simply because of a single attraction, such as the Eden Project in Cornwall or the Great Wall in China

destructive waves – a plunging wave with a short wavelength, high frequency and high crest which breaks so that the wave crashes downwards and erodes the beach

destructive plate boundary – where two *tectonic plates meet and where the denser plate is being destroyed by diving under the less dense plate and being converted to *magma

developed – used to describe a country which has many industries and a complicated economic system

developing – used to describe a country which is poor and trying to make its industry and economic system more advanced

digital divide – the gap between those with regular, effective access to digital and information technology, and those without it

disaster – when a natural hazard has serious effects such as a large loss of life or property

discharge – the quantity of water flowing through any cross-section of a stream or river in unit time

diversify – to spread industrial commitment over a large range of activities so that there is no overdependence on one activity alone

downsizing – a) reducing the number of people who work in a business in order to reduce costs, b) selling your house and moving to a cheaper one in order to save money

drought – a long continuous period of dry weather, caused in Britain by the persistence of warm anticyclones and in the tropics by the failure of the Inter-tropical convergence zone (ITCZ) to move sufficiently far from the equator

dynamic system – a social or geophysical structure that is constantly changing

E

eco-approaches – environmentally friendly ways of addressing a problem

economic migrants – people from a poor area who move to a richer area in search of a better life

edge cities – cities which have grown up on the periphery of older cities, to which new industries and services have moved, away from the old CBD

enclosed or semi-enclosed sea – a gulf, basin or sea connected to another sea or the ocean by a narrow outlet

eustatic changes – a world-wide change of sea level which may be caused by the growth and decay of ice sheets

eutrophic fertile, productive – eutrophication is the process by which ecosystems become more fertile environments as detergents, sewage and artificial fertilizers flow in

exception policy – building on land where it would not normally be permitted, in rural villages where housing needs have been proven

expansion of suburbs and villages – increasing job opportunities in cities lead to people looking for affordable and pleasant housing in settlements within commuter access of the city

exponential – a rate of increase which becomes faster and faster

eye – the calm area at the centre of a hurricane or tropical cyclone

F

fair trade – an organized social and economic movement with regard to trade between developed and developing countries which promotes the payment of a fair price and higher social and environmental standards in areas related to the production of a wide variety of goods

fault line – a fractured surface in the Earth's crust where rocks have travelled relative to each other

fetch – the distance that a sea wave has travelled from its beginning to the coast where it breaks

First World – Western Europe, Japan, Australasia and North America. These were the first areas to industrialise. Also known as the '*developed world'

fjord – a long narrow arm of the sea which is the result of the 'drowning' of a glaciated valley

footloose – industry that can be sited in any of a number of places, often because transport costs are unimportant

formal economy – all kinds of employment which offer regular wages and hours and which carry with them employment rights on which income tax is paid

Fortress Europe – the term sometimes given to the concept of the EU's efforts to keep non-EU goods, businesses and nationals out of the Union's twenty-seven member states

free trade – trade between countries which takes place completely free of restrictions

freeze-thaw – the weathering of rocks which occurs when water which has penetrated joints and cracks freezes and expands

front – the border zone between two air masses which contrast, usually in temperature

fuelwood – wood that is used to produce heat or power. It is normally used in the absence of more efficient fuel sources by poor communities and often involves chopping down scarce tree reserves

G

geology – the scientific study of the Earth, including the origin and history of the rocks and soils of which the Earth is made

geomorphological – to do with the nature and history of landforms and the processes which create them

glaciologist – a scientist who studies glaciers

global conveyor belt – another term for the thermolhaline circulation

greenfield sites – sites which have never been developed or used for an urban use, or are on land that has been brought into active or beneficial use for agriculture or forestry, i.e. fully restored derelict land

greenhouse effect – the warming of the atmosphere as some of its gases absorb the heat given out by the Earth.

grey pound – a term used in the UK to refer to the economic power of older, retired people

greying – a term used to mean the increase in the proportion of older people in a population

H

hazard – an unexpected threat to humans and/or their property. By this definition the Indian monsoon is not a hazard but its failure is

Heavily Indebted Poor Nations (HIPC) – a group of 37 least developed countries with the highest levels of poverty and debt, which are eligible for special assistance from the International Monetary Fund and the *World Bank

hydraulic action – the force of the water within a stream or river

hydrograph – a graph of *discharge or of the level of water in a river over a period of time

I

illegal migrant – an illegal migrant is a foreigner who either has illegally crossed an international political border, be it by land, sea, or air, or a foreigner who has entered a country legally but then overstays his/her visa in order to live and/or work in the country

image – the impression that is given to the public

immigration – the inward flow of people into a country

index of deprivation – see *deprivation

industrial development – the growth of manufacturing industry in an area

informal economy – employment which is not formally recognised. Workers in the informal economy generally do not have contracts, fixed hours or employment benefits

infrastructure – the framework of communication networks, health centres, administration, and power supply necessary for economic development

inner-city zone of transition – according to concentric zone theory this is the crowded , multi-occupied zone of a city first invaded by migrants

inputs – something that is put into a system or project

intangible costs and benefits – costs and benefits that cannot be measured but which are important, such the good and bad effects of a project on people

integrated coastal management – an approach which sees the coastal zone as an interactive and dynamic complex of sub–systems. Human activities in one sub–system may adversely affect other sub-systems. For this reason, it considers that various parts of the coastal zone cannot be considered in isolation

Intergovernmental Panel on Climate Change (IPCC) – a United Nations organization which looks at the evidence and publishes reports on the human impact on climate change

isostatic changes – the rise or fall of the Earth's continental crust, often in response to the melting or accumulation of glacial ice

K

key players – the most important people or organizations involved in a project

L

lag time – the interval between an event and the time when its effects are apparent

lahar – a flow of volcanic debris, either dry or mixed with water as a mud flow

landfill – a site for the disposal of waste materials by burial

legacy – a situation that will exist in the future because of events and actions that take place in the present

longshore drift – the movement of sand and shingle along the coast

low pressure area – another term for a depression

M

magma – the molten rock found below the Earth's surface which can give rise to igneous rocks

magma chamber – a large underground pool of molten rock lying under the surface of the Earth's crust.

manufacturing industry – the mechanized and usually large-scale processing of materials into finished or partly finished products

market-led regeneration – the encouragement of private investment through planning, transport and land policies as well as substantial public investment

mega-delta – where the mouths of several large rivers emerge close together, e.g. the Bay of Bengal

Milankovitch cycles – three interacting astronomical cycles in the Earth's orbit around the Sun, believed to affect long-term climatic change

Millennium Development Goals – eight goals that 192 United Nations member states have agreed to try to achieve by the year 2015 to help raise development standards

minority group – a subordinate group whose members have significantly less control or power over their lives than members of a dominant or majority group

mitigate – to make something less harmful or serious

multiple hazards – where a region suffers from a number of different natural or man-made *hazards, such as unusual tropical storms, tsunamis and rising sea levels, which make life difficult for people living there

N

North Atlantic Drift – a warm ocean current, driven by prevailing south-westerly winds from Florida to north-west Europe, bringing warmer conditions than would otherwise be expected at those latitudes

O

occluded front – where a *cold front catches up with a leading *warm front and lifts it up or where a warm front catches up with a leading cold front and slides over it

OPEC (Organization of the Petroleum Exporting Countries) – an international organisation of oil producing countries which is concerned with the best means of safeguarding their interests, individually and collectively

output – something that comes out of a system

P

pandemic – a disease that spreads over a whole country or across the whole world

periphery – the area most remote from the wealthy 'core' of a country or region

permafrost – areas of rock and soil where temperatures have been below freezing point for at least two years. Permafrost does not have to contain ice

plant succession – the gradual evolution of a series of plants within a given area. This series of plant communities occurs in a roughly predictable order

positive ice albedo effect – when melting snow exposes more dark ground of lower albedo which in turn causes more snow to melt

positive multiplier effect – the positive economic or environmental consequences of an action, e.g. If new jobs are created, people then have money to spend in the shops, which means that more shop workers are needed. The shop workers pay their taxes and spend their new-found money, creating yet more jobs in industries

poverty trap – when people are discouraged from taking higher paid work because that would mean that they would lose their benefit payments. Once income tax and other deductions have been allowed for they would end up earning less

primate city – the largest city within a nation, being more than twice as large as the second city. A primate city dominates the country not only in size but also in influence

privatisation – where government-owned businesses are sold to private owners

push factor – in migration, any adverse factor which causes movement away from the place of residence

pyroclastic flows – a dense cloud of lava fragments thrown out by an erupting volcano as result of bursting gas bubbles within the magma

Q

quotas – a fixed level indicating the maximum amount of imported goods or persons which a state will allow in

R

radiation – energy travelling in the form of electromagnetic waves

re-grading – changing the shape of the landscape by moving soil in large quantities or by shaping cliffs

relief – the shape of the Earth's surface. An area of 'high relief' has large differences in the height of the land; 'low relief' indicates little difference in altitude

remittance payments – money that is sent home to families by people working in a foreign country or in a city in their own country

replacement level – the fertility rate needed to maintain population at its existing size by natural change (without allowing for net immigration)

replacement migration – where migrant workers are encouraged to move from states which have a labour surplus to states which have labour or skills shortages. This requires cooperation between states and is best managed within an organisation like the EU

resource-based industries – these include agriculture, energy, forestry, fisheries, recreation, and mining

reverse colonialism – where companies from previously colonised countries buy up companies in previously colonial countries e.g. Tata Steel from India taking over the British/ Dutch steel company Corus in 2006

ripple effect – the effect on house prices as waves of people move from the city to the countryside

rip-rap (also known as revetment, shot rock or rock armour) – rock or other material used to protect shorelines from water erosion

rising limb – that section of a river *hydrograph which covers the beginning of the increased discharge until the maximum flow

rock armour – see *rip-rap

rural housing enabler – a local government officer in the UK dedicated to helping solve the need for affordable housing in rural communities

rurality – the degree to which an area is rural. Rurality can be measured using the Cloke index which employs a number of indicators of rural life

rural-urban migration – the movement of people from the countryside to the cities in search of work

S

sailing and leisure craft – dinghies, yachts and motorboats

saline – salty

salt marsh – mud flats in estuaries and sheltered bays on which vegetation has grown

scenario – a description of how things might happen in the future

scour – to make a passage, hole or mark in the ground or on rocks as a result of movement

sea change – an Australian term for *coastalisation, with the added special sense of people moving to small coastal towns for the improved lifestyle and low house prices

sea floor (or seabed) – the bottom of the ocean

Seasonal Agricultural Workers Scheme (SAWS) – an EU scheme allowing people (mostly students from eastern European countries) to work in other EU countries for limited periods

seasonal workers – a worker who is allowed into a country to work only for a limited period, usually in agriculture

sediment cells – a length of coastline and its associated nearshore area within which the movement of coarse sediment (sand and shingle) is largely self contained. There are 11 sediment cells around England and Wales some of which can be divided into sub-cells

sediment input – the amount of sediment being deposited

service industry – economic activity concerned with the distribution and consumption of goods and services

sewage disposal – the disposal of human excreta and other waterborne waste products from houses, streets, and factories

shanty towns – an area in or near a town where poor people live in houses made of wood, metal or cardboard, often without access to electricity or running water

sheltered accommodation – a term used for self-contained homes specially designed for the elderly. The aim is to provide independent secure accommodation with additional social and domestic facilities

shoreline management plans – plans that take into consideration an entire *sediment cell or sub-cell

slumping – a mass movement where rock and soil move downwards along a concave face

social cohesion – the linking together of people who are tied by one or more specific types of interdependency, such as values, visions, financial ties, friendship and kinship

social exclusion – what can happen when people or areas suffer from a combination of linked problems such as unemployment, poor skills, low incomes, poor housing, high crime, bad health and family breakdown

social landlord – a local authority or a housing association which provides affordable housing to lower income earners on a not-for-profit basis

social segregation – segregation on the basis of class or economic status in which an underclass develops which is separated from the rest of the population

spatial distribution – the spread of varying observations (such as income) within a population across an area

structural adjustment – introducing changes to a nation's economy, such as currency devaluation, promotion of exports and cuts in public services. These changes are usually made in order to qualify for a loan from the IMF

subduction – the transformation into *magma as a dense *tectonic plate dives under a less dense plate at a *destructive plate boundary

subsidy – money that is paid by a government or organization to reduce costs so that prices can be kept low

sustainability – meeting the needs of the present generation without compromising the ability of future generations to fulfil their own needs

swell – a long series of ocean waves, generally produced by wind, and lasting after the wind has ceased

T

tangible costs and benefits – costs and benefits that can be measured, such as things that cost and earn money

tariffs – a list of duties or customs to be paid on imports

tectonic – to do with the processes acting to shape the Earth's crust

tectonic plate – a rigid segment of the Earth's crust which can 'float' across the heavier, semi-molten rock below

terminal groyne syndrome – beach erosion occurring just after the last of a series of groynes

Third World – an outdated term for the poor or developing countries of Africa, Asia, and Latin America

three Ds – jobs which are difficult, dirty and dangerous

topography – the physical features of an area of land, especially the position of its rivers and mountains

Total Fertility Rate (TFR) – the average number of children who would be born per woman if she were to live to the end of her childbearing years and follow normal patterns of fertility

tourist enclave – an area set aside for tourists where they have little or no contact with the society they are visiting. It may have no benefit for the local economy as all its goods and services are brought in from outside

trade winds – tropical easterly winds blowing towards the equator from the subtropical anticyclones at a fairly constant speed

tree line – the line beyond which trees will not grow

tundra – the barren plains of northern Canada, Alaska and Siberia where both temperature and rainfall are low

U

undeveloped – a country which does not have modern industries and has a low standard of living

urban shadow – the effect on a settlement of being close to a city where residents will often commute to the city, turning it into a dormitory settlement

V

vent – the opening in the crust through which volcanic material flows

volcanic emissions – the materials given out when a volcano erupts. These include gas (mainly sulfur dioxide), lava and ash

vulnerable – weak and easily hurt

W

warm front – a *front where the warm air rises over the cold air

World Bank – a bank that is effectively controlled by subscriptions from rich countries which provides aid to the developing world

Index

A

Africa,
    debt 41, 43, 78-81
    global warming 41-3
    globalisation 78-81
    slavery 71
albedo 38
anticyclones 133
AONB (Area of Outstanding Natural Beauty)
    218, 219, 226
Atlantic storms 13
Australia,
    bushfires 20-1
    coastalisation 176
    drought 158-64
        climate change 162-4
    Murray-Darling Basin (MDB) 158, 164, 165
    Sydney Olympics 262-5
    water resources 165-7

B

Bangladesh, flooding 60-1
BedZED energy conservation 59
Boscastle flood 138-47, 149
    causes 142-5
    timeline 140-1
Boscombe, artificial reef 185
Botswana 230-9
    diamond mining 230, 231, 234
    ecotourism 239
    Gaborone's urban poor 236-7
    HIV/Aids 231, 232, 233, 238
    inequality 231, 233, 234, 235
    land rights 232
    National Development Plans 233
    population 230, 238
    remote area dwellers 232, 233, 234, 235
    rural poor 233
    San Bushmen 232, 233, 234-5
    Self Help Housing Association 237
    urbanisation 230, 231
    women 231, 232, 233, 235, 238
Bournemouth, coastalisation 178-9
economic/population boom 180
    future growth 181
    Burgess and Hoyt, urban land use 241

C

California, multiple hazards 27-9
carbon offsetting 50-3, 58, 129
Caribbean,
    banana wars 128
    colonialism 71
    tourism 124-5
China, superpower 86-91
    Chongqing 88
    as consumer 87
    economic growth 118
    energy resources 86
    Export Processing Zones 87, 89
    exports 86, 89
    industrialisation 88-9
    manufacturing 76, 86-9

population 87, 89
    Shanghai 87, 118, 119
'Chindia' 76
CHP (Combined Heat and Power) 56-7, 59
climate change 30-7
    the Arctic 36-40
    Greenland 34-5
    world oceans 36-7
    see also extreme weather; global warming
coastal erosion 192-9
    zoning 201
    see also Holderness, coastal erosion
coastal flooding, UK 204-9
    see also Thames estuary; Thames Flood
    Barrier
coasts 174-5
    coastalisation 176-9
    industrial use/development 186-91
    UK 174-5, 178-9
    see also Jurassic Coast; Southampton
    Water
colonialism 70-1, 72, 82, 83
core and periphery theory 71, 72, 271
Cornwall
    deprivation 270-3
    Eden Project 274-5
    Newquay Airport 276
    Objective One 277-9
    population 273
    regeneration/rebranding 276-9
    tourism 271, 276
cost-benefit analysis 200, 202, 256
Cuba, effects of tourism 124-5

D

debt,
    Africa 41, 43, 78-81
    environmental consequences 122
    Honduras 122, 137
    Indonesia 63
    Nicaragua 122
    poorest countries 78-9
    Senegal 76
'Debt for Nature Swaps' 123
deforestation 122-3, 137
deindustrialisation 113, 246
depressions 143
deprivation,
    audit 243
    indicators 240
    multiple index 273
    UK cities 242, 250, 253
disasters 9-11
Disney, global brand 66-7
Dorset, tourism 184-5
    see also Jurassic Coast
drought 158-64

E

earthquakes 15
    California 27-8
    Indonesia 62-3
    Kobe 10, 14

East Anglia,
    AONB 219, 226
    future development 228-9
    housing 216-17, 218, 219
    income 215, 218, 219
    inequalities 214-19
    Jaywick, cheapest streets 214, 220
    population 215, 218
    ripple effect 216-17
    SnOasis 228-9
    social exclusion 220, 225
    Southwold and Walberswick 221
    suburbanisation 216
    sustainable communities 222-6
        housing 223
        jobs 225
        Key Service Centres 224
        landscape 226
        services 224
        transport 223
    tourism 214, 225
ecological footprints 126-7
economic development 72-5
    global differences 74-5
    Rostow' model 72, 73
Eden Project 274-5
effluent 187
El Niño 27, 29, 63, 162, 163
energy use 56-9
Environment Agency 146, 147, 148
environmental,
    impact assessment (EIA) 202
    magnetism 212
    refugees 6, 49, 60
erosion processes 192-7
ethical shopping 128
ethnic enclaves 114, 241
ETS (European Emissions Trading Scheme) 52
European Union,
    economies compared 97
    enlargement 96-7, 100
    migration 96-101
    population balance 98-9
eutrophication 189
extreme weather 132-7
    see also Boscastle; hurricanes

F

fair trade/free trade 128
farming (UK), in crisis 280
    eco-approaches 281
    EU subsidies 280
    rebranding 281
    tourism 281
fertility rates 94, 105
Fisher-Clark model of employment 71
flood risk 136-7
    human activity 145
    insurance 149
    London 23, 204, 206-7, 208-9
    rising sea levels 205
    urbanisation 136-7
flooding 13

Bangladesh 60-1
Boscastle 138-49
coastal 204-9
defences 146, 148-9
Florida 177
forecasting 147
Gloucestershire 168-71
managment of 146-9, 170-1
footloose industries 180, 251

G
'G7' 91
GDP (Gross Domestic Product) 75
George, Susan 78-9, 117
global,
    brands 66
    connections, world cities 111
    conscience 127
    differences 74-5
        North-South divide 74
        three worlds 74
    groupings 76-7
    supply chains 126
    tourism 124-5
    trade 76-7
        China 86-9
        environmental impact 86, 89
        human cost 86, 89
    village 69, 111
global warming 6, 32-3, 44-5, 48-9, 153
    Africa 41-3
    the Arctic 36, 38-40
    Bangladesh 60-1
    dealing with 50-5
    future scenarios 46-7
    Greenland 34-5
    sea level 48, 205
    tipping point 48
    see also climate change
globalisation 68-9, 72, 127
    Africa 78-81
    environmental impacts 122-3
    internet use 90, 93
    Mumbai 116
    social consequences 124
    sustainability 126, 127, 128-9
GNP (Gross National Product) 75
greenhouse effect 30-1
greenhouse gases 7, 31
    reduction of 50, 51-3, 58-9
groynes, erosion managment 198-9
Guatemala, colonialism 72

H
hazards 8, 10
    Dregg's model 9
    global 10, 12-13
    managment of 63
    tectonic 25, 27
vulnerability to 9, 14-19, 60-3
HDI (Human Development Index) 75
HIPC (Heavily Indebted Poor Countries) 78, 79

HIV/AIDS,
    Botswana 231, 232, 233, 238
    Uganda 283
Holderness, coastal erosion 192-203
    cliffs 195-7
    economic impacts 193
    longshore drift 194, 195
    management of 198-203
        beach nourishment 201
        coastal zoning 201
        hard engineering 198-200
        retreat/relocation 201
        sea walls 200
        shore line management plans 202
        soft engineering 198, 201-2
    slumping 197
    wave energy (fetch) 194
housing 216-17, 218, 219
    affordable 216-17, 223, 259, 262
    dumb-bell market 219
    rural sustainability 222-3
HSI (Human Suffering Index) 75
Hurricane Katrina 10, 150-3
    causes 152-3
    effects of 150
    evacuation 157
    flood defences 152, 156
    hurricane warnings 157
hurricanes 150-7
    management of 156-7
hydrographs 144
hyper-urbanisation 117, 230, 231

I
illegal immigration by sea 100-1
IMF (International Monetary Fund) 76, 77, 78, 79
India, economic development 76, 90-1
    Bangalore,hi-tech city 90
    see also Mumbai, megacity
Indonesia, multiple hazards 62-3
inequalities, UK 212-13
    East Anglia 214-19
    London 240-5
    North-South divide 212
IPCC (Intergovernmental Panel on Climate Change) 35, 46, 48, 60, 135

J
jet stream 170
Jurassic Coast 182-5
    Kimmeridge Bay 182, 184
    Studland Bay 183, 184
    tourism 184-5
    World Heritage Status 182, 183, 184, 185
'just-in-time' production 66, 83

K
Kenya 81
Kiribati 6-7
Krakatau 63
Kyoto Protocol 55, 127

L
La Niña 26, 27, 29, 163, 170
Lampedusa, Italy 100, 101
London,
    dominating UK 212
    flood risk 23, 204, 206-7
London, East London dockland,
    decline 252
    deprivation-Canning Town 254, 255
    local population 254-5
    regeneration-Canary Wharf 253-4
London, inequalities 240-5
    deprivation index 240, 242-3
    ethnic enclaves 241
    Hackney 242, 243-4
        gentrification 244-5
        New Deal 244
    Hampstead Garden Suburb 242, 245
    regeneration strategies 242-4
    social exclusion 241
    urban land zones 241
London, Olympics 256-61
    affordable housing 259
    cost-benefit analysis 256
    key players 257
    regeneration, Stratford 261
    site, Lea Valley 257, 258
    transport regeneration 260-1
longshore drift 194, 195
Los Angeles, megacity 110, 111, 112-15
    energy use 113
    ethnic enclaves 114
    land use 114
    pollution 112, 113
    problems 113
    suburban sprawl 112-13
    sustainable living 115

M
Manchester, re-imaging 266-7
megacities 110-22
megapolis 111
Mexico City 70
migration 92-101
    Chinese immigrants 101
    economic 97
    European Union 96-101
        population 99
        foreign national enclaves 95
        illegal migrants 100-1
        international 92-3
            host/source nations 92, 93
        into Spain 94-5, 177
        into UK 96
    Mumbai 117
    net migration 98-9
    quotas 100
    remittances 93, 98
    replacement migration 98
    Seasonal Agricultural Workers Scheme (SAWS) 95
Milankovitch cycles 32
Mississippi wetlands 153, 156

multiple hazards 7
Bangladesh 60-1
California 27-9
Indonesia 62-3
Philippines 24-6
multiplier effect 218, 275
Mumbai, megacity 116-21
Dharavi 118, 119
dual societies 116, 118, 119
globalisation 116, 121
Navi Mumbai 120
population growth 117
rural-urban migration 117
vision for the future 119-21

N

negative multiplier effect 39, 260
North Atlantic Drift 36, 48
nuclear power, UK 57-8

O

Objective One (EU funding) 277-9
OECD (Organisation for Economic Co-operation and Development) 77
OPEC (Organisation of Petroleum Exporting Countries) 77, 79
outsourcing 90

P

Pacific Ocean, climatic variation 162-3
Pathfinder Projects 247
Philippines, multiple hazards 24-6
post-industrial economy 251
post-production countryside 270
PPP (Purchasing Power Parity) 75
PQLI (Physical Quality of Life Index) 75
primary employment 270

R

rebranding 250-1, 262
British seaside 268-9
Cornwall 276-9
Doncaster airport 250
farming 280-1
LEDCs 282-5
Manchester 266-7
tourism 251
regeneration 253-4
Jaywick 220
London docklands 253-4
Nightingale Estate 243
Sydney Olympics 262-5
transport, East London 260-1
Rostow, development model 72, 73
rural deprivation 270-1
Rural Housing Enabler 219
rural-urban continuum 214
rurality 227
Cloke index 227

S

Saffir-Simpson scale 155
salt marshes 187, 208-9
Abbots Hall Farm 209

San Andreas fault 27
SAPs (Structural Adjustment Packages) 79
Senegal, unfair trade 76
SnOasis, Great Blakenham 228-9
the Solent 186, 187, 188
Southampton Water , industry 186-91
Dibden Bay protests 190-1
environmental impact 187-9
Fawley oil refinery 186, 187, 188
metal pollution 187, 189
salt marsh (SSSI) 187
sewage/waste 188, 189
Southern Oscillation Index (SOI) 163
Spain,
coastalisation 177
colonisation 70, 71, 94
foreign national enclaves 95
illegal migrants 100, 101
inward migration 94-5
population 94
supermarket supply 85
SSSI (Sites of Special Scientific Interest) 187, 191
The Stern Review 49, 50
Sub-Saharan Africa 74, 75, 81
subsistence farming 283
SUDS(sustainable urban drainage systems) 171
supermarket power 83-5, 280
environmental impact 85
pesticide use 84, 85
sustainability 263-5
sustainable,
communities
Dharavi 118
East Anglia 222-6
development 127-9
urban communities 246-7
Pathfinder Projects 247
Sydney,
Olympics 262-5
affordable housing 262, 264
contaminated site 262, 263
'Green Olympics' 262
solar suburb 264
sustainable development 262-5
synoptic charts 133, 144

T

Tanzania 80
Tenochtitlan 70
Thames estuary 204, 206, 208-9
Thames Flood Barrier 23, 204, 206
Thames Gateway development 206-7
thermohaline circulation 36, 48
TNCs (transnational corporations) 69, 72, 76, 77
China 87
development of 82-3
finance and investment 83, 87
globalisation 82-3
India 90
tornadoes, UK 22
tourism,

Arctic 40
Boxing Day tsunami 19
Caribbean 124-5
Cornwall 271
destination tourism 276
ecotourism 239
farming (UK) 281
global business 124-5
Indonesia 63
rebranding places 251
trading blocs 76-7
tsunami 17
Boxing Day (2004) 10, 12, 16-17
Banda Aceh 62
early warning 62
Sri Lanka 18-19
tourism 19

U

Uganda,
Equatorial College School 284
poverty trap 283
rebranding/regeneration 282-5
remittance payments 283
Village Phone Initiative 285
women and girls 283
UK population change 102-9
ageing 103, 108-9
baby booms 104
birth rate 104-5
ethnic structure 106-7
fertility 105
health and housing 108, 109
migration 106-7
pensions crisis 109
UNEP (United Nations Environment Programme) 41, 46
urbanisation 110-11, 230-1

V

volcanic activity 33
Indonesia 63
Philippines 24-5

W

the Walker cell 162
Walton-on-the-Naze 268-9
waste disposal 129
WHO (World Health Organisation) 108
WMO (World Meteorological Organisation) 46
workers exploited 66-7, 84, 85, 88, 89, 90-1
World Bank 76, 77, 78, 79
World cities 111
WTO (World Trade Organisation) 77, 128

Z

Zambia 80